Transition in Power

Transition in Power

Technological "Warfare" and the Shift from British to American Hegemony since 1919

Peter J. Hugill

LEXINGTON BOOKS

Lanham • Boulder • New York • London

Published by Lexington Books
An imprint of The Rowman & Littlefield Publishing Group, Inc.
4501 Forbes Boulevard, Suite 200, Lanham, Maryland 20706
www.rowman.com

Unit A, Whitacre Mews, 26-34 Stannary Street, London SE11 4AB

British Library Cataloguing in Publication Information Available

Library of Congress Cataloging-in-Publication Data Is Available

ISBN 978-1-4985-4422-1 (cloth : alk. paper)
ISBN 978-1-4985-4423-8 (electronic)

∞™ The paper used in this publication meets the minimum requirements of American National Standard for Information SciencesPermanence of Paper for Printed Library Materials, ANSI/NISO Z39.48-1992.

Printed in the United States of America

Contents

Acknowledgments

I thank several colleagues at Texas A&M University for comments on various parts of the manuscript as I developed it: Brett Cooke of the Department of Modern Languages; Larry Napper of the Bush School International Affairs Program; Jonathan Smith and Vatche Tchakerian of the Department of Geography; and Dick Startzman of the Department of Petroleum Engineering. I also thank some thirty-odd colleagues from many different disciplines with whom I have critically read books and interacted over the past forty years at Texas A&M University in an informal and lively Friday afternoon seminar known locally as the NAR (not a semi-nar but the world's only full NAR). The NAR has given me a much wider multi and interdisciplinary background and it is, from my perspective, what a true university should be about. I thank Alan Dobson of Swansea University for his comments on aviation at several transatlantic studies association conferences, Simon Rofe of the Department of Politics at the School of Oriental and African Studies, University of London, for his comments on President Woodrow Wilson, and Fiona Venn of the Department of Politics at the University of Essex for helping me understand Middle Eastern oil diplomacy a little better. I am also indebted to Curtin University of Technology in Perth, Western Australia, the John Curtin Prime Ministerial Library, Texas A&M University, and the Bush School of International Affairs Program for their support of the travel and research needed for chapter 8. I thank David Black and Roy Jones at Curtin University for their critical and insightful comments. None of the colleagues named bears any responsibility for any errors I have subsequently made.

Chapter 1

Transitions

The Nature of Transition Struggles and the Background to the Anglo-American Struggle

In much of the literature that addresses transitions in world power such transitions are treated:

(a) as fairly straightforward matters of relative economic and military power;
(b) as if the state actors involved are relatively homogenous; and
(c) as if only two main actors were involved.

Neither (a), (b), nor (c) is true. As this book will show, the transition between British and American hegemony was the most drawn out struggle in the history of the world-system as well as the most complex, involving multiple actors in what we now call a multi-polar world. Analyzing that struggle is complicated by the fact that many historians, especially diplomatic historians, see its most active period, between 1919 and 1947, as a period of rapprochement between Britain and America that began in the 1890s and the emergence of what is usually called "the special relationship." In fact the struggle began as early as 1861 and, if anything, intensified between 1919 and 1947.

The book has three organizing principles. The first is to elaborate a model of transitions based on earlier work (Hugill 2009a). This requires commentary on far more than just America and Britain, so that the central focus on that transition will occasionally digress into consideration of events in other states and polities. Complicating such commentary is that many states have more than one polity embedded in them. This was in particular true of America, with at least three. The second is to build a cultural component into a hitherto largely materialist model. And the third is to argue that the last major transition, between Britain and America, was driven largely by technological change. In that regard the book is informed by the remarkable statement made by President Woodrow Wilson while attending the Versailles Conference in 1919 that, in the coming world, there would be three technological arenas of competition between America and Britain: in international transportation, in international communications, and in petroleum. The book is structured on the basis of Wilson's analysis. Whether this was Wilson's own analysis or derived from his advisers is unclear. Wilson himself had little technical education, his Ph.D. was in Political Science from the Johns Hopkins University in 1886, and he also studied history, economics, and German. One of his main advisers was Rear Admiral William H. G. Bullard, director of naval communications from 1919 to 1921, who almost certainly framed Wilson's understanding of international communication.

The struggle over international transportation is divided into chapters on the struggles over maritime transportation and aviation. International communications is divided into struggles over telecommunications and broadcasting. Petroleum sits by itself as part of what one contemporary observer called *The Oil War* (Mohr 1926), a concept that more recent observers reject (Bromley 1991), although one well supported by accounts from the 1920s. The book concludes by examining the shift of Australia from its dependence on Britain as late as 1941 to the ANZUS Treaty of 1951 when it accepted American hegemony.

The term world-system is here used generically in the lower case as a way of describing quickly the increasingly interlocked and globalizing system of economic, cultural, and political development that has characterized the world-economy since the 1430s (Hugill 1993), without implying much adherence to World-System theory in the upper case as espoused by Immanuel Wallerstein (1974), although some of Wallerstein's later work provides a useful framework (1984). The relative economic and military power of the competing states and polities in the Britain to American transition was certainly important, but such materialist arguments are by no means determinative: cultural power also played a strong role. Complicating such materialist and cultural analyses is that a wide range of types of state actors were involved, although it makes sense to treat them as falling into two main types: trading states and territorial states (Hugill 2009a). But perhaps the most serious problem is that the length and complexity of the Britain to America transition makes it difficult to see the transition as the single process it really was. Although the process began with the American declaration of economic "war" on Britain represented by the Morrill Tariff of 1861 (Hugill 2009b) the transition's most active period was from 1919 to 1947, when America finally shifted to seeing itself as a world rather than as a regional or a hemispheric power (Meinig 2004).

CRITICAL QUESTIONS

The most critical questions throughout the transitional struggle were:

(a) what actually was the nature of the struggle for hegemony: was it in the main a material struggle or were there powerful cultural components?
(b) if, when, and how America could and would supersede Britain?
(c) how both countries thought about the transition, which both clearly recognized was occurring?
(d) what role would be played by the other serious challenger, Germany, as well as a series of lesser actors and sometimes challengers: China, France, Italy, Japan, and Russia?
(e) and what forces underlay and drove the transition: in particular geopolitical; technological; and cultural?

The nature of the struggle for hegemony is perhaps the hardest question to pin down, and its cultural component harder still. Previous interpretations of the hegemonic struggle tend to be heavily materialist (see, for example, Wallerstein 1974; Modelski 1978; Hugill 1993). I am indebted to Jonathan M. Smith, one of my colleagues in the Geography Department at Texas A&M University, for helping me attempt to define the cultural component.

As Wilbur Zelinsky pointed out many years ago, American culture is driven by notions of messianic perfectionism that arose from the settling of New England. This derived in large part from the Calvinist fervor brought by many English settlers to that part of the New World and their attempts to create a perfect community, the classic "City on a Hill" of their religious imagination. Zelinsky also propounded "the doctrine of first effective settlement" to explain the disproportionate influence such beliefs would have on American society going forward (Zelinsky 1973). But such Calvinist ideals were by no means lacking in England itself, since many of the Calvinists remained in England. Their ideals drove Cromwell's forces in the three English civil wars that stretched from 1642 to 1651, and had a heavy and continuing influence on English society despite the move away from the excesses of Puritanism at the Restoration of the monarchy in 1660.

At what might be described as the deep cultural level, the transition from British to American hegemony was understood as part of a "flight of the divine spirit" by Americans who saw themselves as founding a "New" England purified of an archaic aristocracy and an established church. John Dewey expressed this cultural interpretation of hegemonic shifts well in his *German Philosophy and Politics* when he wrote:

> Philosophical justifications of war follow inevitably from a philosophy of history composed in nationalistic terms. History is the movement, the march of God on earth through time. Only one nation at a time can be the latest and hence the fullest representation of God. The movement of God in history is thus particularly manifest in those changes by which the unique place passes from one nation to another. War is the signally visible occurrence of such a flight of the divine spirit on its onward movement. (Dewey 1915, 118)

In Britain a great many Calvinists, Whigs, and radicals embraced this "divine spirit." They admired, envied, and, as far as possible given the complexity and constraints of British society at the time, emulated New England's ability to purify itself of Old England's shortcomings (Jonathan M. Smith, personal communication, December 2014). The passage of the "unique place"—in the terms used here, the shift in hegemony—was thus very peaceful on a cultural level, even if filled with alarums and excursions on the level of material struggles. Even those material struggles were predominately economic in character, with military alarums pretty much gone from the scene following the War of 1812. They popped up in the American imagination from time to time, most notably in misplaced northern fears of Britain supporting the Confederacy on materialistic grounds in the American Civil War to ensure its cotton supply. From a cultural perspective, however, given the triumph of the antislavery movement in Britain, driven strongly by the rise of Presbyterianism in Scotland and Methodism in England, the drive by such "Saints" of the Church of England as William Wilberforce to embrace a strong moral stance against slavery, and the creation of the Clapham Group, such fears seem inexplicable. The main British diplomat in Washington throughout the Civil War period, Lord Lyons (served 1858–1865), had no sympathy whatsoever for the emissaries of the Confederacy who constantly importuned him, and his extensive correspondence with Whitehall available in the British National Archives makes this completely clear.

The most active phase of the Britain to America transition began in 1919. With World War I (WWI) over and Germany seemingly defeated it was obvious that Britain would seek to restore the primacy in world export markets it had been steadily losing to America and Germany from the 1870s. President Wilson both saw this clearly and followed an entirely materialistic, economic interpretation of the struggle, commenting "we are on the eve of a commercial war of the severest sort, and I am afraid that Great Britain will prove capable of as great commercial savagery as Germany has displayed for so many years in her competitive methods" (footnote 17 in DeNovo 1956).

The climax of the struggle over the three technological arenas identified by Wilson at Versailles occurred between 1944 and 1947. At the popular level, as well as that of academics using a "wars of hegemonic transition" model,

American hegemony emerged from military victory in 1945 in a two-front world war, primarily against Germany but also against Japan. Economic historians lean to hegemony emerging from the forging of the Bretton Woods financial agreement in 1944, after which the American dollar replaced the British pound as the global reserve currency (Steil 2013). But the real struggle between Britain and America did not involve military action and the loss of status by the pound sterling was evident as early as 1919 (Tooze 2014). Only the accumulation of American fiscal mistakes that led to the economic collapse of the Great Depression prevented the dollar triumphing earlier (for a cogent account of this failure see Ahamed 2009).

In practical terms American hegemony emerged from the complex set of cultural, policy, and technological struggles foreseen in part by Wilson at Versailles. These were the struggle over marine transportation (technology and policy, chapter 2), the struggle over international air transportation settled by the Bermuda Agreement of 1946 (policy and technology, chapter 3), the struggle over international telecommunications settled in Atlantic City in 1947 (policy, chapter 4), the less easily defined and dated struggle over broadcasting that was not really settled until the 1960s (cultural, chapter 5), and the struggle for the petroleum resources of the Middle East effectively settled by President Franklin Roosevelt's agreements with King Ibn Saud of Saudi Arabia in 1945 (technology and policy, chapter 6). By 1947 American hegemony was clearly established in all but marine transportation, where American technological and policy missteps were even more legion than those of Britain, and international telecommunications, where a declining Britain was able to muster enough international support to fight off the worst of the American challenge. In some ways these struggles continued after 1947, especially over the sale of aviation technology (Engel 2007), but by that point there was no longer any question about American hegemony.

By the standards of conventionally structured histories a period that extends from 1861 to 1947 fits into no conventional "era." As for complexity, the relatively straightforward economic struggle between two great trading states begun in 1861 was generally masked by the much bloodier politico-military struggles, often geopolitical in nature, between trading and territorial states and between territorial states that, conventionally, we define as WWI and World War II (WWII), and which can also be called the German wars. It is fitting, however, that the period from the American challenge of 1861 to hegemonic victory by 1947 effectively begins and ends with military conflicts.

In the first of those conflicts, its Civil War, America finally defined itself both politically as an industrial state focused on the North and geopolitically as a federal union of states that controlled the midsection of the North American landmass and would brook no secession. The Civil War was also the first really modern war: it saw, among other horrors, the first use of machine guns, barbed wire, trenches, submarines, and heavily gunned ironclad battleships. In that war many northern politicians still regarded Britain as a serious potential military opponent, believing Britain tied culturally and economically to the Confederacy through its need for cotton.

In the aftermath of the Civil War Britain embarked on a process of appeasement of America (Kennedy 1983b). Britain settled the *Alabama* claims, signed the Treaty of Washington in 1871, and accepted American claims over Canadian in settling arguments over the Bering Sea, fishery rights, and the international border. The treaty was the first step toward the rapprochement of the 1890s. When American Marines deposed the Hawaiian monarchy in 1893 Britain made no attempt to enforce its 1843 treaty supporting Hawaiian sovereignty. In 1898 Admiral Dewey was given enough warning by the British to coal his warships in Hong Kong before war was officially declared and he proceeded to the Battle of Manila Bay. Without such warning the provisions of the Treaty of Washington would have prevented Dewey from coaling. Britain alone of the world powers politically and publicly supported the Americans in the resulting Spanish–American War. And in 1901 Britain backed away from the Clayton–Bulwer Treaty of 1850 requiring an internationally controlled canal across the narrow isthmus separating the Atlantic and Pacific oceans and signed the Hay–Pauncefote Treaty that allowed America to build Panama Canal under American control.

In the last of those wars, the long war of the twentieth century—conventionally WWI and WWII—that began in 1914 and, separated by an armistice, ended in 1945, the declining hegemon, Britain, fought side by side with the rising hegemon, America, as allied trading states with and against an assortment of other states, predominately territorial with one exception, who also fought among themselves, several of whom changed sides. In WWI American support of Britain was slow to come, in part because of American reluctance to enter the war, in part because of America's internal cultural struggle. Although many Americans favored Britain there were also strong pro-German feelings within an America settled heavily by Germans, and strong anti-English feelings among America's Irish. In addition Wilson was no fan of Britain (Tooze 2014). Even so, the only consistency in the long war of the twentieth century was that Britain and America were allied against Germany. France operated as a trading state allied with Britain and America in the first of those wars. After incompetently opposing Germany in 1940, however, she sat out the second phase of the war as a conquered power in a not totally unwilling alliance with Germany (Blouet 2011). Italy, Japan, and Russia were, at various times and sometimes in the same war, willingly allied with states that at other times were their opponents.

Such transitional periods look confusing when they are occurring, and are characterized as periods of multi-polarity. Given the essentially self-preserving nature of power in the world-system, Britain continued to appear powerful on the world stage after WWI albeit, as one historian has it, as a "weary titan" (Friedberg 1988). In retrospect, however, some of the challengers recognized the essentially Potemkin village nature of British power by the 1920s, none better than the Japanese. Japanese exposure to Britain was recent compared to America and Germany, so Japan was less impressed by British power. Although Japan had embraced a German vision of how to organize its army after the German defeat of France in the Franco-Prussian War of 1870, German occupation of Kiautschou Bay in China between 1898 and 1914 as the base for its Asiatic squadron tended to push Japan away from Germany, and Japan fought WWI as part of the western alliance, occupying Kiautschou Bay from 1914 to 1922 (Miyake 1977). Kaiser Wilhelm's inflammatory comments about "the yellow peril" also did not help relations. But the virulent stand taken against free Japanese migration to the west by Australia, pressed to do by America at Versailles, and the discriminatory treatment of Japan under the terms of the Washington Naval Treaty of 1922 pushed Japan away from its WWI allies. Although Japan spent much of the 1920s and 1930s in geopolitical isolation, Japan was also increasingly influenced by a German *geopolitik* that emphasized the inevitability of German rise and British decline as well as by the sentiments in favor of a German–Japanese alliance expressed by the author of that *geopolitik*, Karl Haushofer (Hayes 2004; Herwig 2016).

Certain forces run through this British–American transition process, inform this book, and are embedded from chapter 2 to chapter 6. The most significant force has been whether a polity acts predominantly as a trading or a territorial state at specific historical moments, which determines its geopolitical role. However, this and other forces were heavily affected by the massive time-space compression made possible by rapidly evolving new systems of transportation and telecommunications (Warf 2008). This rapid acceleration of the rate of change in important geopolitical, technological, and cultural systems was devastatingly difficult for the relatively simple political and economic structures of the mid- to late-1800s to manage, and it forced those structures to evolve rapidly and often painfully. Some structures just collapsed.

Other than state type, three forces were dominant and all three underwent constant and rapid change: technology; math and science; and demography. Important lesser changes came in cultural attitudes to ethnic, racial, class, and gender differences. The changes in technology can be summed up as the shift, in different regions, on different schedules, and with different levels of human input, from the Eotechnic via the Paleotechnic and Neotechnic (the terms are from Mumford 1934) to the beginnings of the Infotechnic, although it was the changes in transportation and telecommunications technologies that had the most geopolitical impact (Hugill 1993; 1999a). The states that adapted best to and best managed these changes did best overall. The states that failed to do so usually failed disastrously.

First, the world-system of 1861 was, with one important exception, still mired in the Eotechnic: driven by ancient technology, as the Greek prefix *Eo*-indicates. It was organically powered, with small increments of power from falling water and wind, the last being most important for a global transportation system based on wind-driven ships. The exception was the hegemonic power, Britain, which had achieved that hegemony, in part, by its pioneering embrace of technological innovation, math and science, and thus the Paleotechnic: conventionally, the steam-powered factories of the Industrial Revolution and the application of steam to both waterborne and overland transportation. Cultural forces, such as a form of property rights that tended toward representative government, also played a part, as authors such as Ferguson make clear (2011). Other powers, especially would-be hegemons, had to struggle to catch up, but catch up they did, by embracing the emerging math and science-based technologies of the Neotechnic from the 1870s on, just as Britain started to fall behind. In 1861 the most powerful explosive was gunpowder: by 1947 the first nuclear weapons had been used. By 1947 states in Europe, America, and Asia had fought two global wars using such powerful Neotechnic technologies as telecommunications, organic chemistry, electricity, radar, and petroleum, to say nothing of such military devices as machine guns, big gun battleships, aircraft, aircraft carriers, submarines, tanks, nuclear weapons, and the like. On the cultural front some states had begun to encourage women to enter the labor force en masse and treat groups who had been regarded as second-class citizens more equally. Others, however, had defined some second-class citizens as subhuman using pseudoscientific racism or as political enemies of the state and systematically attempted to exterminate them.

Second, underlying these technical and cultural changes, not initially responsible for many of them, but ultimately the driving force behind their developed forms even when misused, was the force of scientific change. Although European polities were already moving toward science much earlier (Ferguson 2011), between 1861 and 1947 the world went from being largely ignorant of math and science and empirical in its approach to changes in technology and culture to one in which such change was driven by math and science and thus seemed at least to be more predictable. The key math and science driven technologies were organic chemistry, electronics, fluid—and thus aero—dynamics, and nuclear physics. These Neotechnic technologies did not develop at all evenly in the world-system and the states that did best at educating as many as possible of their citizens in math and science, whatever their ethnicity, race, class, or gender, tended to gain at least temporary advantages over those that lagged. Attempts to apply science to cultural areas for which it was not designed, as in the case of social Darwinism, could also produce disastrous problems through the application of what was called scientific racism. Germany suffered most from this.

Third came a more subtle change, in demography. Birth and death rates were everywhere high in 1861, life expectancy moderate, and education levels generally low, more notably so for some groups, especially women, than others. Birth and death rates fell at different rates in different regions, and life expectancy and education levels rose, though unevenly, especially for women. In 1861 Europe and European-derived cultures were not much different demographically from other major world regions. By 1947 the differences in life expectancy, education levels, and the relative status of different groups, especially women, who generally had attained voting rights after WWI, between most European and all non-European cultures with the marginal exception of Japan were remarkable.

If, in retrospect, these three major changes, other lesser ones, and the way they were managed were dominant in shaping the transition, they were not always seen as such as the transition unfolded. Thc morc traditional historical forces of dynastic change, military activity, the maneuvering of political elites for power, and the agency of powerful individuals did not stop operating, nor have historians stopped writing about them. Those forces, however, now had to deal with the technological, scientific, and demographic changes that were beginning to assume much more centrality to the world-system and to the process of hegemonic transition. The polities that adapted to, brought under control, and discovered how to direct and create these new forces did best. In the long run these polities were in the trading, not the territorial states.

This book is about America's particular success in managing those changes in its drive to succeed Britain as global hegemon. America's cultural self-image generally denies any such drive for hegemony, but it is self-evident in American culture at almost all levels: from the ideals of American exceptionalism in the late 1800s to the neoconservative behavior of the late 1900s; from the religious ideals of the various "Great Awakenings" of the nineteenth century to the swirl of mission activity that characterizes much American religion today; and from the boy's adventure novels that became popular in the late Victorian period to such modern popular culture movie icons as "Rambo."

TRANSITION THEORY AND STATE TYPE

States tend to fall into two main types, trading and territorial, although when there are multiple polities embedded within states they are often in conflict for control of the levers of state power. The first person to argue that multiple polities were embedded in one state was the founder of British geopolitics, Halford Mackinder, when he noted that, after the Congress of Vienna of 1815, Germany came to contain two polities that pursued conflicting geostrategies: "Prussia obtained a detached territory in the old Germany of the west, which territory was divided into the two provinces of the Rhineland and Westphalia." The emerging German state was thus divided between its "liberal Rhineland"—liberal in the sense of classic "Manchester liberalism" and thus a trading state—and "the conservative Brandenburg of Prussia," a classic territorial state. The primary aim of the Prussian Junker class was to derive income from land and to expand east into the Slavic-dominated Heartland, control of which was identified by Mackinder in his seminal article of 1904 as central to global hegemony, to obtain more land. In the west, Mackinder noted, "the Teutonic element along the Rhine . . . [has] no deep-seated hostility to the French," whereas in "East Europe there are also two principal elements, the Teutonic and the Slav, but no equilibrium has been established between them." Although this last quote can be interpreted as shot through with the sort of notions of Anglo-Saxon superiority that were common in 1919, it also accurately reflects the geopolitics of the situation and the different geostrategies being pursued. Mackinder summed up the impact of this when he noted that, at the outbreak of WWI, "Berlin had not decided between her political objectives—Hamburg and overseas dominion, or Bagdad [*sic*] and the Heartland" (Mackinder 1904; 1919, 122–123, 131–132, 154; Hugill 2005).

The first thorough account of a state hoist between two competing polities came from the American historian Edward Whiting Fox, who reinterpreted French history as the history of "Two Frances," a coastal trading polity looking out into the Atlantic that found itself warring with a territorial polity bent on expansion eastward. In Fox's estimation this split, more than any other factor, accounted for French failure in the long series of wars with Britain that occupied the period from the late 1600s to 1815. French defeat in the Napoleonic Wars was largely because of the inability to focus on defeating its trading state enemy, Britain, while focusing on its principal territorial state enemies, Prussia and Russia. In the end, however the Russian winter, Prussian resurgence, and British ability, because it could generate wealth though trade, to pay or loan money to its allies to help prosecute the war, combined to bring down Napoleon and the French territorial polity (Fox 1971; Hugill 2005).

Recasting the long, complex transition between Britain and America in such theoretical terms requires at least some comment on the history and nature of transitions and on the nature of British and American polities in the transition period. In 1978 Modelski identified five hegemons in the world-system since the onset of the European expansions in the 1430s: Portugal, Holland, Britain 1, Britain 2, and America (Modelski 1978). Scholars such as Wallerstein have argued for only three hegemons in this period—Holland, Britain, and America—but where Wallerstein's world-system model provides a useful model for how hegemony is won and lost it lacks a causal mechanism and ties hegemony too narrowly to economic policies. Wallerstein argues that protectionism is the policy of challengers from the semi-periphery, not the policy of a hegemon ensconced in the core, and that once a challenger "wins" it assumes the core policy of free trade (Wallerstein 1974; 1984; Hugill 2009a). Some of Wallerstein's followers have argued that all transitions are necessarily accompanied by major wars between a declining hegemon and a rising challenger (Goldstein 1988; Taylor 1985). Wallerstein's model excludes the transition from Holland to Britain 1 since Britain did not adopt free trade in the late 1600s but waited to do so until after the transition to Britain 2 in 1815. Taylor's model excludes it because Holland and Britain never fought a major war and both either avoided or stayed relatively clear of the Thirty Years War, the major war of the Holland–Britain 1 transitional period.

Yet the Holland–Britain 1 transition is a good model for the Britain 2–America transition for precisely the reason that the two states (a) were trading states and (b) did *not* fight a major war, although America did switch from protection to free trade in the 1940s after achieving hegemony, thus meeting the needs of Wallerstein's model. Both Holland and Britain were states dominated by trading polities and their struggle, pretty much over after 1688 when the Dutch national debt was moved to London, was economic rather than military. The British and the Dutch did, it is true, fight three almost entirely naval wars in the 1600s, but these were wars about trade that hinged on control of merchant shipping. Such wars, though brutal for the sailors involved, were more concerned with taking merchant ships as prizes than running up the body count or obtaining territory. Indeed, at the end of the Second Dutch War in 1667 the victorious Dutch actually ceded their North American colony of New Amsterdam to the British.

The first of the three Anglo-Dutch Wars of the 1600s (1652–1654) had broken out over Britain's adoption of a particular form of protection under the guise of the Navigation Acts, requiring that only British ships carry goods out of British ports. The Dutch argued for a policy of *mare liberum*, "free seas," in which ships should be free to drop off and collect goods in any port regardless of their place of registration. In other respects the Dutch and British economies of the period were protected through the general policy of mercantilism, which the British began to abandon in the late 1700s under the influence of the writings of first Adam Smith (1776), then David Ricardo (1817). When America adopted a policy of protection in 1861 it was in a very different economic period. Britain had repealed the Navigation Acts and had become aggressively liberal, mercantilism was dead, and Ricardo's doctrine of "comparative advantage" was everything. Although by the late

1800s some American politicians were drifting into more aggressive forms of protectionism American protection was generally only about tariffs to protect infant industries from foreign competition, a concept embraced by Smith in *The Wealth of Nations* (1776). Internally America was completely liberal, with free trade between states enshrined in its constitution.

One must also examine the cultural issue. Following the Reformation the main countries of Europe were all heirs in various ways to the notion of "messianic perfectionism" that centered on Calvinist belief. In this sense a world-system model that includes Portugal might appear problematic, and a material explanation for Portuguese success rooted in material innovations in navigation and ship design might well be paramount (Hugill 1993). But Portugal was also driven, as later was Spain, by the *Reconquista*, a Christian messianic drive to remove Islam from Europe. Holland, Britain, and America, the subsequent hegemons, fit the drive for "messianic perfectionism" precisely. The principal challenger powers, France, Germany, and even Russia have also understood themselves in these messianic terms. Russia has seen itself very clearly as the "New Rome," and the purity of its Orthodox vision of the world has driven it to attempt to root out the "heresies" of Catholicism and Protestantism.

TYPES OF TRANSITIONAL STRUGGLE

Transitional struggles work out differently between different types of states. There are three main types of struggle: Type I struggles are between trading and territorial states; type II struggles are between trading states; and type III struggles are between territorial states. Hegemonic transitions are rarely straightforward type I, II, or III struggles. They are usually complex mixes of two or even three types of struggle. Type II struggles are largely economic, involving trade wars, tariffs, and highly focused military action aimed at gaining trade advantages such as capturing a larger merchant fleet from an opponent or gaining control of strategic geographic choke points. They are about controlling the flows of goods and raw materials that generate wealth and about the critical nodes through which such flows move and where they can be relatively easily interdicted. Type III struggles are basically struggles over control of territory and routinely eventuate in serious military conflict and great loss of life. Type I struggles can combine elements of both, but the trading states involved rarely seek any territorial advantage other than control of choke points.

There have thus far been four hegemonic transitions in the world-economy, all involving a type I military struggle. The states in italics and brackets were major actors in the military struggles in question. The states underlined were the principal territorial state challengers in that transition.

1. Portugal-Spain [*Britain*] Holland
2. Holland [*France*] Britain I
3. Britain I [*France/Prussia/Russia*] Britain II
4. Britain II [*France/Germany/Japan/Russia-USSR*] America

In each case, however, the type I struggle has been the most obvious and the one most written about. Transition 1 is the most problematic and the most obviously a type I struggle, since Spain's interests in the New World mark it out as a territorial state. Portugal, forcibly aligned with Spain by royal marriage, was less of a territorial state, despite its territorial interests in Brazil. In transition 2 the "Wars of Louis XIV" were a type I struggle initially between Holland and France aligned with Britain, but ultimately between France and Holland aligned with Britain. They tended to obscure the type II struggle between Holland and Britain. Transition 3 was effectively between a commercial Britain and an industrial Britain that embraced first the Eotechnic, then the Paleotechnic forms of the Industrial Revolution (Hugill 1993). As such it was technically a type II transition if not much of a struggle, and in the history books it is the massive challenge from France and the type I struggle that caused that has been of most interest.

Transition 4, which is what concerns this book, was the most complex hegemonic transition yet in the world-economy, embodying all three types of struggle. Deep down it was a type II struggle between Britain and America. It was also a type I struggle in which Britain and America united as trading states against Germany in WWI and WWII, and against a France clandestinely aligned with Germany and Japan in WWII. Russia was caught between pillar and post, allying with everyone at various points between 1914 and 1945. No predominately Type III transition exists, but transition 4 contained two Type III struggles. In WWII, which in many ways began with Japan's invasion of Manchuria in 1931, Japan became caught up in a Type III struggle with Russia in Manchuria. The definitive Type III struggle was, however, the intensely vicious struggle between Russia and Germany in both WWI and WWII (Hugill 2009a). The nature of that struggle in WWI has been far less written about than that of WWII, although recent work by Norrell makes the continuities between WWI and WWII on the eastern front very clear (2018).

POLITIES WITHIN STATES

The issue of polities within states adds a further layer of complexity. What somewhat confuses any account of the transition from British to American hegemony is that America contained within it three major polities, Britain only one. In terms of Fox's model, Britain was purely a trading state, dominated totally by a seaborne trading polity in full control of its internal politics, thus its geopolitics and geostrategies. The American state that broke away from Britain in 1776 proclaimed itself a republic and in its new institutions attempted to restrain private interests for fear they would corrupt public virtue. But as Agnew notes, cultural forces triumphed: "the essentially liberal political economy inherited from the past proved more significant in integrating the vast new country than did the institutions of republicanism" (2005, 15). Although the new state had clear trading state origins, it soon broke into three polities: a seaborne trading polity based on the Atlantic seaboard; an Atlantic to Great Lakes inland trading and manufacturing polity; and a southern aristocratic slave owning polity at serious odds with the other two. The first two of these were culturally similar, the third was not. This would result in very serious internal tensions, prevent any coherent control of its internal politics until the Civil War, and delay America's rise to hegemony.

The Atlantic Seaboard Seaborne Trading Polity

Initially, from the end of the revolution through the mid-1840s, the Atlantic seaboard region from Portland, Maine, to Baltimore, Maryland, acted as a continuation of the seaborne trading polity inherited direct from Britain. Settlers here had focused on the numerous natural harbors and exploited the region's excellent natural forest resources to build by 1776 at least a third of the ships trading around the first British Empire (Albion 1926, 246). However, the ships of the American republic that emerged in 1783 were, under Britain's Navigation Acts, denied the right to carry goods out of British ports. But ten years later Britain was again at war with France in the French Revolutionary (1793–1802) and Napoleonic Wars (1803–1815). Many British merchant ship captains vanished into the British navy where prize money was to be made, and New England's shipowners were thus routinely licensed to carry goods out of British ports. But in 1815 peace broke out, the Navigation Acts were reasserted until their repeal in 1849, American ship captains had to seek new areas to trade with, such as China and the Pacific basin, and some Americans began to look to their continental interior for markets, not across the seas of the world.

Despite its loss of its privileged position within the British Empire the Atlantic seaboard polity remaining focused on sea trade, but the seaboard cities involved were, in Vance's terms, "misplaced colonial cities" that needed either their trade with Britain or some alternative to prosper (Vance 1990, 267). The China-Pacific trade was nowhere near enough. A secondary trading polity thus developed as another alternative. This looked to trade with the continental interior, which area the Atlantic seaboard polity sought to open up in the late 1820s via canals. Two of these were completed though others were planned. New York State's Erie Canal linked New York City to the Great Lakes, entering them above Niagara Falls at Lake Erie. The route went north up the Hudson River, thence west through the Mohawk Gap onto the Lake Ontario shore plain, and thus into a rapidly burgeoning maritime trade on the Great Lakes, effectively America's inland sea. The Pennsylvania Mainline Canal linked Philadelphia to the Great Lakes via a collection of railroads, canals, and an inclined plane over the Alleghenies. It is best described as a whole Heath Robinson mess, and it was a spectacular and expensive failure. Canal boats had to be dragged up this last by a steam engine installed at the summit, which was thus monumentally expensive to fuel. As a consequence, only New York State prospered on the interior trade, a trade that was still fundamentally waterborne in nature along the highly profitable Erie Canal and into the Great Lakes.

In the early 1850s, however, a series of trans-Appalachian railroads, with trains hauled by a new generation of much more powerful American steam locomotives, bolstered the Erie Canal. These connected New York (via the New York Central Railroad), Philadelphia (via the Pennsylvania Railroad), and Baltimore (via the Baltimore and Ohio Railroad) to Chicago and thus to the Great Lakes. This inland, continental trading polity finally corrected the geographic problem of three of the "misplaced colonial cities" and also began to stress domestic manufacturing (Hugill 1993).

John Quincy Adams prefigured the emergence of this inland trading polity in a speech at Quincy Massachusetts in 1802 when he borrowed a line from bishop Berkeley and commented that "westward the star of empire takes its course," in the process giving New York its tagline as "The Empire State." New York prospered most because it had both canal and rail connections inland as well as excellent port facilities for the Atlantic trade. The intersection of the seaboard and inland trading polities with the republican institutions of the American state is crucial, since the republican ideal was based on "limiting government powers and seeing the United States as a fixed territorial enterprise without 'interests' beyond its immediate geographical confines." The problem was that trading polities need constantly expanding markets to ensure the continual increase in profits necessary to the functioning of a capitalist economy and "in practice the republican model has always failed to control the expansionist impulse" (Agnew 2005, 15).

These two dominant American trading polities are well represented in the pages of *Hunt's Merchants' Magazine and Commercial Review*, published monthly in New York from 1839 until the outbreak of the Civil War in 1861. In the early 1840s *Hunt*'s was still almost entirely concerned with the seaboard polity operating out of the Atlantic coast, describing appropriate maritime markets for American goods and sources of imported goods and raw materials. The entire table of contents for volume 1 of *Hunt*'s references only one inland American city, and that is simply a short table entitled "The Trade of Buffalo," most of which was passing along the Erie Canal and out into the Atlantic at New York (*Hunt's* 1839:1, 90). There are numerous short articles on the Atlantic trade and eight dealing with various aspects of the Atlantic seaboard cities' trade with Asia that reached into the Pacific via the Straits of Magellan and the Hawaiian Islands. Two short articles also describe world atlases. *An Illustrated Atlas of the United States and Adjacent Countries* was, according to *Hunt*'s, "recently adopted by the Secretary of the Navy of the United States, as one of the works to be furnished to our public ships" (*Hunt's* 1839:1, 375–376). The second, *A Pictorial Geography of the World*, is "particularly entitled to our attention as a storehouse of valuable and accurate commercial intelligence. Geography is one of the first studies that should engage the attention of merchants" (*Hunt's* 1839:1, 463–464). But such early American atlases depicted a global, maritime geography, not an inland territorial one.

THE ATLANTIC TO GREAT LAKES INLAND TRADING AND MANUFACTURING POLITY

Hunt's began to define what would emerge as the inland trading polity when they began a series of major articles on western transportation in 1846 with "Commerce of the Great Western Lakes: Increase of Lake Marine in 1845" (*Hunt's* 1846, 15:4, 348–359). They then followed this up with a series of articles on the burgeoning railroad system of the northern states. In 1847 they began a long running, numbered series on "Commercial Cities and Towns of the United States." Although the first of these dealt with the Great Lakes' port city of Buffalo the second covered Rochester (*Hunt's* 1847, 16:6, 596–602; 17:1, 46–54). Third came St Louis, fourth Nantucket with its important whale fishery, then Toledo fifth, Pittsburgh sixth, and Chicago seventh (*Hunt's* 1847, 17:2, 167–174; 17:4, 368–377; 17:5, 489–491; 17:6, 587–594; 18:2, 164–171). As an occasional nod to the seaboard polity they published a much less regular series on "Commercial Cities of Europe." The first of these, on the French port city of Le Havre, appeared in 1848 (*Hunt's* 1848, 18:1, 43–48). The series on American cities of the continental trading polity petered out around the outbreak of the Civil War with numbers seventy-six, Fort Wayne, Indiana, and seventy-seven, Dover, New Hampshire (*Hunt's* 1860, 42:6, 697–703; 43:2, 199–204). *Hunt's* would vanish during the Civil War.

The success of the three railroads linking seaboard cities to Chicago caused more railroads into the interior to follow. American trade shifted away from an overarching focus on the seas to a continental pattern that would predominate for almost the next hundred years. This shift was part technological, part political. At the level of technology, in the late 1830s American railroad engineers began to produce a much more efficient and powerful style of locomotive than those pioneered in Britain and imported into America in the 1820s during the earliest phase of American railroad development. This had its wheels arranged in the 4-4-0 pattern—four wheels up front on a pony truck, four driven wheels, and no wheels under the firebox. The particular innovation of this *American* class locomotive was the use of three point compensated suspension. The pony truck was connected to the locomotive by a single pivot and acted as the first point of contact. Each pair of driven wheels pivoted in the middle as the other two points. Effectively the locomotive became a mobile tripod, the three points of contact of which could move up and down. This flexibility allowed *American* class locomotives to run well on the cheaply laid track that characterized early American railroads, capital being in very short supply. Earlier British influenced designs with rigid frames and no suspension derailed much too easily. In the late 1840s this 4-4-0 design was refined with a longer wheelbase that both rode better at speeds and that allowed larger fireboxes. Larger fireboxes gave more power to haul heavier trains as well as allowing higher speeds. By the early 1850s such locomotives were capable of hauling decent sized trains from the Atlantic seaboard cities across the Appalachians and into the continental interior, a feat well beyond earlier locomotives (Hugill 1993).

At the political level the shift was made possible by the repeal by Britain in 1846 of its last tariff, the Corn Laws. These made imported wheat expensive, protected the landowners of Britain, and restricted imports from America to that area of Upstate New York that could ship its wheat across Lake Ontario into Upper Canada and enter Britain as if it were Canadian wheat. The new railroads completed across the Appalachians between 1851 and 1852 opened the wheat lands of the American west to the British market. This fundamentally changed the trade pattern of America, making Chicago a vital inland focal point and bringing America's inland trading polity into preeminence (Hugill 1993, 176–178).

The passage of the Morrill Tariff in 1861 and northern victory in the Civil War secured the preeminence of this inland trading and manufacturing polity within America for two generations. It was self-sufficient in energy and raw materials. Continuing immigration made it a massively expanding market. Secure behind a tariff wall that generally increased in height for the hundred years after 1861, and thus released from excessive competition from older and cheaper producers such as Britain, the region's factories burgeoned. It became America's manufacturing core. This region saw its trade as internal, preferred a policy of tariff protection to free trade, and generally supported a Republican Party that made protectionism a central part of its party platform from 1861 on.

However, this inland trading polity followed the seaboard polity in some very important cultural respects. Although neither polity was a particularly democratic or classless society in a modern sense their shared cultural values of hard work and education attached to Protestantism allowed individuals relatively good upward social mobility. Most citizens in the two trading polities, including women, were literate, and were encouraged to be so by a general preference for heavily Protestant, often Calvinist religious forms that emphasized moral responsibility, thrift, and education, inasmuch as godly individuals needed to be able to read the Bible for themselves, not rely on its interpretation by a minister. Wealth and status might intersect with land ownership in some areas (Hugill 1995), but they were not tied to it as they were, for example, in the American South. The seaboard polity had established numerous colleges early in its history and the inland polity continued that tradition. Initially such colleges tended to be associated with particular religious groups, but in the aftermath of the Civil War both the seaboard and inland polities established through the Morrill Act of 1862 a system of wholly secular colleges focused on teaching the useful arts, initially mostly agriculture and engineering, often referred to as agricultural and mechanical colleges. With the restoration of the Union in 1865 and the end of Reconstruction in 1877 such colleges were also established in all the former Confederate States. Although the American state lagged behind Wilhelmine Germany in the development of sophisticated graduate schools and research universities in the run-up to WWI, once the problem posed by the better German universities was identified American universities caught up rapidly in the 1920s and 30s. In addition, although women were clearly not well served by American higher education in the period covered by this book, in the two trading polities they almost always had schooling through twelfth grade, enough to be fully literate and thus to often hold responsible jobs.

America's republican institutions served America well in the period through the 1890s, as the expansionist impulse was effectively contained within the geographic limits of what became the lower forty-eight states. But when the frontier "closed" at the end of the 1800s (Turner 1893) it proved difficult and then impossible to redirect the expansionist impulse externally in the European fashion: America's republican institutions thus effectively prevented the imperial solution to overaccumulation used by various European states from the 1880s on. By the 1920s the manufacturing core was beginning to suffer from serious overaccumulation caused by overproduction relative to a saturated domestic market. Despite considerable foreign-direct investment by American companies seeking to sell their product in Europe, American companies could not easily penetrate tariff-protected markets elsewhere.

Singer, Ford, and McCormick were merely three of the American companies that established overseas factories to solve this problem. As Harvey puts it, "the formation of closed empires after the First World War almost certainly played a role in the inability to solve the over-accumulation problem of the 1930s" (2003, 140). Harvey notes that there were ultimately four solutions: New Deal spending, which was too little and thus only a partial solution; the much more massive spending of WWII; post-WWII spending on infrastructure and education; and the postwar return to global free markets heralded at Bretton Woods in 1944.

The other economic impetus was that by the late 1800s and certainly the early 1900s America was recognizing that it could construct a culturally based economic imperium that would not depend on the political control of territory. This new imperium would be based on "cheap mass products, . . . dazzling technology, and . . . alluring mass culture" (Rosenberg 1982, 6). America was the first great marketer of cheap mass products through Woolworth's innovation of the "Five and Ten Cent Store." Dazzling technology came on line with such things as Singer's sewing machines, Ford's Model T, and McCormick's harvester. Alluring mass culture was on display in Hollywood films and, later, recorded music on discs and broadcast radio. Such forces ensured that "American expansion . . . [would be] characterized not by the *acquisition* of new territories but by their *penetration*" (Huntington 1973, 344, italics in the original). This would be the focus of American policy after WWII.

The Southern Aristocratic Slave Owning Polity

America contained, of course, a third major polity, the aristocratic slave owning polity of a South that would focus after 1815 or so on the massive cultivation and export of cotton. In its aristocratic pretensions and its insistence on deriving income from land this might seem to be a territorial style polity, but slavery made it an odd one, inasmuch as southern land owners saw their wealth and prestige derived as much or more from ownership of slaves than land. Rightly or not, there has also been a sense that southern slave owners "used up" the land by growing crops such as tobacco and cotton that depleted the soil of its nutrients (though see Earle 1988 for a counter). Certainly southern elites spent almost none of their own money or that of the states they controlled on infrastructure or increasing the value of their land in the way that landowners in northern states did (Wright 1978). They failed to build roads and railroads, which required both social organization and a great deal of capital, at anywhere near northern levels, preferring privately held steamboats operating on the free right-of-way represented by rivers.

Wallerstein's World-System model (1974) would identify the South as part of the periphery on the basis of its controlled cost labor systems, slavery before the Civil War and sharecropping thereafter. As a periphery the South did not support protectionism: it made its living growing and exporting cotton, not manufacturing. Its main export market was free trade Britain, and it wanted the cost of its imports from Britain kept low. Although the Constitution of the Confederate States of America generally followed that of the United States of America it contained two major deviations: one, naturally enough, enshrined slavery, but Clause 8 went so far as to explicitly ban protectionism.

But however much Wallerstein's argument makes sense, the South was an odd periphery as well as a conflicted polity. If the South had been an independent state or part of a formal Empire Wallerstein's rules would reasonably apply, but the South arose from, failed to secede from, and was ultimately embedded back into an American republic that was clearly derived from a British trading state just at the point at which that British state began to make the transition from mercantilism to free trade in the late 1700s. Adam Smith and David Ricardo gained a powerful hold on American economic reasoning, and the preference of such as Hamilton for "the protection of infant manufactures" was perfectly respectable in Smithian terms, even if Hamilton could not persuade a Congress dominated by southern interests to pass Protectionist legislation. In any case, the Democratic politicians who represented the former Confederate States polity after the end of the Civil War continued that polity's long commitment to free trade. However tempting in some of its aspects, Wallerstein's World-System model is too crude to be applied straightforwardly to the complex republic with its multiple polities that arose in America after 1776.

As Levine's recent work (2013) makes clear the South was little more than an unusually unpleasant version of the French *ancien régime*. It was riddled with aristocratic pretensions and, apart from slavery, suffered from a rigid class structure that helped bring it down. As the Civil War wore on white Southerners who lacked slaves and land increasingly saw the war as one in which they were being asked to die to protect the property rights of an *ancien régime* they had no good reason to love. The South educated only its aristocratic elites, and then generally in such a way as to give them an excessive sense of entitlement. In the 1850 census over 90 percent of the population of Mississippi was enslaved, unskilled workers, or whites who recorded themselves as farmers but owned no land. Almost all of these would have been illiterate or barely literate. The comparable figure for New York State that year was less than 60 percent, many of whom were literate (Hugill 1988b, 117–118). The South did its utmost to prevent slave literacy and neither encouraged nor paid attention to developing literacy among poor whites who lacked land and slaves. Baptism, the most popular religious form among all the poorer classes of the South, depended on ministers being "called," not educated as they had to be in the scholarly religious forms that predominated in the northern trading polities.

Before the Civil War the structure of American government allowed the South to exert strong political control over the American state via its control of the Senate, a control that it sought assiduously to continue by trying to restrict the entry of more free states into the Union than slave through such devices as the Missouri Compromise. The result was that American internal politics ensured that no consistent geopolitical strategy could ever be followed. The continuing power of the Senate to make foreign policy even after the Civil War ensured that American geopolitics would remain inconsistent until the rise of a more powerful presidency via Progressivism and the accession and election of Theodore Roosevelt.

As the North grew by immigration, especially from the 1840s on, the South increasingly lost control of the democratically structured House of Representatives. The accession of several new and potential new states in the late 1840s following the settlement of the Oregon question, the annexation of Texas, and the Treaty of Guadalupe Hidalgo that settled the Mexican War began to unbalance the Senate in favor of the North and thus threatened the South's ability to use the Senate to curb the House. Although the crisis of 1850 resulted in the "Great Compromise" between the three polities the compromise was no solution. No solution could have been found that would have been acceptable to the slave owning aristocrats who controlled the southern polity and who even dreamed of establishing more slave states than free in order to fully control the Senate. The "Great Compromise" merely delayed the onset of the Civil War for a decade (Bordewich 2012). Put simply, for the trading polities of the North to successfully challenge Britain they had no choice but to bring the wayward aristocratic polity of the cotton South to heel, and in the end the only way to achieve that was through Civil War. However much the Civil War was about slavery, it was also an important step toward American hegemony, if unrecognized as such at the time.

The two trading polities did not always see eye-to-eye on economic policy. The protectionism of the manufacturing core never had complete support in an Atlantic seaboard polity that made its living by sea trading. Even when the seaboard polity was represented in Congress by Republicans, as it always was after the Civil War, it continued to prefer open markets. It was the inland polity that was most resistant to free trade, although its resistance was ultimately worn down by the overaccumulation crisis of the 1920s, the Depression, and its inability to penetrate tariff-protected markets elsewhere. In many ways the inland polity most clearly embraced America's republican institutions, but it was these institutions that prevented the use of classical imperial tools and pushed America into developing mechanisms to exercise power without exercising territorial control.

The American return to global free markets was not easy. With the political collapse of the Republican Party in the Great Depression and the election of a Democratic national government in 1932 came the political opportunity to move away from protectionism and back toward free trade. By the 1920s the American economy had elaborated to embrace both agricultural and industrial production. In addition, America's traditional foreign trade as a producer of agricultural raw materials, most notably cotton and wheat, had begun by the 1890s to suffer from competition from lower cost producers in Latin America and Australasia. The farm subsidies favored by the Democrats from 1932 on effectively lowered the price of American agricultural goods on the world market so that American farmers could again compete. This was not free trade, but neither was it the traditional protectionism of tariffs.

In the industrial sphere, by the teens American factories had begun to wholeheartedly embrace the principles of scientific management and of time and motion study espoused by Frederick W. Taylor and Frank B. Gilbreth. Put into practice on a grand scale by Andrew Carnegie at U.S. Steel, Henry Ford, and Alfred P. Sloan at General Motors, and diffused to lesser manufacturers by publication and example, the application of these principles made American factories the most efficient in the world by the 1920s, but they still needed a playing field leveled by free trade to expand their market share in the rest of world (Taylor 1911; Gilbreth 1911; Wilkins & Hill 1964; Sloan 1964). In the absence of free trade, companies such as Singer, Ford, McCormick, and General Motors routinely set up manufacturing plants to produce such things as sewing machines, automobiles, and harvesters outside America in the 1920s to avoid local tariffs. But this was not enough to solve the problem of overaccumulation. In the end America had to find an alternative to imperialism that did not radically violate its republican institutions. As Agnew notes this produced a unique American hegemony that required not only a culturally based economic imperium but also "constructing alliances (such as the North Atlantic Treaty Organization [NATO]), building international institutions (such as the United Nations [UN] system, the International Monetary Fund [IMF], World Bank, the World Trade Organization [WTO], etc.), and using economic and military leverage (such as the U.S. dollar and the threat of nuclear weapons)" (Agnew 2005, 15).

BRITISH ACCEPTANCE OF THE AMERICAN DRIVE FOR HEGEMONY

The conventional view of British–American relations by American historians is that they were, at best, problematic throughout the long nineteenth century, a century that, for America, could be reasonably said to have begun in 1776.

In this reading negative American feelings about Britain were reinforced by the War of 1812 and continued until the very end of the 1800s, fading slowly between the rapprochement of the late 1890s and the effective end of the long nineteenth century when WWI broke out, at which point the two powers entered into a hitherto unimaginable military alliance that was the starting point of a "special relationship" that many believe continues through the present. This is, of course, much too simple a view of a turbulent relationship in which any possibility of a cultural or military struggle faded quite early but in which an economic struggle continued throughout and even intensified as American power increased. That economic struggle was the real transition, but the very early fading away of the possibility for cultural struggle and the somewhat later fading of the possibility for military struggle also marked a steadily increasing British acceptance of America's drive for hegemony.

Many authors have noted that negativity toward Britain was exacerbated by American domestic politics, which tended to be driven most heavily by the migration to America of a very vocal and very anti-English Irish community from the late 1840s on. But before the emergence of the Irish problem there were clearly moments of considerable British–American tension over the Canadian boundary. Canadian politics were heavily polarized against America by the American invasions of the War of 1812, the war on the Great Lakes, and the looting and burning of York—modern Toronto—which prompted the British to sack Washington and burn the White House in reprisal. Wrapped up in the Canadian boundary issue was the issue of the appropriate western boundary between the two countries, with the "Oregon Question" seeming to loom large as a potential casus belli in the 1840s until settled peacefully under the Polk administration in 1846. Even then the two countries managed to confront each other over their exact boundary in British Columbia's Straits of Juan de Fuca, between the mainland and Vancouver Island. The result was the bizarre "Pig War" of 1859, the only casualty being a British pig barbecued by American marines.

Conventionally most historians relate the beginnings of Anglo-American rapprochement and thus the modern "special relationship" to the late 1890s, although it really depends on the Treaty of Washington of 1871 for its origins. This conventional interpretation links rapprochement to Imperial Germany's naval ambitions and the passing of the five German Naval Laws between 1898 and 1912, the most worrisome of which was the second of 1900, which together with the 1898 law approved the construction of a High Seas Fleet intended to ensure that Britain would not "risk" going to war with Germany. There are certainly good grounds for accepting that the potential challenge posed by the High Seas Fleet caused Britain to abandon both its near-century long policy of "splendid isolation" and the use of the "two-power standard" by which to measure its naval supremacy.

In support of this focus on the German Naval Laws is the fact that, starting in 1901 Britain clearly restructured its geosecurity arrangements with the ultimate goal of concentrating its far-flung fleet back into the North Sea to oppose the building High Seas Fleet. In 1901 the Hay–Pauncefote Treaty with America nullified the Clayton–Bulwer Treaty of 1850 and gave America the right to build a canal across the Isthmus of Panama free of international control. An American navy free to move between the Atlantic and the Pacific reduced the need for British ships to patrol Pacific waters. In 1902 Britain entered into a formal naval alliance with Japan that allowed British possessions in the Pacific to be defended by a Japanese navy that would show its mettle when it destroyed a Russian fleet in the Russo–Japanese War of 1904–1905. 1904's *Entente Cordiale* saw Britain moving closer to its former enemy, France, the second ranked naval power of the period. Finally, in 1907 Britain effectively ended the "Great Game" with Russia by agreeing to a solidifying of boundaries in disputed areas of central Asia—Afghanistan and Tibet—as well as to the de facto partition of Persia into British and Russian spheres of interest and a neutral zone.

There are several periods after 1861 where Britain either accommodated the rise of American power or ceded power to America that are crucial to the peaceful nature of the hegemonic transition and which helped ensure that Wilson's prediction of "commercial savagery" was little more than political hyperbole, however accurately he outlined the nature of the technological and commercial struggle that defined that period of the transition that followed WWI. There were two important periods of this British accommodation: the American Civil War and its aftermath; and the buildup to and aftermath of the Spanish–American War.

The American Civil War and Its Aftermath

When the Civil War broke out many in the secessionist Confederate States of America and some in the North believed that Britain might well join the war on the side of the South to protect its access to the American cotton supply upon which the cotton manufacturing towns of Lancashire seemingly depended. Britain certainly felt a cold wind from the loss of the American supply, but its warehouses were jammed with the fruits of the last two cotton harvests before the Civil War broke out, by far the two best years yet. British cotton factors simply stopped reexporting cotton so it was Continental European mills that were devastated. And efforts to increase cotton exports from Egypt were also successful (Farnie 1979). In addition, there were many cultural reasons why such a materialist interpretation was not likely as well as incorrect, and there is no evidence in the British official records that any alliance with the Confederate States was even vaguely considered. Indeed, the diplomatic correspondence from the British envoy in Washington during much of the war, Lord Lyons, can be broken into three main parts: a deepening annoyance at the importuning of the Confederacy; frustration with his inability to fully trust Lincoln's secretary of state, William H. Seward, who Lyons suspected—correctly as it turned out—of having designs on British North America; and annoyance at the behavior of the more hawkish and anti-British members of Lincoln's government, such as Charles Sumner, the powerful and influential chair of the Senate Foreign Relations Committee.

Despite Lyons continual attempts to reassure Seward of British neutrality the North responded at times as if British involvement was possible. The single biggest fear was clearly of British naval involvement, especially in the aftermath of the *Trent* Affair of November 8, 1861, in which two Confederate diplomats who were on their way to seek diplomatic recognition of the Confederacy by Britain were removed from the British mail packet RMS *Trent* on the high seas by an American warship, the USS *San Jacinto*. The American claim was that the diplomats were "contraband of war" but this was a clear assault on British sovereignty and a substantial exchange of diplomatic notes ensued, forcing the North to release the diplomats, although no apology was ever issued. The naval problem for the North was that in 1861 Britain possessed the two most powerful warships on the planet, the fast, maneuverable, steam-powered, iron hulled, and heavily armed frigates, *Warrior* and *Black Prince*. Had either of these ships turned up in a northern harbor in support of the Confederacy the Union Navy's wooden warships would have been utterly defenseless. Even the crude, slow, hardly maneuverable, and hastily built Confederate ironclad *Virginia* demonstrated this at Hampton Roads in her first combat deployment on March 8, 1862. Only the arrival of the radical new Union ironclad, *Monitor*, the next day prevented further damage to the North's wooden warships and forced the clumsy *Virginia* to retreat. As recent research has shown, however, *Monitor* was as much a message to the British to keep their ironclads out of northern ports as it was the means of destruction of the *Virginia* (Fuller 2008). *Monitor* also represented the first of a radically new class of steam ironclad warship, henceforth called monitors, armed with very large and powerful guns mounted in turrets, in the case of *Monitor* two 11" Dahlgren guns. Such weaponry, which rapidly grew in size, was more than a match for the much lighter broadside armament of such ships as *Warrior*. Rotating turrets meant that a warship did not have to maneuver alongside an enemy: they could open fire whatever their position.

By far the most serious crisis that arose in relations between Britain and America in the entire transition from 1861 to 1947 was the row over the *Alabama* claims. In the early part of the Civil War the Confederate States purchased a number of commerce raiders from British shipyards, chief of which was the Confederate States' Ship (CSS) *Alabama*. The North argued that this sale was in violation of the neutrality laws of the period. The British Lord Chief Justice at the time denied the violation, but comments by Lyons to Seward indicate that the British Foreign Office had attempted to prevent the departure of the *Alabama* and was massively embarrassed by its escape. Queen Victoria herself signed a letter severely criticizing the Scottish yard that built *Alabama*. Throughout the war the *Alabama* issue featured prominently in Lyons' correspondence with Whitehall. *Alabama* alone captured or burned sixty-five Union merchant vessels and sank one Union warship. Other Confederate cruisers built in British yards were similarly effective: CSS *Shenandoah* captured or sank another thirty-eight Union ships. At the war's end the Union demanded compensation both for direct and indirect damages, the usual figure for direct damages being $18 million, the value of the ships and their cargoes.

There were also claims for indirect damages, the most extreme of which came from Senator Sumner, who argued that the *Alabama* and her sisters had extended the war by two years. The Senate Foreign Affairs Committee thus demanded $2.125 billion or, alternatively, the ceding of Canada. Secretary of State Seward took a middle ground. When he negotiated the Alaska Purchase of 1867, he saw it as a first step in a plan to gain control of the entire northwest Pacific Coast. Seward came to believe that Britain might be maneuvered into accepting the annexation of some, but not all, of Canada—British Columbia, the Red River Colony (Manitoba), and Nova Scotia were on his shopping list—in exchange for America's dropping the *Alabama* claims. This agitation for things Canadian peaked in 1870 with American expansionists, Canadian separatists, and British anti-imperialists seeming to combine forces, but was stalled by, among other things, a growing Canadian nationalist sentiment in British Columbia that emerged from the Act of Dominion of 1867 that had made Canada a virtually sovereign nation.

Initially Britain simply rejected the *Alabama* claims outright. But in 1866 America dispatched one of its newest and most powerful monitors, *Miantonomoh*, armed with four massive 15" Dahlgren guns in two turrets, across the Atlantic. *Miantonomoh* was one of four very similar vessels. *Monadnock* and *Agamenticus* were the first two completed, and *Monadnock* herself was sent round Cape Horn as a harbor defense vessel for San Francisco in late 1865. *Miantonomoh* and *Tonawanda* were slightly modified *Monadnocks* and all four ships were referred to as the *Miantonomoh* class. As contemporary accounts in such newspapers as the London *Times* made clear, the British had no ship capable of matching the *Miantonomoh* class ships, even *Warrior*. It would take the British until 1871 to launch *Devastation*, a classic "answer ship" capable of defeating the *Miantonomoh* class ships, and even longer to complete *Devastation's* sister, *Thunderer* and her half sister, the first *Dreadnought*. By that time the *Miantonomohs*, hastily built in the heat of war using green lumber in their frames, had been broken up and America had returned its focus to continental, not maritime, expansion (Sandler 1979; Fuller 2008; Hugill 2009c). But in the late 1860s the *Miantonomohs* still represented a powerful threat. Britain backed off its rejection of the *Alabama* claims and agreed to let them go to the first ever international court of arbitration as part of a comprehensive deal between Britain and America that, formalized in the Treaty of Washington of 1871, settled a large number of outstanding issues, many of them related to Canadian boundary problems. The Geneva-based court found in favor of America in 1872 and Britain accordingly paid the sum of $15.5 million in compensation for the *Alabama* claims, although money also changed hands in the opposite direction in compensation for some of the Canadian problems.

The peacefully legislated settlement of the *Alabama* claims and the Treaty of Washington set the tone for the resolution of all British–American geopolitical disputes going forward, but it contained the seeds of a major problem. Although Britain refused to admit that it had violated the Neutrality Acts it agreed that, henceforth, it would abide by them. This meant that, once a war had been declared, it would be illegal for a neutral power to sell war materiel to a belligerent power. In the run-up to the Spanish–American War British actions showed that they regarded naval fuel as war materiel (see below). The struggle that developed after 1919 over petroleum had, at its heart, the terms of the Treaty of Washington and Britain's need for a geopolitically secure source of oil for naval fuel, one within the Empire (see chapter 6).

The Buildup to the Spanish–American War

When the Spanish–American War broke out in 1898 American aggression toward Spain was almost universally condemned in Europe. The one country to support America, geostrategically as well as verbally, was Britain. Although it is reasonable to see this as heralding a sea change in American public opinion about Britain, inasmuch as the newspapers that molded American popular opinion fundamentally abandoned their anti-British position in 1898, that sea change clearly had an agent. That agent was the man who, first posted to Washington in 1889, became in 1893 Britain's first ambassador to the United States, Julian Pauncefote. Until 1893 America had stood on the high ground of its Revolutionary roots and refused ambassadors as clearly a mark of monarchical power.

In fact, three main cultural forces operating in the aftermath of the Civil War set the stage for Pauncefote's success and Britain's support of America's increasing ambitions on the global stage. These three cultural forces were followed by a set of events that can be divided into two periods, before and after the outbreak of the Spanish–American War.

The first cultural force was the realization by the American elite in the post-Civil War period, as American society began to polarize on ethnic lines through skyrocketing immigration, that their ethnic identity was English. This helped lead to the idea of the Anglo-Saxon Compact (Hugill 1986; 1989; 1999b). Second was the substantial increase in the number of wealthy Americans and British making transatlantic voyages as fast, reliable, safe steamships emerged in the later 1800s. For Americans this meant "grand tours" of Europe with England almost always the starting point (Hugill 1985). Third was the substantial increase in the number of transatlantic elite marriages. This was in part a result of the first and second forces and tended to take the form of the marriage of young American women from wealthy families to British aristocrats in need of an infusion of wealth—trading dollars for social status (Brandon 1980, though see Montgomery 1989 for a less materialist interpretation). The successful recent British television show, *Downton Abbey*, plays on this history. Winston Churchill's father, Randolph, had similarly married American money, in his case Jenny Jerome from New York.

Three events of significance in British–American relations preceded the Spanish American War: the Samoan Crises, the Behring Sea Arbitration, and the annexation of Hawai'i.

The Samoan Crises of 1887–1889 and 1898–1899

The first of these crises in particular pushed Britain into the realization that German companies in the Samoan islands were attempting to enslave the Samoans under the guise of supporting one side against the other in the First Samoan Civil War, and that support for the British resistance to that enslavement came strongly from America. In itself this might seem a very minor backwater of imperial history but the King of Hawai'i, David Kalakaua (ruled 1874–1891) offered to send military assistance to King Malietoa of Samoa, who was supported by America. The German chancellor, Bismarck, then threatened to invade Hawai'i, pushing the American secretary of state, James G. Blaine, into a strongly anti-German stance (Kennedy 1974). After the second crisis, and in the midst of the chaos of the Spanish–American War, the Samoan islands were divided between Germany and America with Britain receiving "the cession of German rights over Tonga and certain of the Solomon Islands" (Mowat 1929, 224).

The Behring Sea Arbitration, 1889–1893

This arbitration was a complex argument that grew out of the American purchase of Alaska and concerned the rights of Canadians to kill seals in what Britain and Canada claimed were international waters. America's claim was that, under the terms by which Alaska had been acquired, the Behring Sea was a "closed sea" and not international waters. After some mis-starts the Americans finally allowed it to go to an international court of arbitration. Some British observers were concerned about this, pointing to the British loss of the *Alabama* claims at such a court. In the event the court found in Britain and Canada's favor and America paid nearly half a million dollars in compensation for the Canadian fishing vessels they had seized. Some nationalist Americans objected to this, but it greatly improved British opinions of America and was generally accepted by the more responsible of America's elites as marking an important step forward in relations between Britain and America. As his biographer notes, it was the first major arbitration in which Pauncefote demonstrated his legal and diplomatic skills in successfully arguing Britain's case in America (Mowat 1929, 138–150).

The Hawaiian Annexation, 1893–1898

In 1843 Britain had briefly annexed the Kingdom of Hawai'i, but had returned it almost immediately to independence following American protests. The American note of June 13, 1843, lays out the centrality of the Hawaiian chain to developing American geostrategic interests in the Pacific basin and of American geostrategic thought with regard to territorial expansion along the Pacific Coast of North America. The acting American secretary of state at the time, Hugh Legaré, wrote "it is impossible to overrate the importance of the Hawaiian group as a stage in the long voyage between Asia and America. . . . It seems doubtful whether even the undisputed possession of the Oregon Territory and the use of the Columbia River or indeed anything short of the acquisition of California (if that were possible) would be sufficient indemnity to us for the loss of these harbors" (BNA/FO58/259.4.c). The British wrote back on August 15, 1843, that "the occupation of the islands by Lord George Paulet was a fortunate event, inasmuch as it prevented them from being taken possession of by a French squadron, which (it is said) was on its way for that purpose. Had France got possession of the islands, she would certainly have retained them" (BNA/FO58/259.4.d). On November 28, 1843, Britain and France jointly agreed "to consider the Sandwich Islands as an independent State, and never to take possession either directly, or under the title of Protectorate, or under any other form, of any part of the territory of which they are composed" (BNA/FO93/33/40A). The British offered America a chance to sign this treaty, but America demurred. In 1893 the American planter elite, concerned that the producers of beet sugar in the mainland states were about to rig American tariffs to freeze out Hawaiian cane sugar, deposed Queen Lili'uokalani, believing that the time was ripe for annexation. A secondary issue was that the lease on Pearl Harbor as a coaling station was up in 1894, although that had far more impact in 1898 when very real fears arose that the lack of a secure coaling station in mid-Pacific would be both fatal to American geopolitical ambitions in Asia and play into the hands of an aggressively expansionist Japan (Russ 1992, 217, 301). Unfortunately for Sanford Dole and his fellow planters Benjamin Harrison failed to hold the presidency in the 1892 election and was replaced by Grover Cleveland, a Democrat. The Democrats had no desire for foreign expansion, and the planters were forced to declare the Hawaiian Republic with Dole as its president. The Republic lasted until 1898 at which point the Republicans once again held the presidency and the events of the Spanish–American War persuaded Congress that annexation had become a geostrategic necessity. Even so, throughout the period 1893 to 1898 Britain clearly recognized American primacy over Hawai‘i.

With regard to documentation of British involvement in the Hawaiian issue the British Foreign Office records are regrettably incomplete for early 1893, there having been problems with the encryption and decryption of diplomatic telegrams at the time of the initial landing of American marines in 1893, the deposing of Queen Lili‘uokalani on January 17, and the subsequent creation of the Hawaiian Republic that spring. The documents in the British National Archives contain one cryptic reference by Pauncefote. On May 6, 1893, he wrote to Whitehall "Your Lordships telegram no. 23, Hawaii. The answer to your Lordships question is yes, but I venture to suggest that I be allowed to inform the United States Govt of the fact, and of the reason, in order to avoid the risk of misconstruction" (BNA/FO5/2189). Pauncefote's widow destroyed his personal papers after his death, the copy of the telegram sent to him has gone, and no copy of telegram 23 exists at the British National Archives. The tone of all Pauncefote's correspondence to Whitehall is highly positive with regard to America's increasing role in global affairs and there is no mention of the treaty of 1843.

The Spanish–American War

There were two key areas of support by Britain for America in the Spanish–American War. The first was the way in which the American Asiatic squadron re-coaled at Hong Kong. The second was the tacit support by the British of American action at the Battle of Manila Bay. The re-coaling is of interest since the British admiral at Hong Kong, learning that the Americans were about to declare war, went out of his way to warn the American commodore, George Dewey, that, in order not to violate the Treaty of Washington, he needed to acquire his coal supplies before war was declared. Commodore Dewey immediately purchased the British collier *Nanshan* complete with 3,300 tons of coal.

Second, at the Battle of Manila Bay, May 1, 1898, Captain Chichester, in command of the British squadron, placed his squadron between the German and American squadrons during the bombardment to prevent any German interference (Mowat 1929, 214). Dewey then lost his temper with the commander of the German Asiatic squadron, Otto von Diederichs, whose instructions from Berlin were ambiguous and whose warships at Manila Bay apparently ignored the American blockade. Dewey went so far as to threaten to fire on German warships for doing so (Ganz 1977, 128).

Kaiser Wilhelm was obsessed with vacuuming up as much of the collapsing Spanish empire as was possible: "I am determined, when the opportunity arises, to purchase or simply to take the Philippines from Spain" (quoted in Herwig 1976, 25). Even Queen Victoria seems to have agreed with her foolish grandchild on this occasion, condemning the American action in Cuba as "monstrous" (Herwig 1976, 26). The British Foreign Office paid no notice to their monarch and in Washington Pauncefote's instructions were clearly to support the American actions against Spain in every possible way. Wilhelm also wanted a naval base in the Caribbean and Admiral Eduard von Knorr, the commanding admiral of the Imperial German Navy, suggested one of the Dutch or Danish islands. All such suggestions fell apart for three main reasons: America's imperial ambitions; British support of such ambitions; and the manifest weakness of the Imperial German Navy. For Wilhelm the last was perhaps the central issue. Without a High Seas Fleet and a "system of German naval bases girdling the globe" Britain and America would always have the upper hand (Herwig 1976, 35). Wilhelm summarized the issue in 1899 when he told German chancellor Bernhard von Bülow, a committed imperialist, that "there would be no repeat of the present misery once the fleet was built. 'After twenty years, when it is ready, I will adopt a different tone'" (Herwig 1976, 34).

The Aftermath of the War

In the aftermath of the Spanish–American War relations between Britain and America continued to improve rapidly, with Pauncefote taking a major leadership role. In multiple areas Pauncefote moved Britain and America closer together, though nowhere more so than his negotiation of the Hay–Pauncefote Treaty of 1901. Pauncefote was, in almost all ways, the ideal man for the job. His first great diplomatic triumph had been in settling the questions surrounding Britain's purchase in 1875 of the shares of the Suez Canal owned by the Egyptian Khedive and amounting to 40 percent of the total, which resulted in Britain and France becoming joint owners. After several fairly serious international incidents and much concern over the extent to which the canal was international waters Pauncefote's diplomacy was central to the settlement reached in 1885. Pauncefote's "perfect knowledge of French, his complete equipment in international law, combined with an un-lawyer like elasticity, his moderation and straightforwardness, all made him excellent at dealing with the French delegates" (Mowat 1929, 85). Such qualities would later stand him in good stead in a Washington where many politicians were also lawyers by training.

On the American side the role of John Hay cannot be underestimated. From 1897 to 1898 he was the American ambassador to the Court of St. James, and he was as effective in improving the opinion of America in London as Pauncefote was in improving the opinion of Britain in Washington. Just after the Spanish–American War broke out Hay was recalled to Washington as secretary of state. Before leaving London he wrote to Senator Henry Cabot Lodge, the majority leader of the Senate, that British elite opinion of America had become so positive that "if we wanted it—which, of course, we do not—we could have the practical assistance of the British Navy" against Spain (Mowat 1929, 205).

More than any other British action the Hay–Pauncefote Treaty indicated the extent to which Britain was willing to accommodate America's geopolitical rise. Hay–Pauncefote resolved one of America's most serious geostrategic problems, being bordered by two oceans in an era dominated by the navalist geostrategic thinking of Admiral Alfred Thayer Mahan (Mahan 1890). Britain had long understood such geostrategic thinking: Mahan's book derived its main lessons from Dutch and British hegemonies founded on the use of naval power by those trading polities. In an age we describe as dominated by the "New Imperialism" in the aftermath of the Berlin Conference of 1884–1885 navies were clearly crucial to global power projection. By 1900 America's critical geopolitical boundaries were no longer in the Americas: they were with Europe and Asia, both a major ocean away. Absent a canal under American control, America would have needed to build and maintain two battlefleets in order to secure its interests in each ocean, an appallingly expensive solution. Such a canal had to be under American control if American warships were to be freely able to pass between the Pacific and the Atlantic in time of war. A canal under international control could have too easily been closed.

Pauncefote's role in the shift in British–American relations that resulted in Hay–Pauncefote was crucial. Upon his arrival in Washington in 1889 he found a city that suited him. His "manner of carrying out his diplomatic duties, a little formal but transparently honest, suited the political atmosphere of Washington, while his quiet and unaffected social agreeableness made him fit easily into Washington society" (Mowat 1929, 122). That social agreeableness brought him very close to many in the social elite of Washington, including Henry Adams. Adams and Hay were extremely close and had built their houses next door to each other on Lafayette Square in Washington: "Hay had no ally, abroad or at home except Pauncefote, and Adams always maintained Pauncefote alone pulled him through" (Mowat 1929, 130).

Wilson's "Rejection" of Rapprochement

So the question becomes, why was Woodrow Wilson in 1919 so convinced that "we are on the eve of a commercial war of the severest sort" with Britain? There is nothing in Wilson's academic career to make it obvious why he reached such a conclusion. Wilson admired Britain's parliamentary system of government, arguing for it in his book *Congressional Government* (1885) and against the complexity and dysfunction of the American system of checks and balances. But after he entered politics his anti-imperialist ideology took over. Wilson was also from the Commonwealth of Virginia, the first Southerner to become president since the disaster of the Civil War, a racist who believed fervently in white supremacy, and a president elected in part to roll back the nascent American Empire begun in 1898. Britain ruled a complex multiracial Empire that Wilson distrusted on the grounds of both anti-imperialism and that Empire's relative tolerance of racial, religious, and cultural differences. For Americans such as Wilson the system of Empires by which the world was ruled in 1914 created conflicting spheres of interest. America had risen by rejecting imperialism not once but twice and by creating a society predicated upon continually expanding its commerce and industry. That demanded order, not conflict or revolution.

Wilson's experience of the chaos wrought by the Mexican Revolution led him to be terrified of the prospect of more revolutions. He was particularly concerned by the apparent shoring up by the British of Mexico's Huerta government. The agent of this was the notoriously anti-American envoy to Mexico, Sir Lionel Cardin (Gardner 1984, 54). At Versailles it seemed to Wilson that Britain's prime minister, David Lloyd George, was much too willing to accommodate the Bolsheviks (Gardner 1984, 326). Wilson also believed that nationalism was all-important for self-determination and imperial structures suppressed nationalism (Tooze 2014). Finally Wilson had overseen the passage of the Federal Reserve Act in 1913 that raised New York's banks level with those of London for the first time. In foreseeing "commercial war" he must have been heavily influenced by New York's bankers.

Wilson's comments at Versailles provide the structure of the shift from British to American hegemony between 1919 and 1947. It is a shift that should, at least theoretically, have occurred in 1919 or so, not 1947, but it was a complex and multifaceted shift that was inadequately supported by an America that lacked the political will to step into British shoes. WWI had been, had it not, "the War to end all Wars?" In any case, of the powers that fought the war, by 1919 Germany was defeated and in economic ruin, France essentially bankrupt, Russia in Red Revolution and Civil War, no one really knew much about Japan, China was a chaotic mess, and Italy simply didn't matter. Only America and the British Empire were "left standing," and the latter would reach its maximum territorial extent with the "mandates" negotiated at Versailles. For Wilson, the British Empire still looked powerful and was the obvious problem for the continued expansion of American influence. He defined the conflict he saw coming in terms obvious at the time: in struggles over international transportation, international communications, and petroleum.

WWI had already brought these three struggles into stark relief. Britain and France needed American raw materials, especially petroleum after the collapse of Russia, to fight the war, and these had to be shipped. By the time America entered the war Britain had placed such massive orders for American built ships that they owned, on paper at least, the output of American slipways for several years. Wilson commandeered those slipways but did nothing to increase output (Tooze 2014). The British ability to control global telecommunications was obvious, as demonstrated by the British release of the Zimmerman telegram that helped give Wilson little option but to declare war on Germany. At Versailles Wilson instructed Rear Admiral Bullard to return to America and set up what would become the Radio Corporation of America (RCA) as a wireless alternative to Britain's submarine cable network (Hugill 1999a). Finally, WWI was the first oil war. In its aftermath American companies such as Standard Oil were on the lookout for supplies of oil outside America, the Middle East was the obvious target, and Britain and France had agreed to carve up the Middle East for their own ends by means of the Sykes–Picot(–Sazonov) Agreement of 1916.

Chapter 2

The Struggle over the Evolution of the Military and Merchant Marines

Throughout the period 1861 to 1947 global power hinged heavily on maritime power, although air power would become increasingly important in the 1930s and come into its own in WWII. Maritime power, which was almost exclusively held by trading states, was structured in two ways. Trading states grew wealthy on the flows of raw materials and food in and manufactured goods out that could be carried in their large merchant marines. Their navies protected that trade in times of war or interdicted that of opponents. The first trading states, Holland, then Britain, had learned by the 1600s that a standing navy was a necessity for the routine protection of merchant fleets against piracy, although until that fleet was large and powerful enough to encompass the globe some merchant ships, such as East Indiamen, continued to go armed since they sailed in dangerous waters. By 1861, however, the day of the armed merchantman was long over and the British navy was well able to protect the global flows of raw materials, food, and labor that made Britain the world's wealthiest state. In 1892 the American admiral, Alfred Thayer Mahan, examining the naval histories of Holland, Britain, France, and Spain, expounded what we can describe in modern terms as the first reasonably comprehensive theory of how a state achieves hegemony in the global system (Mahan 1892; Modelski & Thompson 1988). The Portuguese duke of Goa and second viceroy of India, Alfonso du Albuquerque, had said in the early 1500s that who rules the seas of the world rules the world, but that hardly amounted to a theory. For Mahan, a hegemonic state needed the world's most powerful navy, and it needed to structure its warfare along the lines of blockade of the opponent's ports, raids upon the opponent's commerce, and the destruction of the opponent's battlefleet in a climactic battle. Although Mahan did not say so directly, the clear implication of his work was that hegemony would also hinge on possession of the world's largest

Table 2.1 Tonnage of Merchant Shipping

Year	British Empire		American Ships in Foreign Trade		American as
	Steam Power	*Total*	*Steam Power*	*Total*	*% of British*
1850	187,631	4,232,962	44,942	1,585,711	37.5
1860	500,144	5,710,968	97,296	2,546,237	44.6
1861	561,023	5,895,369	102,608	2,642,648	44.8
1862	597,932	6,041,358	113,998	2,291,251	37.9
1863	657,026	6,624,403	133,214	2,026,114	30.6
1864	769,398	7,103,264	106,519	1,581,894	22.3
1865	902,052	7,322,604	98,008	1,602,583	21.9
1870	1,202,134	7,149,134	192,544	1,516,800	21.2

Source: Lindsay (1876, 619).

merchant marine, the trade of which would generate the state income to pay for the battlefleet needed to protect it and to act as the training ground for the large number of sailors such a fleet would need.

The maritime history of the American–British struggle is thus a central part of the long struggle for hegemony between the two powers that stretched from 1861 to 1947. A pivotal period in the maritime struggle was clearly the American Civil War. There were three consequences of this (table 2.1). The first was simply the loss of northern ships to the Confederate States navy which created the major international incident known as the *Alabama* claims: America demanded payment for the massive damages wrought by Confederate cruisers built in British yards (Hugill 2009c). This claim went to the first major international court and resulted in the payment to America by Britain of $15.5 million, basically the value of the northern merchant ships lost to such ships as CSS *Alabama* and the other Confederate States' cruisers, built in what America successfully claimed was a violation of the Neutrality Acts. The second was the substantial increase in the size of the British merchant fleet during the Civil War. The third was the disproportionate increase in the relative size of the steam-powered portion of the British merchant fleet over that of Americas.

In 1850 British imperial shipping amounted to well over 4 million tons. At 1.6 million tons American shipping engaged in foreign trade amounted to 37.5 percent of British. A further part of total American merchant shipping, not included in table 2.1, was in the coastal and internal trades on America's river and canal systems and on the Great Lakes. This always exceeded the tonnage in foreign trade, sometimes very substantially, being well over double in 1865. By 1860, although British tonnage had increased to nearly 6 million tons, the relative size of the American foreign trade fleet had increased to 44.6 percent of that of the British, and it continued to increase in 1861. From that point on, however, the size of the foreign trade component of the northern merchant fleet began to suffer from the serious depredations of such Confederate States' cruisers as CSS *Alabama* and her sisters. By the end of the Civil War absolute American tonnage was down almost to 1850 levels. Second, during the war the absolute tonnage of British merchant ships continued to increase, with over 1.5 million tons added between 1860 and 1865, at which point the American tonnage was reduced to just on 22 percent of British. Third was the much faster shift of the British merchant fleet to steam power than the American fleet. In 1850 just under 4.5 percent of British merchant tonnage was steam-powered. Ten years later that percentage had about doubled to just under 9 percent. But that rate of growth continued, and by 1865 the figure was over 12 percent and by 1870 it was at almost 17 percent. To some extent America caught up once the war ended, but in 1870 steam vessels accounted for not quite 13 percent of the American total. Steam, of course, made for much more efficient use of tonnage, since a steamship could make a significantly larger number of voyages per year than a sailing ship, and with vastly greater regularity, so that while British total tonnage ceased to increase between 1865 and 1870 annual carrying capacity increased substantially.

Since the Confederate ships that caused the damage were all built in British yards this could be seen as a British assault on American merchant shipping. Such thinking may well have lain behind the *Alabama* claims, although there is no evidence that it did, and no evidence in the diplomatic record that the British were other than substantially embarrassed by the whole set of events and deeply annoyed at the Confederate States as a result.

Britain's advantages by the mid-1800s were substantial: it could legitimately claim to be "the workshop of the world." Napoleon had earlier called the British a "nation of shopkeepers" without seeming to fully understand that the shopkeepers sold what their compatriots in the workshops manufactured. In the early 1800s Britain came to dominate not only global manufacturing but also the global carrying trade, in which industrial products, emigrant labor, and, later, coal were shipped out and raw materials and foodstuffs were shipped back in to the British Isles from almost every corner of the globe. Britain's share of global manufacturing slipped as the 1800s came to a close and Britain's role in the global economy shifted to more of the carrying trade: "World Economics is tending to slowly force us to close down some of our shops but to increase at the same time the number of our carriers' carts" (Hardy 1931, 1).

CHANGES IN MERCHANT SHIP TECHNOLOGY

In merchant shipping the technologies of construction and propulsion evolved with frightening speed, in particular between roughly 1900 and 1920. Over the longer period 1861 to 1947 wood construction gave way to wrought iron, then steel, then much better steel, the steels of the late 1800s and early 1900s becoming somewhat brittle in cold waters. The tragic loss of the Royal Mail Steamer (RMS) *Titanic* may well have been in part caused by the brittleness of its hull when it collided with the fatal iceberg. In propulsion technologies sail gave way to a variety of increasingly complex steam engines that in turn gave way to a variety of forms of the compression ignition internal combustion engine patented by Rudolf Diesel on the basis of his thorough theoretical understanding of thermodynamics in 1898. Paddle wheels gave way to screws driven direct by reciprocating steam engines, then driven via gearing or electric motors by much faster rotating steam turbines, then direct again by diesel engines. Direct drive was always more efficient. To complicate matters further, the new propulsion systems required new energy sources, first wood, then coal, then oil, whether to generate steam or burned direct in diesels. The geographic distribution of these fuels was not the same. Most countries that had gone through the first industrial revolution had adequate access to wood as well as domestic reserves of coal, but oil was much less common. In terms of the comments made by Woodrow Wilson at Versailles in 1919 international maritime transportation and petroleum were intimately related arenas of struggle, even more so as oil replaced coal and diesel engine motor ships proliferated.

Steam power was initially adopted more rapidly in merchant shipping than in navies. By the early 1800s the world's fastest growing and most vibrant economies surrounded the North Atlantic basin, creating a lucrative market for the rapid movement of mail and financial information. In 1816 the first packet line, the American Black Ball Line, opened for business on the North Atlantic. Unlike the vast majority of sailing vessels, sailing packets sailed on a set schedule, come rain or come shine. Their ships were of the highest quality, built for speed, with their captains and crew selected for their aggression and ability; they quickly captured the lucrative trade in mail and financial information. Like all sailing ships their passage times could vary and however aggressively and well-handled they were always somewhat at the mercy of the weather. Most such ships advertised a three-week passage time, but only rarely achieved it. Steam-powered ships had huge advantages. They were in little or no danger, for example, of being blown on a lee shore, or having to stand off in stormy weather before entering a harbor. Almost as soon as transatlantic range steamships emerged they captured the lucrative packet trade in mail and financial information. They quickly showed their ability to cross the Atlantic in almost any weather in two weeks or less, and with remarkable regularity compared to sailing ships. Their vastly better safety record made them particularly attractive to well-heeled passengers, who paid up to twenty times the fares charged on sailing packets. The initial adoption of steam power for long distance merchant ships was, however, geographically restricted to the all-important Atlantic shuttle, where the need for speed was paramount, regularity and safety were all-important, range requirements were not too great, and a ship could refuel on either side.

Three ships from one British designer, Isambard Kingdom Brunel, epitomize the technological changes in merchant ship design that preceded the American Civil War. The first passenger ship with transatlantic range was Brunel's *Great Western* of 1838. It was wooden-hulled, albeit of the most advanced construction possible with wood, and therefore size limited, although it could carry enough coal to comfortably cross the Atlantic, which it did in fourteen days. As a single example, however advanced, it was not a secure basis for a regular service. In any case Brunel's restless engineering mind had already moved on to bigger and better things, and he was never able to focus on shipping alone: he was as comfortable and successful designing bridges, tunnels, and railroads. Brunel's second great steamship, the screw-driven *Great Britain* of 1843, was wrought iron-hulled and large enough to comfortably carry enough coal to have better than transatlantic range, but like *Great Western* it was a one-off. In Brunel's last great ship, the giant wrought iron *Great Eastern* of 1858, he attempted a steamship so large it would carry enough coal to literally encompass the globe, although like his other ships sail was an auxiliary power source (Hugill 1993, 134). The engines, boilers, and knowledge of fluid dynamics of the time were simply not up to Brunel's vision for *Great Eastern*. William Froude's work on hull shape was not properly understood until the 1870s, Brunel in particular preferring to use John Scott Russell's "waveline" theory, which made *Great Eastern*'s hull shape much less efficient than it could have been. Fuel consumption was, accordingly, too high. Osborne Reynolds did not begin publishing his pioneering theoretical work on fluid dynamics until the early 1870s and then it was only fully adopted by ship builders in the 1890s. As ships grew in size such understanding was crucial to increases in speed and propulsion efficiency. *Great Eastern* was a leap too far in the 1860s, no similar size ship being built until the early 1900s. It was thus a desperate failure as a passenger ship, although its immense size gave it an important second career in the 1860s and 1870s as the world's first cable-laying ship (Hugill 1999a, 29, 33). None of Brunel's ships made profits so despite Brunel's immense technological virtuosity it was the Canadian entrepreneur, Samuel Cunard, who first understood how to make money out of Atlantic steamships. Cunard emphasized safety and built a fleet of simple, sturdy ships of similar design and performance to *Great Western*. He was thus able to offer a much more frequent service, and in 1840 he obtained the all-important British government mail contract.

The Adoption of Steam Power by Navies

Navies were initially much more reluctant to adopt steam power than merchant shipowners, since steam ships needed geostrategically secure coaling ports. With no need for fuel, sailing ships had effectively unlimited range, although their wooden hulls needed expensive copperbottoms or regular and significant attention to prevent fouling and serious loss of speed. In this respect, and even though the French and British navies experimented with steam power and developed steam-powered ironclads just before the American Civil War, such ships were no more technologically advanced than *Great Britain* from nearly twenty years earlier. But war, as it often is, was the crucial technology forcer. In *Monitor*, commissioned in 1862 just before the crucial Battle of Hampton Roads, the Union Navy pioneered a powerful, big gun warship powered solely by steam, albeit with short range. Both the French *Gloire*, commissioned in 1860, and the British *Warrior*, commissioned in 1861 were, effectively, global range sailing ships with steam auxiliary engines to allow them to maneuver freely in combat situations. Their total broadside weight was high, but by no means comparable in hitting power with that of the two 11" Dahlgren guns carried by *Monitor*. Despite the many shortcomings that eventuated from its hasty design and construction, *Monitor* was a radical new type of warship that gave its name to a whole new class of warships and, effectively, ushered in the age of the big gun battleship.

There was never any real naval conflict between Britain and America once the War of 1812 had ended, although *Monitor* can be construed as a response to American naval fears of Britain at the time the Civil War broke out (Fuller 2008). Certainly the republic of the North, the Union, was gravely concerned about Britain's potential to intervene militarily in support of the cotton South, although in the event such intervention was likely only in the eyes of the radical Republicans or the more deluded representatives of the Confederacy. The main Union worry was the theoretical possibility of a British steam-powered ironclad, *Warrior*, or her sister, *Black Prince*, entering a northern harbor on behalf of the cotton South and laying waste to a union navy composed entirely of wooden warships. British documents show no such interest in intervening on the side of the Confederacy: indeed the correspondence between Lord Lyons, the British envoy in Washington from 1858 to 1865, and the British Foreign Office shows increasing impatience with representatives of the Confederacy and no willingness at all to support it much past the time of the Trent Affair of 1861. However, the exceedingly brief career of the Confederate ironclad, *Virginia*, hastily and incompetently converted from the Federal warship *Merrimack* after the Union had burned her to the waterline to prevent her falling into Confederate hands, showed at the Battle of Hampton Roads that a well-handled, heavily armed British ironclad would certainly have been capable of immense damage to the northern warships harbored there. In the event, *Monitor* arrived in time to prevent *Virginia* from more than a single day of destruction, and *Virginia* was forced to withdraw from Hampton Roads. In this respect, *Monitor* was seen by the Union as much a message to Britain to stay out of the conflict as she was a threat to the Confederacy (Fuller 2008).

By the end of the Civil War monitor class ships had matured into vastly more powerful and capable warships, such as the USS *Miantonomoh*, commissioned in 1865. *Miantonomoh*, with its four powerful 15" Dahlgren guns in two turrets, caused a panic when it appeared in Britain in 1866, pushing the British government into reconsidering its initial refusal to entertain the *Alabama* claims. The claims eventually went to international arbitration in

Geneva and resulted in Britain paying America $15.5 million and signing the Treaty of Washington of 1871 in which Britain effectively agreed to abide in future by the Neutrality Acts. *Miantonomoh* also pushed the Royal Navy (RN) into creating in Her Majesty's Ship (HMS) *Devastation* a classic "answer ship" (Hugill 2009a). *Devastation* was commissioned in 1873 as the first all steam warship with transatlantic range. Ericsson, the designer of *Monitor*, said of *Devastation* that she would be able to "steam up the Hudson in spite of our batteries and our monitors, and dictate terms off Castle Garden" (quoted in Sandler 1979, 243).

After *Devastation*, her sister *Thunderer*, commissioned in 1877, and her half-sister *Dreadnought*, commissioned in 1879, concerns over coaling stations pushed the British navy to switch back to a variety of experiments combining sail and steam propulsion. Some experiments, such as the wildly unstable HMS *Captain* that combined turrets with sails and steam, ended with tragic loss of life, and the late Victorian British navies were aptly described as "fleets of samples" since no obvious technological solution to the range problem emerged quickly (Daveluy 1910, 255). By the late 1800s, however, the development of the much more energy efficient triple-expansion engine initially used in merchant steam ships and, alongside it, the emergence of a global network of coaling stations, allowed the RN to move back to a complete dependence on steam power. Removing sails as a means of propulsion literally "cleared the decks" for a return to turret mounted guns of the sort favored on such pioneering ships as *Monitor*, *Miantonomoh*, and *Devastation*.

A global network of coaling stations and the triple-expansion engine set the stage for the emergence of *Royal Sovereign*, commissioned in 1892, the first of a class of eight ships that were the world's first globally capable, steam-powered, heavily armored, turreted battleships. Guns, armor, weapons, weapons mounts, and power systems all improved rapidly from the 1880s as the Neotechnic period of the second industrial revolution developed. The *Royal Sovereigns* exemplified all these improvements. Better metallurgy allowed both better armor and better weapons, guns that could use much more powerful explosives without being too heavy or themselves exploding. Top-heavy turrets derived from ships like *Monitor* gave way to barbette mountings that lowered metacentric height and thus increased safety in heavy seas. The revolution in organic chemistry pioneered in Imperial Germany in the late 1800s allowed far more powerful propellants and explosives. The range of the most powerful naval guns rose quickly from two to three miles to five, then ten, then twenty, and ultimately, by WWII, thirty. Such ranges required the development of whole new weapons systems, including sophisticated rangefinders and the beginnings of what we can now see as primitive ancestors of computers, designed to predict where a shell traveling twenty to thirty miles fired from a ship moving at twenty or more knots could expect to meet another ship moving at twenty or more knots. By WWII the need to fire at targets concealed by smoke, bad weather, or darkness had pushed navies into the innovation of radar direction of gunfire. Better, multi-tube boilers allowed much more steam to be generated to drive triple and then quadruple-expansion engines that radically increased engine power over the ships of the 1860s and 1870s. By the early 1900s steam turbines allowed even more power, and oil fuel made that power far more usable since delivering it to the boilers did not require human labor that quickly tired of stoking massive boilers with coal when a ship needed to be kept at combat speed.

At the first genuinely modern naval battle, at Tsushima Straits in the Russo-Japanese War in May 1905, both fleets opened fire at a range of five miles. Superior Japanese range finding and tactics, combined with the high quality of a navy built largely in Britain and a flagship, *Mikasa*, that was a modified and updated *Royal Sovereign* class ship, brought an easy and convincing Japanese victory. The lessons of Tsushima Straits spread rapidly, since every naval power had observers present. Although a competent ship, *Mikasa* suffered from mixed armament and its range finding could not accurately keep pace with the shell falls from guns of different sizes. It was obviously better to fit a battleship with only one size of guns so that range finding only had to deal with one prediction, a conclusion that led quite clearly to HMS *Dreadnought* of 1906 with its armament of all 12" guns as well as to new classes of dreadnought battleships in all the world's navies.

Yet too much of the focus in the literature is on dreadnought battleships. Many other radical new forms of naval warfare had emerged by WWI. Most of these had their origins in the American Civil War, as whole new classes of ships were developed to defend harbors, attack merchant ships, or attack existing battlefleets, increasingly driven by the changes in technology exemplified by the Neotechnic. Several important new classes of ships were developed between the American Civil War and WWI: monitors, submersibles, torpedo boats, and torpedo boat destroyers, as well as dreadnought battleships. Monitors and torpedo boats proved themselves in the American Civil War, which also saw the first successful use of submersibles, albeit at high cost to their crews. The high-speed of torpedo boats necessitated a larger, more powerful, better-armed boat to destroy them, the torpedo boat destroyer, later simply just called a destroyer. The immediately post-Civil War British response to American monitors embodied in *Devastation*, *Thunderer*, and *Dreadnought* prefigured the all-big gun dreadnought battleship that emerged when the second HMS *Dreadnought* was commissioned in 1906.

The last addition to this group of new warship types of the period 1861 to 1947, the aircraft carrier, was developed during WWI, although it did not come into its own until WWII, at which point it displaced the battleship from its position of chief arbiter of global power. In many ways, and despite the importance of the nuclear submarine, the aircraft carrier still plays this role to this day. Although the first aircraft carrier was British, it was the American and Japanese navies that developed the concept of carrier aviation in a serious way between 1920 and 1941. The British clearly believed they would be able to rely on land bases for their air power and tended to relegate carriers to a secondary role, even relying on the land-based Royal Air Force (RAF) to procure planes to serve on British carriers, a nearly fatal error since the requirements of the RAF were antithetical to those of the RN. America and Japan knew they were likely to confront each other on the essentially trackless wastes of the Pacific, an ocean covering half the planet, and that carrier air power would thus be crucial. Between 1920 and 1941, when hot war finally erupted at Pearl Harbor, both America and Japan put a huge amount of effort into developing effective carriers, airplanes, weapons, and tactics. The lesson of WWII, most forcibly brought home at the climactic Battle of Midway, was that it was the aircraft, weapons, and tactics that mattered most, not the ships (see chapter 3).

In the struggles between Britain and America that are central to this book, the three outlined by Wilson at Versailles, naval struggles are, however, irrelevant, if interesting. In the German Wars Britain and America would fight side by side. Despite occasional glitches their two navies cooperated extensively. In WWII, for example, British naval aviation was fundamentally saved by the provision of excellent American naval aircraft to serve on British carriers in lieu of the ineffective types that had been procured for the RN by the RAF. The American navy in its turn profited greatly from British radar, especially at such crucial battles as Midway. There was, quite simply, no naval struggle between the two trading states.

The Real Struggle: Merchant Shipping

The struggle between America and Britain over merchant shipping was, however, far more substantial, interesting, and complex. It extended throughout the period 1861–1947, and although there were periods of American equality, Britain was usually dominant, a fact recognized by Wilson at Versailles. In the 1880s the development in Britain of cheap steel and improved triple-compound steam engines allowed the construction of merchant ships that could be operated as or more cheaply than the best sailing ships of the period, and which reached their destinations much more reliably and on schedule.

Despite the advantages of steam propulsion to shipping its problems were manifold. The first application of steam to merchant shipping was on the rivers of the world, where there was plenty of fresh water for boilers and easy access to supplies of wood for fuel along the river's banks. Such use exploded on the rivers of the American South after the application in 1817 by Henry Shreve of high-pressure engines to the stern wheeled keelboat on the Mississippi. The use of steam for open-ocean navigation required a way of recycling fresh water, more fuel-efficient engines, and hulls large enough to carry copious amounts of fuel. *Great Western* was capable of making the Atlantic voyage in 1838 in part because it pioneered the use of Hall's surface condenser and Brunel built her hull using the system of diagonal iron riders developed for British warships by Robert Seppings, which made it possible to build a wooden-hulled ship to a much more generous scale than if construction had been in wood alone. Coal fuel reduced the amount of bunker space needed.

Engines and boilers remained a problem. The working pressure in marine engines doubled in the 1840s and doubled again in the 1850s to around thirty pounds per square inch (psi), but then skyrocketed past 200 psi as rapidly improving steel quality allowed much better boilers by the 1880s. Working the steam expansively became the norm in the 1860s, tripling power output at the same time it reduced fuel consumption by a third. Compounding the engine, which allowed the steam to expand in a second, then a third, and ultimately a fourth cylinder, gave even greater fuel savings, but its engineering complexity made the period of adoption stretch from a first use of compounding in 1854 to full adoption only in the mid-1880s. By the late 1880s the steel-hulled triple-compound steam engine vessel could compete on cost terms with the finest steel-hulled clipper ships, a remarkable achievement. Even so, the triple-compound engine was still a relatively inefficient energy converter requiring frequent bunkering and thus the establishment of a global chain of coaling stations.

By the late 1880s the British had created just such a chain of coaling stations, supplied by colliers from Britain. Whereas in 1840 coal accounted for some 1.4 million tons of a total British overseas trade of 20 million tons by 1887 it was close to 50 million out of a total of 140 million tons (Craig 1980, 18). This coal trade, mostly out of the coalfields of South Wales, increased the profit to Britain of a global trading system that was moving rapidly to dependence upon merchant ships built almost exclusively in British shipyards, most of them lined along the River Clyde around the city of Glasgow. Such ships were not really possible until several innovations combined to allow them in the 1880s. The Thomas and Gilchrist technology made mild steel much cheaper and was finally adopted in shipping in 1879 with the construction of *Rotomahana.* Over the next ten years annual tonnage produced along the Clyde jumped from 18,000 to over 320,000 tons. Despite experiments in the 1870s the first successful triple-expansion engined deep-sea ship was not launched until 1882 (Hall 1998, 348–377).

Financial aspects of the rapidly increasing global trade were also important and two forms of financial organization of the shipping industry developed. As late as the early 1880s the ownership of many British merchant ships still operated much as it had in the Dutch Republic of the 1500s, on the "64th" system. Most 64ths were owned by fairly small investors and ship operators themselves, so that the 64th system effectively spread risk. Few of these ships operated on fixed schedules or routes and are thus referred to as "tramps." The return on capital for such ships in the 1870s was routinely 20 percent–25 percent and could reach as high as 40 percent. 64ths sold easily and were routinely mortgaged. Risk was unlimited in such a system, so investors routinely spread their investment over several ships, although "managing owners" routinely owned 4/64ths and many owners had a considerable sense of identity with the ships in which they had partial ownership. But as ship size and cost increased the tramp system began to give way to ships running on fixed routes and schedules, called "liners." Liners needed much higher levels of capitalization and were better managed by much larger, more formal corporations. As British laws changed in the late 1800s these were better operated as limited liability joint stock companies and used insurance via such groups as Lloyds to cover their risks (Craig 1980, 40–41). In the case of either tramps or liners, high returns encouraged many British investors to enter the market, and it was generally easy to finance the expansion of the global carrying trade in Britain in a way that was not easily matched in an America where capital was less easily available and investments closer to home tended to be preferred. For example, a typical wealthy American family in the post-Civil War period kept its investments in land, especially land for urban growth, energy, industrialization and, as they increased in importance, railroads (Hugill 1995, 90–93).

Steam turbines, developed late in the 1800s, promised even greater efficiency than reciprocating engines, but ran at very much higher rotational speeds. Without gearing that meant that screw propellers ran much faster than was efficient. In any case, speed was never a priority in general purpose cargo ships, where fuel economy was important to maximize profit. For example, a typical 6,000 ton triple-expansion engined steamer of 1904 would consume 20 tons of coal per day at a speed of 9 knots, 25 at 10 knots, and 32 to 35 tons at 11 knots (Craig 1980, 34). Ungeared turbines worked best in large ships that needed a great deal of power, such as high-speed passenger liners or capital ships: they were a costly and complex solution that only made sense when the need for speed was paramount. To achieve maximum efficiency turbines had to be either equipped with heavy, complex, and expensive gear trains to reduce the speed of their propellers or used to generate electricity that in turn drove electric motors that ran at the appropriate speed for efficient propulsion. Neither was a technologically easy solution. Geared turbines were not adopted for the basic merchant ship until the teens of the new century. Turboelectric drives, though excellent in theory, were never easy to apply in the damp environment of steamships, and tended to be limited to a few classes of highly specialized vessels, such as very powerful tugs where very large variations in power demand were normal. Since tugs were constantly in port the high maintenance levels needed by turboelectric drives were more easily taken care of.

Like sailing ships, wood or coal-fueled steamships required a substantial number of crew to deal with propulsion. Steamships needed many stokers to keep their boilers fed. More crew meant more space was needed to accommodate them and more provisions to feed them, which meant that less cargo could be carried. The use of oil fuel allowed a radical reduction in the number of stokers, and thus marks the last major improvement in the efficiency of the steam-powered merchant ship. For example, when the British tanker *Strombus*, built in 1900, was converted to oil fuel "six men were able to do the work of the 26 normally employed in the stokehold." In addition fuel consumption dropped from 42–43 tons of Welsh coal per day to 28–30 tons of oil (Craig 1980, 17). Despite such advantages, oil turned out to be a far more useful fuel when burned directly in propulsion than when burned under boilers to generate steam. The last great jump in the fuel efficiency of the merchant ship thus came with the application of Rudolph Diesel's revolutionary engine.

For almost all merchant vessels the diesel or heavy oil engine was a far better solution than the steam turbine. It was simple, easy to maintain if difficult to manufacture, and much more efficient than any steam engine. In addition, oil fuel has a higher energy density than coal, so range was greater for a given weight of fuel, and bunkering was radically easier than bunkering a coal-fired ship. The first effective seagoing diesel engine ship was the Danish *Selandia* of 1912, built for the Far Eastern trade and powered by two 1,250 horsepower engines. She "embodied so many novel features that she might well be regarded as one of the most original ships ever built" (Craig 1980, 17). Diesels were, at least initially, adopted for range: "the need to provide space for considerable quantities of coal . . . had an adverse effect on the earning capacity of ships making lengthy passages" (Maclennan 1994, 158). Diesel engines also turned slowly enough that they needed no reduction gearing and could drive their propulsion screws direct.

Merchant Shipping and the Oil "Problem"

President Wilson's evaluation of the global technological situation at Versailles was prophetic, and two of his three issues, international transportation and petroleum supply, turned out to be inextricably interlinked. The economic struggle between Britain and America that had begun with the

Morrill Tariff of 1861 and that was seemingly damped down by the period of rapprochement leading up to WWI broke out with renewed vigor after Versailles, at which point only two powerful actors seemed left standing on the global stage: an up-and-coming America and a British Empire acquiring significant new territories under League of Nation's mandate and not yet obviously in decline, however "obvious" that decline has been in retrospect. Wilson had always seen maritime transportation as crucial to the struggle over international transportation, stating at the beginning of the war "we have grossly erred in the way we have stunted and hindered the development of our merchant marine . . . and now when we need ships, we have not got them" (quoted in Denny 1930, 350).

Unfortunately Wilson's understanding of the ways in which international transportation would develop from 1919 to 1947 failed in four crucial areas. First, he lacked understanding of the remarkable technological changes then underway in the field of marine propulsion systems as triple-expansion steam engines gave way to steam turbines and, far more significantly, to diesel engines. In today's world merchant ships are almost exclusively powered by diesel engines (Smil 2010). Although many modern warships now use mixed propulsion systems with oil-fueled gas turbines for combat speed and diesels for cruising, only a very few passenger ships on time critical routes such as the Atlantic use, like warships, gas turbines alongside their diesels to provide high-speed when needed.

Wilson's second failure was that, in 1920, he signed into law the protectionist Jones Act which required that all goods transported by water between American ports had to be in American flagged, constructed, owned, and crewed ships. Such water routes were usually short-haul, around the American coast, through the Panama Canal, and throughout the Caribbean, with long runs only to Hawaii and the Philippines, routes where refueling was easy and the long range of diesels scarcely mattered. The Jones Act could not protect American shipping on long hauls in international waters where the diesel engine came into its own. By encouraging American shipowners to stay with steam technology it actually worked against America's long-term interests. America "is a great maritime nation within her own tariff walls. Outside she can never be" (Hardy 1931, 74).

Wilson's third failure was that he completely failed to see the rapidly increasing importance of aerial transportation. As it turned out this was the component of international transportation in which, by 1947, America would convincingly out-perform Britain (Hugill 2009a).

But Wilson's fourth and greatest failure was really in his failure to grasp the intimate connection that had emerged between international transportation technology and petroleum supply in the first twenty years of the century.

In 1919 he believed America would continue to dominate the arena of world petroleum supply just as it had since the late 1860s. In fact, by the late teens many observers had begun to believe that the global supply of petroleum was insecure and were theorizing that American "oil exhaustion" was on the horizon, especially in an America that, according to Canadian banker E. Mackay Edgar in an influential comment, had "recklessly and in sixty years run through a legacy that, properly conserved, should have lasted her for at least a century and a half" (Edgar 1919). To some observers at the time the link between the struggles over maritime transportation and petroleum supply was clear: "Britain waged an exhausting war with the main object of ruining German shipping for ever. . . . Yet now, from the very war which destroyed that competitor, a new one has arisen, twice as formidable as the old, for, in addition to a superiority in tonnage, it enjoys the practical monopoly of a fuel of which Britain has none" (Delaisi 1922, 21).

Global Oil Supply Issues, 1901–1945

This section encompasses three main time periods: from 1901 to around 1919, from 1919 to around 1931, and from 1931 to 1945. It summarizes briefly the much more complete account in chapter 6.

In the first period American production predominated, Russia and Indonesia were important secondary sources, and most contemporaries believed that the Middle East promised a huge though unproven bonanza. The Bolshevik Revolution withdrew Russian oil from the world market after 1917.

In the second period, fears of American "oil exhaustion" arose, especially as the automobile continued to accelerate domestic demand for what at the time seemed a limited and increasingly uncertain supply. Throughout the 1920s there was an intense policy struggle over access to the Middle East, which precluded much production but seemed to be resolved largely in Britain's favor both by a series of treaties settling the disposition of the former Ottoman possessions in the region and by the Red-Line and Achnacarry agreements of 1928 between the major oil producers (Yergin 1991). The main American company to sign these agreements was Standard Oil of New Jersey.

In the third period the discovery of major new fields in Texas allowed American domestic production to skyrocket, serious production began in the Middle East, and new technologies exploited by a company that was not a party to the Red-Line Agreement gave an American company access to Saudi Arabia's massive reserves. This amounted by 1945 to a massive American end run around Britain's previous and largely policy driven dominance of Middle Eastern oil reserves.

Oil Issues, 1901–1919

By 1901 all the major industrial states had recognized that petroleum was the coming source of power, especially for their navies, but only America and Russia had proven domestic supplies. Through exploitation of its resources in Pennsylvania, California, Oklahoma, and Texas, and through its invocation of the Monroe doctrine and the Roosevelt corollary to control the increasingly important resources of Mexico and Venezuela, America completely dominated the global oil supply at the moment WWI broke out. Only the Russian fields around Baku seemed competitive, and only briefly, around 1900. Indonesia, controlled by the Dutch, was a third, much smaller supplier, but oil from the region was shipped by Shell, a British company that got its start shipping exotic seashells for European collectors. The other powers, in particular Britain and Germany, thus began to search for geostrategically secure sources of oil.

Well before WWI broke out Germany and Britain began to focus on the likely oil reserves of the Middle East that neither America nor Russia could control. The area was already geostrategically important to Britain since the main maritime route from Europe to Asia ran through it via the Suez Canal. Before oil was discovered in the region Germany had begun to contest control of the region with Britain. This contestation was sharpened by the increasing incompetence and corruption of a collapsing Ottoman Empire that, nominally at least, controlled all of the Middle East except Persia, itself an area of contestation between Britain and Russia as part of the "Great Game." Persia was effectively removed from that contestation when Russia and Britain agreed to the partition of the country into a Russian "sphere of influence," a British sphere, and a neutral zone, in 1907.

The collapsing Ottoman Empire was another matter. By 1900 it was pretty obvious that several Ottoman provinces held significant amounts of oil. British military officers reported on this to the Royal Geographical Society as early as 1886 (Ardagh 1886; Stewart 1886). By 1901 a German technical commission would describe Mesopotamia as a "lake of petroleum" which it would be advisable to develop the break the stranglehold on world supplies held by the Americans (Earle 1923, 15). In 1903 Germany negotiated a ninety-nine-year lease of mineral rights along the line of the Berlin to Baghdad Railway (BBR) concession in Ottoman territory, though the BBR never fulfilled its obligations to the Ottoman government and thus was technically not entitled to those mineral rights. These rights focused on the three Ottoman provinces that made up Mesopotamia: Mosul, Baghdad, and Basra, modern Iraq. As the Ottoman Empire collapsed and was superseded by the Turkish state, and because its rights to exploit the region's minerals under the 1903 concession were in dispute, the German interest in BBR evolved into a partnership with Britain in the Turkish Petroleum Company (TPC) of 1912, but the Turks had not approved any renewal of the 1903 concession before war began in 1914 (Earle 1924).

Britain's interests were more convoluted. They began with the D'Arcy concession of 1901 from the Shah of Persia, which evolved after the 1907 rapprochement with Russia into the Anglo-Persian Oil Company (APOC) of 1908, then ran through the effective merger of APOC with TPC and BBR in 1912 into the secret agreement signed by the British, French, and Russian Foreign Ministers, Sykes, Picot, and Sazonov, in 1916 dividing the former Ottoman provinces among Britain, France, and Russia when the war ended. When Russia left the war the Sykes-Picot-Sazanoff agreement became Sykes-Picot, and it amounted to a French agreement to cede to Britain the Ottoman provinces of Baghdad and Basra and a significant part of Mosul, including all its oil resources, expanded just before Versailles to all of Mosul. This resulted in the creation of the state of Iraq under League of Nations mandate after the war was over and, after Britain took the TPC as a war prize, the alignment of APOC with TPC.

The massive strike at Spindletop in Texas in 1901 secured America's position as the dominant global supplier, a position it would hold through 1945 and beyond. Before Spindletop the early American oilfields were, for the most part, marketed by Standard Oil and its associated companies. But Standard Oil failed to see Spindletop's potential and, in fact, thought of it as a fluke. The company's geologists, all trained as hard-rock mining engineers, believed oil was only found trapped in the folds of porous sandstones or in fractured rocks such as limestones. Spindletop's oil was trapped in a salt dome and not until the 1930s did geologists at Louisiana State University (Louisiana State University Geology online) and companies such as the California Standard Oil Company (CASOC), later renamed Standard Oil of California (SOCAL), begin to understand such structures. CASOC had been broken away from Standard Oil proper by antitrust legislation in 1911 and had gone very much its own way.

Oil Issues, 1919–1931

Following French complaints that they had been denied Mesopotamian oil, especially after their cession of Mosul to Britain at Versailles, the British agreed at San Remo in 1920 to cut the French in for 25 percent of the TPC. An important consequence of San Remo was that Britain and France between them effectively excluded American companies from the Middle East. This pushed Standard Oil of New Jersey into fomenting a storm of American diplomatic protest that eventually forced Britain into giving American signatories to the Red-Line Agreement of 1928 a further 25 percent share in TPC.

That still left Britain with rights to all of Persia's oil via APOC and half of Iraq's via TPC, which changed its name to the Iraq Petroleum Company (IPC) in 1929. By 1931 it seemed that such policies would allow Britain to control much of the oil of the Middle East.

One of the main problems in the early period of oil exploitation, basically through 1931, was that there was no proration policy to protect landowners from reservoir depletion by their neighbors. Massive numbers of wells could be drilled in close geographical proximity based simply on possession of title to the land. This first came to a head in America when the output from Spindletop began to drop precipitately in the late teens from massive over-drilling and serious worries about "oil exhaustion" began. In 1919 the Texas Railroad Commission, which came to control oil and gas production in the state after 1917 via its power to regulate pipelines as common carriers, began to control well spacing to reduce fire hazards, but it did not adopt its crucial proration policy until August 1930. Even so, more generous well spacing combined with deeper and more careful drilling allowed Spindletop's output to recover somewhat by the mid-1920s but the growth of the American supply was far from certain until the huge East Texas field was discovered in October 1930. The other major Texas field, in the Permian basin of West Texas, had been explored in the late 1920s, but the cost of building the pipelines needed to connect West Texas with the pipeline hub in Houston was initially prohibitive (Texas Handbook Online). Texas began serious proration in 1931 (Texas Railroad Commission Online).

The history of American oil output from 1901 through 1931 is preeminently the history of Spindletop and is thus one of considerable concern over the amount of oil available and the rate at which it was being used up as automobile related demand rose rapidly in the late teens and twenties. There was no sense of how large the country's reserves might be, merely figures for production and consumption. After 1931, a combination of the discovery of the East Texas field, easily connected to Houston, increased exploitation of the Permian basin field, conservationist policies, especially that of proration, better survey technologies, and better geological understanding developed in the universities of the region, ended fears of American domestic shortages by the mid-1930s. New wells and new production were added faster than old wells were being closed.

Oil Issues, 1931–1945

In this period American domestic production surged and the struggle over access to Middle Eastern oil came down to a strong British rearguard action fought primarily on the grounds of policy and a remarkable American end-run around British interests on the basis of technology developed in American fields. SOCAL, later the Arab American Oil Company (Aramco), now Chevron, recognized that oil was present in Bahrain in salt domes of the sort it was familiar with in California and which one of its partners, the Texas Company, later Texaco, had come to understand empirically at Spindletop. SOCAL cleverly circumvented the Red-Line and Achnacarry agreements, recognized that Saudi Arabia was a massive potential site for salt dome deposits, and was able to persuade King Ibn Saud, who was no friend of Britain, to lease Saudi Arabia's potential oil fields to them. Like Standard Oil of New Jersey APOC and IPC had no understanding of salt dome geology, believing oil was exclusively found trapped in porous rock formations. Shell, one of the main constituent companies of IPC, refused to participate in any lease negotiations in Saudi Arabia since it "did not regard the prospects as favourable from either a geological or a political point of view" (Fitzgerald 1991, 458). SOCAL's final success was in policy. They were able to get President Franklin D. Roosevelt to visit Saudi Arabia on his return from the Yalta Conference in 1945, cementing a close relationship between the Saudi monarchy, America, and Aramco that guaranteed America access to what would prove to be the Middle East's most prolific oilfields.

Ideas and Reality: Oil Fuel, Naval Power, and the Merchant Marine

Tied up with the "oil exhaustion" argument of the 1920s was the realization that oil-fueling of ships had become central to the development and retention of global power at both a military and an economic level. By the 1920s three main strands in the innovation of oil-fueling were apparent. The first was in naval vessels, where the advantages of oil fuel were quickly seen as outweighing the costs and problems. The second was in oil-fueling of merchant steamships. The third was in the adoption of a radically new form of power plant, the diesel engine, which offered very substantial advantages to merchant shipowners, in particular in range. There was also early interest by the Imperial German navy in using diesels to extend the cruising range of its otherwise steam-powered warships. As it turned out this effort was much too far ahead of the materials technology of the period and resulted in failure, although the knowledge gained helped greatly in designing less powerful engines for submarines and then for merchant ships once the war was over (Landahn 1925).

Ideas and Realities in the Innovation of Oil-Fueled Shipping

The innovation of oil fuels for ships was clearly revolutionary, but the technologically oriented response of the engineers was different from the response of those concerned with geopolitics. The engineers were very clear on the technological advantages of oil fuel, but also on its ability to destabilize the shipping industry:

> The old stable situation, where the problems were mainly non-engineering in character, and competition depended on crew wages and size within a given size of ship, on shipping management, and possibly on government cooperation, has for the moment entirely disappeared, and while some of the old problems remain, new ones have come forward and are to-day dominant. The new ones are the result of (a) the use of oil as fuel in competition with coal; (b) the steam turbine; and (c) the internal combustion engine. (Lucke 1921, 14)

The geopolitical responses focused more on the issue of the supply of oil. As demand rose there were fears that supply might falter, fears expressed by supporters of the "oil exhaustion" thesis. Delaisi (1922) and de la Tramerye (1924) both argued that the British had robbed France of possible oil rights in Iraq she should have had under Sykes-Picot (-Sazonov) but was maneuvered out of at Versailles and, later, at San Remo. Both authors linked hegemony explicitly to a nation's ability to control the oil supply and to the strength of its merchant marine.

> The nation which controls this precious fuel will soon see the wealth of the rest of the world flowing towards it. The ships of other nations will soon be unable to sail without recourse to its stores of oil. . . . [T]he nation which obtains the world's carrying trade takes toll from all those whose goods its carries, and so has abundant capital. . . . At one stroke the controlling centre of the world's credit is displaced. This is what happened already in the eighteenth century when, with the development of British shipping, it passed from Amsterdam to London. (de la Tramerye 1924, 10–11, based on Delaisi 1922, 2)

Diesel engines were not initially without drawbacks that deterred many potential owners: "the heavy prime cost of diesels, the high maintenance charges, and, above all, the fact that the power/weight ratio of diesel engines was for long inferior to that of steam, since there was still much scope for improvement in the older mode of propulsion" (Craig 1980, 54). De la Tramerye's argument focused on the technological advantages of the diesel-engined merchant ship (table 2.2). His horsepower rating is clearly a typological error. No diesel that powerful had run by 1924 and a typical steam-powered merchant ship of the period used about 2,000 horsepower. The two engines used in *Selandia* in 1912 generated just over 2,000 horsepower in total, so one should assume de la Tramerye meant 2,100 horsepower.

For oil-firing of steam ships, there would be a reduction

> by 70–80 percent of the stokers since a single man can look after several boilers. [For diesels the labor savings would be even greater.] The fueling of the ship

Table 2.2 Diesel versus Steam Power in Merchant Ships

Power Plant	*Diesel*	*Steam*
Horsepower	21,000 [*sic*]	21,000 [*sic*]
Weight of engine & accessories	1,000 tons	3,400 tons
Space required	5,300 cubic meters	10,000 cubic meters
Daily fuel consumption	100 tons heavy oil	360 tons coal
Consumption for a voyage of fifteen days	1,500 tons	5,400 tons
Bunker space for a voyage of fifteen days	1,700 cubic meters	7,000 cubic meters
Total space required for engine and fuel	7,000 cubic meters	17,000 cubic meters

Source: de la Tramerye (1924, 15).

> is effected cleanly and quietly in a few hours. Hundreds of tons of oil can be pumped into the cisterns in a negligible time, and that even at sea and in heavy weather. . . . In the latest Cunard and White Star liners the economy of space thus realized has been as much as 33 percent. (de la Tramerye 1924, 16–17)

The popular press had begun to point out the technical advantages of the diesel engine motor ship much earlier than those worried about "oil exhaustion." The first successful oceangoing motor ship was, appropriately, *Vulcanus*, an oil tanker built in 1910 for the Anglo-Saxon Petroleum Company. In 1911 the Canadian *Toiler*, built as a Great Lakes freighter, was the first diesel engine ship to cross the Atlantic. *Toiler* burned $15.60 worth of oil a day compared to the $28 worth of coal a similar size coal-burning steamship would have needed. In addition, she saved $5 a day "on stoker's wages, besides the cost of their food, and [gave] an increase in revenue earning space of 120 tons on weight saved by the absence of boilers and coal bunkers." Such ships would use far less fuel for the same amount of work, have more usable cargo space, need less labor, and be much more rapidly refueled (Carter 1911, 592–594). The first really notable motor ship was, however, *Selandia,* which was built for the Far Eastern run where the range made possible between refueling mattered. *Selandia* made her maiden voyage in 1912, and marked "the most striking advance at this time" (Smith 1938, 330).

Oil and Naval Power

As early as 1908, in *Oil Fuel and the Empire*, J. D. Henry, British author and founder of the journal *Petroleum World*, noted that "recognising that in swift ships oil saves time and labor, if it does not save money, the admiralty has decided to use oil fuel in every type of warship, and it is this fact which gives the imperial aspects of the subject such tremendous importance" (Henry 1908, 16). In the run-up to WWI navalists such as Winston Churchill began to see oil-firing of steam turbine engine warships as a highly desirable technology that would substantially increase the fighting capacity of naval vessels by allowing higher speeds and radically easier refueling, plus lowered weights through lighter fuel and a substantial reduction in stokehold crew size, thus allowing more armor plate and bigger guns. Churchill pushed the acquisition of the 15" gunned *Queen Elizabeth* class battleships, which when they entered service in 1915 were by far the most heavily gunned, fastest, and best protected ships in any navy. In many ways they were not improved on until the advent of the American *Iowa* class 16" gunned ships in the early 1940s. To make sure the Royal Navy would have the fuel for its ships, in 1913 Churchill oversaw the acquisition of APOC by the British state. Such policies would then push Britain into the acquisition of Iraq as a source of oil that would be securely within the Empire.

The diesel was a natural fit for the German U-boat, and many British merchant ships were lost to them during WWI. Ironically, large numbers of surplus German U-boat engines found their way into merchant ships at war's end (Hugill 1993, 149). The diesel revolution also seemed initially poised to spill over into other parts of the military arena. British admiral Jackie Fisher, who as first lord of the admiralty had radically transformed and improved the navy in the first decade of the 1900s, argued in 1912 after being informed about *Selandia* that diesel battleships would be far more efficient than oil-burning steamers, make even less smoke, and have the added advantage of needing no funnels—one of the design features of *Selandia*—so their gun turrets could turn a full 360 degrees (Mohr 1926, 120). In the event, the British never went down this technological route for their warships.

On the German side in the experimental period Alfred von Tirpitz, whose influence on the Imperial German navy was as great as that of Fisher on Britain's RN, invested in diesel power for *Prinzregent Luitpold*, one of the five *Kaiser* class dreadnoughts of the 1912 program, but only as a long distance cruising engine replacing her third turbine. In the event the 12,000 horsepower engine commissioned was never installed. *Prinzregent Luitpold* fought at the Battle of Jutland in 1916 on the power of the two turbines that were installed and was scuttled in 1919 at Scapa Flow (Gibbons 1983, 193). Two powerful experimental diesels were commissioned for *Prinzregent Luitpold*, one from *Maschinenfabrik Augsburg Nürnberg* (MAN) and one from the Krupp *Germaniawerft*. An experimental single cylinder version of the Krupp engine first ran in 1911 and justified construction of a full-size six-cylinder, which actually ran at its rated 12,000 horsepower in 1915 but could not be made reliable (Landahn 1925, 169, 172). Such outputs were well beyond the metallurgy of the time. The smaller U-boat and merchant ship engines were not so heavily stressed.

The Germans continued their diesel warship program in the years between WWI and WWII, building the three *Deutschland* class *Panzershiffe* of the

1930s, *Deutschland* (later *Lützow*), *Admiral Graf Spee*, and *Admiral Speer*. The *Deutschlands* were referred to at the time by the British as "pocket battleships," used eight 6,500 horsepower diesels, and were the only major diesel-engined warships to see service in the period. They mounted powerful 11" guns but were really under-armored cruisers, using their diesels to gain great range for their expected wartime role as global commerce raiders. Despite their apparent innovations their speed was not spectacular and they suffered from maintenance problems. A high-speed was claimed and they were certainly fast enough to catch merchant ships. Since they were designed to run on low quality, very cheap fuel, their engines smoked heavily. They were easy for their opponents to track. Good innovations in warfare are quickly copied by potential opponents: the *Deutschlands* never were. The evolution of warship propulsion since WWII has seen more and more warships use diesels for long distance cruising, coupled with gas turbines to achieve high speeds in combat, the basic mixed power idea behind the Imperial German navy's *Prinzregent Luitpold*. Steam remains important in modern warships only in those using nuclear fission as their heat source.

Oil-Firing of Merchant Ships

Before 1914 operators of merchant vessels began to follow a similar line regarding oil-firing for the new steam turbine-powered ships coming to dominate the Atlantic shuttle by virtue of their speed. Such ships, like their warship cousins, used direct drive turbines, which required a great deal of fuel. The high rotational speeds of the turbine did not lend themselves to efficient propeller speeds and design. Smaller merchant ships stuck with the more fuel-efficient triple-expansion reciprocating steam engine, which rotated much more slowly and at a speed that made for efficient propulsion. They also stuck with coal fuel until the disruption of the global economy caused by WWI began to push them toward oil. The massive expansion of steam-powered merchant ships from the 1880s on had created a global network of ports where coal was readily available. The war, however, placed immense demands on the coal mines at a period when union activity and the demand for higher wages and greater safety was growing by leaps and bounds. One of the aftermaths of the war was that from 1918 to 1920 constant strikes at the mines and the docks meant that coal prices rose much faster than oil, which suffered very little from labor unrest (Pobodnik 2006). As a consequence, and even as worries about "oil exhaustion" were beginning to surface, "between 1919 and 1922 the tonnage of steamers fitted for oil-firing rose from 5.3 million to 14.5 million [tons]" (Henning & Trace 1975, 382).

The outbreak of war in 1914 marked something of a technological watershed for merchant shipping. "The geared turbine, the turbo-electric drive, and the diesel engine made their appearance almost simultaneously around 1913 and by 1917 the triple-expansion and quadruple-expansion steam engine began to feel the competition of the diesel engine and the geared turbine" (Chapman 1942, 110). The first burned only oil, the second mostly oil. American shipyards turned out a huge tonnage of basic merchant ships during WWI, the majority using reciprocating engines but increasingly burning oil. By 1920, 56 percent of the American merchant fleet burned coal and 44 percent oil (USDC 1960, 444). When the war was over a steady shift to oil fuel and geared turbines, which slowed propeller rotation to efficient speeds, continued in America, although American shipbuilders generally avoided diesels.

The Diesel-Engined Motor Ship

Before WWI steam engines were clearly important to the adoption of oil as a maritime fuel but in that same period the utility of the diesel also became apparent. Even the earliest diesels were ideal for merchant ships, particularly where long range was needed. *Selandia*'s sister ship, *Fionia*, renamed *Christian X*, was reported on in the *New York Times* upon its arrival in New York on its maiden voyage: "the steamship requires recoaling four times to circle the globe. . . . A fuel consumption of between nine and ten tons for each day . . . makes it possible for the *Christian X* . . . to make a continuous voyage around the world without replenishing her tanks" (Bogert 1912, 4). Motor ships did not initially offer high-speed but they did offer substantial savings in fuel and stokehold labor costs.

Although most diesel engine ships ran at speeds around eleven knots, when WWI was over some companies began to build much larger and faster liners that capitalized on the substantial range advantage of the diesel engine. Motor ships such as *Aorangi*, built in Britain as a first-class passenger and mail liner for the Union Steamship Company of New Zealand, displaced four to five times as much as a conventional cargo ship, had four Fairfield-Sulzer engines generating some 16,000 horsepower in total, and operated at eighteen knots, the normal speed for passenger liners that were not designed for the Atlantic ferry, where much higher speeds were needed. *Aorangi* was built specifically for the Pacific, for "the 'all-red' route between Vancouver, B.C., the Fiji Islands, Honolulu, Auckland, N.Z., and Sydney, N.S.W" (Hardy 1926, 99–101, 202–206, quote from 202). "It is, in fact, largely the tremendous virility of the Pacific Coast, with its ever-increasing markets, which has caused the creation of motorship fleets for transpacific work." Such forces pushed the Japanese into action in 1923, when the Kobe Steel Works expanded their license for submarine engines from Sulzer to allow them to build the slower running engines needed in merchant ships. Japanese owners, very naturally, "are determined to have at least their fair share of the carrying business" across the Pacific (Hardy 1931, 43–46). Such advanced motor ships quickly proved themselves just as popular with passengers as the geared turbine ship, as well as more profitable.

As ships like *Aorangi* demonstrated, WWI proved to be a huge technology forcer in the area of diesel engines. Although the French navy was the first to use diesels in submarines the Germans quickly followed. During the war half of Germany's "submarines ran on MAN's four-stroke diesels, mostly 1,200-hp units" (Smil 2010, 71). At the war's end two forces intruded. First, the Germans had built many more engines than submarines and a large number of such engines became available as war surplus, quickly finding their way to the merchant marine market (Smith 1938, 337). Second, German and other companies were happy to license the new technologies to the rest of the world to help them acquire foreign exchange. By 1939 24.4 percent of the world's merchant ships were powered by diesels. It was, however, the Scandinavian countries and Holland that most rapidly adopted diesels. By 1939 Norway led, with 62.2 percent of its merchant fleet powered by diesels: Denmark was at 52.2, Sweden at 46.6, and Holland at 45.5 percent. By contrast Britain was at 25.6 percent, a figure that is perhaps best explained by a combination of excess capacity in older steamships and cultural preference for familiar coal power. In many ways "Britain became the victim of her long reliance upon her own resources of high quality steam coal and many shipowners for too long pinned their hopes upon an anticipation that the pendulum would swing back in favor of coal once a revival of trade manifested itself" (Craig 1980, 54). Contemporaries noted that "it comes as a little difficult for us [British] to realize that steam and coal are no longer paramount" (Hardy 1931, x). At 26.2 percent of its fleet powered by diesels in 1939 the figure for Germany is much harder to explain, and even Japan surpassed both Britain and Germany with 27.2 percent of its merchant ships diesel powered (Henning & Trace 1975, 354).

In contrast, American shipowners were very much slower to adopt the diesel engine than even the most laggardly Europeans though much faster than they were to adopt oil-fueling of steam ships and abandon coal. By 1940 almost 80 percent of American merchant ships were oil-fueled. Some 69 percent were steamships but only 11 percent used diesels (USDC 1960, 444). Adolphus Busch had licensed Diesel's engine in 1897 and in 1898 he installed one to power his brewery in St Louis, MO (100 Years of Annheuser-Busch Online). The Swiss company, Sulzer, licensed Diesel's engine in 1898 and joined forces with Busch to build engines for the American market. Initially, however, Busch-Sulzer focused on the demand for diesels by local electricity generating stations. When America entered WWI the navy had them develop submarine engines and after the war they moved into marine diesels. Other

American companies, such as the shipbuilding company William Cramp and Company of Philadelphia, also licensed European diesel technology, in Cramps' case that of the Danish company, Burmeister & Wain. Even so, in the mid-1920s, out of fifteen manufacturers of diesel engines worldwide, only two, Bethlehem and Worthington, were American companies building American designed diesels, and neither was successful (Hardy 1926, 308).

A second reason for the slower American adoption of diesels was that the American economy of the early 1920s still lacked some of the skills needed to manufacture and maintain them. Diesels operated at much higher internal pressures and temperatures than even the most sophisticated steam engines, and thus required very high levels of manufacturing skill and excellent metallurgy that increased first cost. Rudolf Diesel himself is believed to have complained that "Americans are obsessed by the principle of first cost." The first major American company to successfully enter the market for diesels without licensing a European design was General Motors, which did so in the early 1930s via its acquisition of the Electro-Motive Corporation (EMC) in 1930, which it renamed Electro-Motive Diesel (EMD) in 1941. By the late 1930s EMC had an efficient diesel engine at work in diesel-electric railroad locomotives, and in WWII EMD would turn its attention to marine diesels (The Electro-Motive Story Online).

CONCLUSION

Commentators in Britain and America by the early 1920s were clear on the relationship between the developing struggles foreseen by Wilson in the arenas of marine transportation and petroleum. They argued that whoever controlled the right mix of marine transportation technology, in particular that of the merchant marine, and petroleum supply would control the emerging postwar international order. However, the ideas that were put around in the early 1920s with regard to the development of petroleum and shipping did not fully mesh with the realities that were apparent by 1945. Four major technological and economic forces intervened to prevent clear American victory in this struggle, although in general America did better than Britain.

First, the "oil exhaustion" thesis of the early 1920s faded rapidly as new American oilfields came on line in the 1930s and oil production finally began in Persia and Iraq. Britain's policy advantages in the Middle East evaporated for three reasons. American companies such as SOCAL and The Texas Company developed practical experience with salt dome oil deposits, while theoretical understanding of these was developed by geologists at Louisiana State University and at the University of Texas' Bureau of Economic Geology. British companies involved in the Middle East failed to understand the importance of salt dome deposits and effectively denied their existence. Finally, sound diplomacy brought Saudi Arabia's massive oil reserves under American control by 1945. Despite fears in the 1920s that America might lose its grip on the global oil supply, from 1930 to 1945 that grip substantially tightened.

Second, although America remained the dominant global supplier of oil, America failed to capitalize on the remarkable increase in its ship building capacity that had been developed in WWI, failed to adopt the diesel engine for its merchant ships when the war was over, and never challenged British dominance of the world's merchant marine in the period between WWI and WWII. In 1894 Britain built 79 percent of the 1,323,538 tons of merchant ships launched that year, America only 5 percent and Germany just under 9. In 1919 7,144,549 tons were launched, with America accounting for 57 percent and Britain for just under 23 percent (Smith 1938, 357–358). During WWI Britain lost the spectacular total of 9 million tons of shipping to German mines, commerce raiders, and U-boats (Craig 1980, 53). This huge loss should have made it easy for American ships to take over the global shipping market, but Wilson's vision of an American merchant marine fell apart for several reasons.

The American merchant marine of 1919 was outdated technologically. Too many merchant ships built quickly as part of the war effort were coal-fired, triple-expansion engine ships, and by 1919 oil-fired, geared turbine engine ships and diesel engine motor ships were rapidly taking their place. The American yards, driven by war and government-funded demand, effectively stopped dead once that wartime demand vanished. Despite all the travails of the Depression of the early 1930s and its huge impact on British shipyards, of the 2,117,924 tons launched worldwide in 1936 Britain built just over 40 percent, Germany 18, but America only just over 5 percent (Smith 1938, 358–359). To make things worse, the fleet of passenger ships that Wilson hoped after 1919 would compete for traffic on the lucrative Atlantic routes dominated by Britain was actually a rather badly mixed bag of ships, many of them German ships interned in American ports in 1914 and seized as war prizes when America entered the war in 1917. Such ships were stylistically and technologically outdated by 1920 and several had been deliberately sabotaged by their crews and required substantial effort and expense to get them back into service. "The ships and shipping of the pre-1914 era are as hopeless to deal with present-day transportation conditions as would be the stage-coaches of the last century to cope with the road traffic of today. . . . The passenger today buys luxury on any ship and his extra passage money is largely for Speed . . . modern cabin-class is easily superior to pre-war first-class" (Hardy 1931, 83, 100, 103). Such upward pressure on luxury was constant. In the Norddeutscher Loyd liner for the North Atlantic trade, *Bremen*, which took the Blue Riband on her maiden voyage in 1929, "third-class staterooms were equal to second-class quarters in older ships" (Baker & Tryckare 1965, 152). As a final nail in the coffin of Wilson's hope for an American Atlantic fleet of passenger ships, almost no well-off passengers were willing to take passage on American ships after the Volstead Act prohibited alcohol on American territory, making American ships legally "dry" from 1920 to late 1933.

Third, the success of the motor ship encouraged by the 1930s significant improvements in the fuel efficiency of the oil-fired steamship that brought its specific fuel consumption much nearer to that of the diesel engine motor ship. The combination of superheated steam, higher boiler pressures, and geared propulsion systems allowed substantial increases in fuel efficiency in steam turbine engine ships. Whereas in 1919 the best such ships consumed one pound of fuel per horsepower hour and diesels 0.42 lbs; by 1940 those figures were 0.60 and 0.38 lbs. respectively (Chapman 1942, 249). The one strength of American merchant shipping in the 1919 to 1939 period was the refinement of the geared turbine, oil-fueled steamship. Unfortunately this was ultimately a technological dead end, and an analysis such as Chapman's based only on engineering criteria was not realistic in the light of the real world of merchant shipping.

> The motorship's great advantage was in cargo carrying capacity . . . this additional cargo capacity accounts for the greater profit earned by motorships in the twenties and early thirties. Note also that whereas the proportion of the motorship's capacity required to be filled in order to break even varied between sixty-seven and ninety-one percent, there were years in which it was impossible, given freight rates ruling, for either the coal- or oil-fired steamships to break even. (Henning & Trace 1975, 375).

Fourth, the price of refined diesel fuel moved against the price of the unrefined bunker C fuel that was burned in oil-fired steamships. By 1941, and corrected "for the differences in specific gravities of the two fuels . . . the ratio of fuel costs per ton . . . =1.64" (Chapman 1942, 250). Given that the ratio of fuel use was 0.60 to 0.38, or 1.58, the oil-fueled turbine engine steam ship held a slight cost advantage. This, the Jones Act, and the short-haul geography of American shipping between American ports encouraged American shipping lines to stay with the geared turbine when they should not have. A diesel engine could run on much cheaper bunker C fuel as well as the more refined and expensive diesel fuel as long as bunker C was preheated so it atomized properly. It was harder to start and needed more maintenance, and American engineering texts of the 1940s were skeptical of its utility and clearly preferred the geared turbine (Chapman 1942, 251).

Diesel fuel also seems to have varied in quality, perhaps because it was hard to distinguish from bunker C and the latter could be passed off as more expensive diesel. The issue of the variable quality of fuel oil concerned some observers in the 1920s and manufacturers of diesel engines seem to have tested them routinely for their ability to run on bunker C fuel. British author Hardy noted that at least one successful British manufacturer, Doxford, ran its engines on bunker C and that "the majority of Diesel engine makers have on trials run their engines on boiler oil, and one often hears complaints from Diesel engine users that so-called Diesel oil which has been supplied to them at bunkering ports is practically boiler oil" (Hardy 1926, 278).

The simple fact is that, despite the initial policy success of Britain in the Middle East, America maintained a huge advantage over Britain between 1919 and 1947 in the struggle over petroleum. Where America signally failed was in its inability to capitalize on the advantages in international merchant ship construction that it should have enjoyed as it emerged from WWI. In both cases technology played a major part. American success in the petroleum struggle was based on the development at SOCAL of the technology to exploit salt-dome deposits of oil and the recognition that Saudi Arabia had huge such reserves. American failure in the struggle over merchant shipping was the failure of American engineers to see the central importance of Diesel's radical new engine and a policy of protectionism that led to a continued preference for oil-fired steamships.

Chapter 3

The Struggle over Air Power and Technology

When President Wilson argued in 1919 that international transportation would be an arena of struggle between America and Britain in the coming world he believed that that arena would be a maritime one and that Britain had a good chance of beating America. In the arena of maritime struggle he was generally correct, albeit not for the correct reasons. The problem was that the main arena turned out to be in aerial transportation, not maritime. There the technologies were new, they diffused easily, and there was little attachment to traditional ways of doing things.

THEORIES OF AIR POWER

The rapid development of aviation just before WWI made it obvious that air power was going in future to be crucial to both the commercial and military power of states. Four theories were developed to deal with the military and political aspects of air power, one dealing with its tactical use to support armies in the field, two with its strategic use, and one combining elements of both. The aftermath of WWI made it obvious that aviation would be important on two more fronts: in terms of the unification or control of a political or an economic space in peacetime. Territorial and trading states embraced these theories and practices differently.

Four Theories and Two Fronts

The Theory of Tactical Air Power

This was central to the British "Plan 1919," which proposed that WWI would be ended by the coordinated advance out of the trenches on the Western

Front of mobile ground forces, tanks and armored cars, which would be supported by ground attack aircraft, the whole air and ground advance being coordinated by wireless (Hugill 1999a). This was the recipe, of course, for the "Blitzkreig" tactics of WWII, used successfully by the Germans, then the Soviet Red Army, and then the American and British armies that invaded Fortress Europe on D-Day. Tactical air power had particular appeal to territorial states. A famous saying attributed to Jimmy Doolittle, one of airpower's great advocates, goes something like "If bombing is compared to milking a cow, then tactical bombing aims to upset the pail of milk. Strategic bombing aims to kill the cow."

The First Theory of Strategic Air Power

This, the most influential, was put forward by the Italian, Giulio Douhet, who proposed that the next war would take the form of a total war from the air that would involve the terror bombing of civilian populations so that their government would be forced to sue for peace.

The Second Theory of Strategic Air Power

Billy Mitchell in America proposed a much more focused approach to strategic air power, the precision bombing of important elements of the enemy's power structure. In WWII this resulted in an American bombing focus on such key elements as oil production, ball bearing production, and aircraft factories. Both theories of strategic air power appealed to trading states, which began to see them as the logical extension of the Mahanian geostrategy of blockade, preventing the civilian population of an opposing state from manning that state's war machine and making it pressure its government into suing for peace.

Air Policing

Air policing emerged in Britain as the fourth theory of air power. The Royal Air Force began to use aircraft to control insurgencies in such problematic parts of the Empire as Iraq and the Northwest Frontier of India in order to enforce the political unity of the Empire. Inasmuch as it avoided land war and would be very much cheaper, it was obviously attractive to trading states.

The Use of Civil Aviation to Promote the Political Unification of Space

Britain moved quickly into this first post-WWI front, the problem being the administration of an enlarged and increasingly complex global Empire. Air travel offered the ability to move senior civil servants much more quickly and efficiently than by sea from the home country to and from distant posts.

The emerging Soviet Union also saw the possibilities of air travel for political integration of a sprawling territorial polity in the 1920s. Even the inefficient and relatively slow civil aircraft that emerged from WWI could, with government subsidies, perform this role.

The Commercial Integration of Large Spaces

This, the second post-WWI front, took longer since it required aircraft that could be operated profitably without government subsidies. Such aircraft were unnecessary for political integration and Britain and the Soviet Union employed a succession of highly unprofitable aircraft, even dirigibles in the case of Britain, in their efforts to achieve political integration in their Empires. Not until the emergence of airplanes like the Douglas DC-2 and DC-3 did America become able to use aircraft to achieve commercial integration of the continental United States.

Practicalities of Air Power

The tactical air power theorized in "Plan 1919" required little more than the sorts of aircraft being procured by 1918. Most were short range aircraft operating from forward airfields which could keep in constant touch by wireless with ground forces and bomb and strafe the opponent's forces just ahead of and to facilitate the advance of its armored divisions and ground troops. Many airplanes from late in WWI could fulfill this role. In WWII the need for tactical air power produced two main sorts of airplane. The first was designed from the outset as a ground attack airplane, the second was adapted from a fighter airplane as a fighter-bomber. The classic German dive-bomber, the Junkers Ju-87 *Stuka*, and the Red Air Force's Ilyushin IL-2 *Stormovik* were custom-designed ground attack airplanes. Both the Nazi and Red armies were embedded in territorial states, the logical geostrategy of which tended to emphasize their need for tactical air power. Stalin famously described the *Stormovik* as just as necessary to the Red Army as bread.

WWII's Allies, classic trading states with a strong preference for strategic air power, preferred to adapt fighter airplanes for tactical purposes. They saw fighters both as defensive interceptors designed to maintain air superiority over the homeland and, eventually, as necessary escorts for their own strategic bombers over enemy territory, but they adapted many as fighter-bombers. Since they had started life as fighters, once they had released their bombs and rockets they could revert to the fighter role. To defend themselves against strategic bombing the British and American air forces procured far more fighter designs than the Germans in the 1930s, and some of these, for various reasons, did not quite measure up as air superiority fighters. If one simply counts those single engine, single seat, piston-engine fighters and fighter prototypes that first flew between 1934 and 1942, and thus had a chance of entering or did enter production in time for the war, the Germans procured seven, thc British eleven, and the Americans twenty-five. The Soviet Union provides an interesting contrast to Germany, presumably because it tended to prefer the defensive, procuring fifteen air superiority fighters, although seven of these were simply increasingly modified versions of two basic designs. In Britain there were only two versions of one design, in Germany three of one, and in America eight of three (data from Green 1970; Gunston 1983; Jones 1975; 1977; Lewis 1979). Failed air superiority fighters were often eminently suited to the role of fighter-bombers. Classic examples from WWII are the American Republic P-47 *Thunderbolt* and the British Hawker *Typhoon*. Unlike the *Stuka* and the *Stormovik*, which needed constant fighter protection, fighter-bombers, although not as effective as pure interceptors in their fighter role, could be reasonably expected to hold their own even in the heavily defended air space over Nazi occupied Europe. Although some twin-engine aircraft were used as tactical airplanes, especially by the Soviets, most tactical ground attack aircraft were single engine and had relatively short ranges. The sorts of aircraft the British used in "air policing" were also ground attack aircraft.

Although the beginnings of strategic bombing can be seen in the events of WWI, the real shift after the war was in the much more thorough development of ever larger, multiengine airplanes for the two main forms of strategic air power: terror bombing of civilians; and precision attacks on industries critical to the war effort. There was one important sub-version, and that emerged in America. America's developing geostrategy in the period between WWI and WWII focused on what became known as "hemispheric defense." This was, in effect, Britain's classic "moat defensive" geostrategy writ large (Hugill 1999a), with the Atlantic and Pacific oceans providing America with far more generous moats. The implication of a policy of "hemispheric defense" was that long-range airplanes would play a major defensive role against naval forces attacking across these oceans, effectively extending America's fortified, defensive coastline several hundred miles out to sea. Two expressions of this ideal were built as part of the United States Army Air Corps (USAAC) XBLR—Experimental Bomber, Long Range—program: the massive XBLR-1, which emerged as the Boeing XB-15 and the equally massive XBLR-2, renumbered as the Douglas XB-19 and actually named the "Hemispheric Defender." Both were seriously under powered and slow. One more design, the Sikorsky XBLR-3, was canceled before it was built (Jones 1980, 36–38, 54–61). The ultimate expression of the technology that would be needed to work this concept of "hemispheric defense" was, in fact, Boeing's smaller, faster B-17 that, in consequence of its intended role, acquired the name

Flying Fortress. In WWII it was pressed into service as a strategic bomber by an air force committed to Mitchell's theories, a role to which it was not particularly well suited since its bomb load was relatively low, designed for it to attack shipping, not cities.

WWII would, of course, turn out to be the proving ground for three of these four theories of air power. "Air policing" faded into the background at this point and would not reemerge until the modern era of cruise missiles, drones, and smart bombs.

It was quickly obvious that all aircraft that might be used for the strategic offense or for hemispheric defense would have to carry reasonable bomb loads and enough fuel to fly long distances. Locating a target was obviously likely to be a problem, whether it be the pin prick of ships on an open ocean, or distant land targets obscured by smoke from factories, cloudy weather, or darkness. Since a substantial amount of the flight time of offensive bombers was likely to be over enemy held and defended territory it became evident that such bombers would need to do three more things. First, as anti-aircraft guns—in German, *flak*, short for *Flugzeugabwehrkanone,* became more and more effective it was clear that strategic bombers would need to fly relatively high to try and avoid it. Alternatively they might fly at night. If they flew high, they would need very accurate bomb aiming equipment. But if they flew at night they would not be able to use the very accurate Norden optical bomb aiming equipment on which America spent so much money between the wars (McFarland 1995). Because America had the Norden bombsight it believed it could bomb accurately from high altitudes in daylight. Without such a bombsight the British moved to night bombing more readily and then developed effective radar bombing systems that allowed them to bomb accurately without visual bomb aiming (Hugill 1999a).

Second, since they would be over enemy territory for some time, both American and British designers believed that their strategic bombers would need to carry a relatively heavy defensive armament to fend off enemy fighters. They thus provided numerous power-operated gun turrets to allow on board gunners to defend their airplanes. Turrets, guns, ammunition, and gunners added weight. Third, some air power theorists argued by the late 1930s that strategic bombers would need to be accompanied by long-range fighters to defend them. The USAAC went so far as to establish a "Fighter Multiplace" category to this purpose and in 1936 procured the Bell XFM-1 *Airacuda* to meet the projected need. It turned out to be thirty miles per hour slower than the B-17s it was supposed to escort, had numerous other problems, and the air force abandoned plans to procure any more such escort fighters in the late 1930s (Pelletier 1992, 19–24). In the event it turned out that this was a mistake, and by the end of the war the escort fighter had come to assume a central role in the strategic bomber offensive. By 1944 the classic escort fighter was a high performance, long-range, single-engine fighter such as the North American P-51 Mustang that could hold its own against almost any German defensive fighter.

What emerged initially out of all this technological and geostrategic ferment, which was stronger in America than Britain and almost absent everywhere else, was thus the classic concept of the sort of strategic bomber that would serve in the opening rounds of WWII: large, powerful, multiengine, high-flying, heavily armed, and "self-defending" by means of numerous machine guns in multiple, powered turrets, and containing a fairly numerous and well-trained crew to operate it all. All this meant weight, weight, and more weight, thus larger and larger airplanes with more and more powerful engines. It also meant substantial training programs for aircrew: as well as gunners this included a flight crew composed of a pilot, copilot, flight engineer, and navigator, around ten in all. By the early 1930s it was thus obvious that there was a great deal of synergy between the demands of strategic military aviation and the needs of commerce for airplanes that could carry large payloads at altitude over substantial distances, such as across the American continent without refueling, across the Atlantic gap, or from America's West Coast to its main geostrategic outpost in the Pacific, Hawai'i. Such commercial airplanes needed the same basic flight crew as strategic bombers.

THE ARENA OF CIVIL AVIATION

By focusing on maritime issues at Versailles President Wilson failed to see the competition in international transportation that was emerging in civil aviation. At Versailles the British even had to advise Wilson that civil aviation was an emerging area of international concern he needed to pay at least some attention to. Given Wilson's illness and death in the aftermath of Versailles, and the shift back in the administrations of presidents Warren G. Harding, Calvin Coolidge, and Herbert Hoover to business oriented Republicans with little or no sense of the complexities of the international arena, or little willingness to deal with them, it is hardly surprising that American aviation policy and technology languished through the 1920s. In 1915 Wilson had at least signed into law the creation of the National Advisory Committee for Aeronautics (NACA). In the aftermath of WWI NACA transferred a huge amount of material on aircraft technology to America, most of it developed in the advanced technological universities of Imperial Germany. This material NACA had translated, and it had copies distributed to America's Morrill Act universities to aid in the development of aeronautical engineering in those universities. Unfortunately, without the sort of robust market for new airplanes at the military or civil level that could only be created at this timeframe by government demand, this wealth of material languished in the dusty stacks of America's still inadequate Morrill Act universities.

In the early 1930s the problems for American aviation set up after Versailles and mitigated only slightly by the deliberations of the Morrow Commission, which opened up something of a market for new military airplanes after 1926, were rapidly remedied. Once President Coolidge and his ilk were gone, the market for military airplanes became more robust, and the procurement of new bomber and fighter aircraft began to accelerate. Coolidge's approach had been both ridiculous and penurious—he once said in response to a war department request for more airplanes "why don't we buy just one airplane and let the pilots take turns flying it?" By the late 1920s, the funding of several chairs in aeronautical engineering by the Guggenheim Foundation and the NACA translation efforts had begun to bring American universities up to the level of German universities in the early teens (Hugill 1993, 278). In the early 1930s America also began to place considerable stress on the problems of how the long-distance civil aviation market was to be organized. There were obviously three major civil air routes where aircraft would be critical in terms of their ability to radically improve the speed of movement of mail and critical economic and political personnel: New York to Chicago, Los Angeles, and London. The problem of organizing the last was that the other powerful actor, Britain, was not about to concede control of that route to America and vice versa. The major policy struggles between Britain and America over this route during the 1930s and 1940s have been well covered in the work of Alan Dobson (Dobson 1991; 2011).

TECHNOLOGY AND POLITICS IN THE ARENA OF AVIATION

In the aftermath of WWI several states emerged as players in the fields of military and civilian aeronautics. In the arena of civil aviation, two were of great significance, two were of lesser significance but important technologically, especially in military technology, and three were little more than interesting bit players. The two main players were America and Britain. The second group contained Germany and Russia. The third consisted of France, Italy, and Japan. Of these the main players were by far the most important, although the others made significant contributions.

In 1913 the British newspaper, the *Daily Mail*, offered a prize of £10,000—$1.16 million in 2017—for the first transatlantic flight to be completed in less than seventy-two consecutive hours. In 1913 London was clearly at the center of the world's major trading state (Hugill 2009a) and still the world's dominant financial center, although America's rapid rise as a challenger trading state was obvious, as was the increasing power of New York as the world's second major financial center. Air connections between the two cities would obviously prove valuable by speeding the connections for mail and high status humans between the two main global financial centers.

But fears of the possibilities of aerial warfare were also burgeoning in 1913. Many of these fears can be traced to H. G. Wells' dark novel, *The War in the Air* (1908), which saw aerial warfare ushering in global war, pandemic disease, and a new dark age. On the positive side novels by Wells and many others prepared the population for the coming of aerial travel as well as aerial warfare (Goldstein 1986; Paris 1992). The outbreak of WWI in 1914 delayed the attempts under way in response to the *Mail's* prize, but in 1919 three successful crossings were made, all in aircraft derived from military experience. The United States Navy (USN) sent three NC class flying boats across from America via the Azores, and one made it. The NC boats were descendants of the Curtiss *America* boat that had been almost ready to make an attempt in 1914, until prevented by the outbreak of WWI. The British rigid airship, R.34, based on downed Zeppelin L33 (LZ76), successfully crossed both ways. Both the NC-4 and the R.34 took over seventy-two hours. Two British fliers, Royal Air Force officers John Alcock and Arthur Whitten Brown, then took the *Daily Mail* prize flying a converted WWI Vickers *Vimy* bomber from west to east to take advantage of the prevailing winds. Several other attempts to claim the *Mail's* prize ended in disaster and sometimes death. From then until the successful opening of the route's first commercial heavier-than-air service in the summer of 1939 it was obvious that closing the Atlantic gap would be technologically extremely challenging however rewarding it might eventually turn out to be economically.

In America WWI had raised a strong military concern with air power. In 1923 the USAAC had flown nonstop from New York to San Diego in just under twenty-seven hours against the planetary headwinds using a single-engine Fokker T-2 (Wehl 1965, 376). They had previously tried to fly west to east to take advantage of the planetary westerlies but the T-2 could not climb high enough to cross the Rockies with enough fuel to reach the east coast. By the early 1920s the burgeoning geostrategic needs of its Pacific interests was pushing the USAAC into trying to connect the Hawaiian Islands to the mainland by air. Such a flight was as long as that across the Atlantic, against the planetary headwinds, and it posed a much more serious navigational problem, how to find a tiny set of islands in a very large ocean. In 1927 the USAAC successfully flew a Fokker C-2, a military version of the three-engine F.VIIB/3M, from the mainland to Oahu, although the plane was little more than a flying gas tank, its normal range being 350 miles (Postma 1979, 70). The USAAC used both a radio compass and an earth inductor compass to successfully find the island chain. That same year the Hawaiian pineapple planter, James Dole, offered a prize of $25,000 for a commercial flight to Honolulu. Of the eight airplanes that began the flight only two made it to Hawai'i. The winner navigated using a sextant and smoke bombs to calculate drift. Three of the entrants crashed and three simply disappeared, almost certainly from navigational problems. No one used the earth inductor compass. Ten people died.

The technological challenges represented by the drive to develop airplanes and then routes for very long-range service can be broken down into a set of nine complex phenomena, the second through sixth of which are heavily interacting, and several of which have important subsets. These nine are:

1. flight control systems;
2. engine reliability;
3. engine performance and the airplane propulsion system;
4. fuel quality;
5. aerodynamic theory;
6. airframe and wing design;
7. take off performance;
8. navigational equipment; and
9. airport infrastructure.

Simply put an airplane had to be controllable, fitted with engines of sufficient power to be able to get its airframe, enough fuel for the crossing, crew, and payload off the earth's surface. Its engines had to run reliably for long enough to get to its destination, and that destination had to be both navigated to successfully and reached reasonably quickly. Given the early realization that air power would be a central part of any future war, innovations occurred rapidly in all those states that saw aerial warfare as likely, and such innovations diffused rapidly. Diffusion occurred via the conventional channels of the scientific literature, via constant movement of skilled engineers between states, and by the clandestine activities of intelligence services. In general the main problem with diffusion seems to have been the nature of the adoption environment. Given that adoption environments are more cultural and economic than technological they are not always easy to analyze though always central to the problem of the acceptance of new ideas (Hugill & Dickson 1988, 271).

Flight Control Systems

The most critical technological development in the earliest period was the development by the Wright brothers of effective three-axis control, demonstrated in their gliders even before their first powered flight of 1903.

The Wrights realized that airplanes had to bank to turn a corner, not merely steer round it as if they were on a flat surface, which is what the early European pioneers of flight were attempting to do (Gibbs-Smith 1970). There were many faults with the Wrights' control system—they used wing warping rather than the hinged, wing mounted control surfaces known as ailerons developed by later fliers, and they mounted their elevator at the front of the airplane, not the back. Their designs were inherently unstable and tricky to fly, but without their insight and innovations controlled flight would not have developed so quickly. Once it was clear what they had accomplished their technology was easily copied and very rapidly improved on.

Engine Reliability

Coming out of WWI few aeroengines ran reliably for long periods since designers were obsessed, rightly given the needs of combat, with keeping weight as low and power output as high as was possible at the relatively low level of technology of the period. Weight concerns tended to make designers prefer air-cooled engines, since these were not encumbered with the extra weight of radiators, hoses, pumps, and coolant, systems that were also vulnerable to combat damage. The most common air-cooled engines used in WWI were rotaries, where the entire engine and propeller rotated as one. Though powerful for their weight, the weight of the rotating engine made for nasty gyroscopic effects. Because of these gyroscopic effects rotaries could not be enlarged much to make them more powerful and power outputs much above 110 horsepower (hp) were rare. They also needed oil that could resist very high temperatures. The best was oil from the tropical castor bean, of which Germany, lacking a tropical Empire, had very limited supplies compared to the Allies. For example, whereas the Fokker E.I had a fuel tank of 70 liters that supplied its 80 hp Oberursel rotary with enough fuel for up to two hours it also needed an oil supply of 16 liters (Weyl 1965, 102). Erratic lubrication from their "total loss" lubrication system also helped make rotaries unreliable. Over the limited geographic area of the Western Front, reliability was far less important than lightness and power, and engines only had to run for a few hours between services or even replacement. Even so, by the end of WWI all sides had begun to focus on high power water-cooled engines that were becoming at least more reliable than the rotaries.

Engine Performance and the Airplane Propulsion System

There were really three interacting components of engine performance that contributed to the airplane's propulsion system. Sheer power output seemed initially to be the main one, and that also depended on fuel quality.

The second was that the way the engine was installed affected overall aerodynamic performance. The lower installed drag of in-line engines seemed, at least to begin with, an advantage over bulkier air-cooled engines. The third was the propeller, effectively a "transmission" system. Early propellers were "one-speed" systems, and because take off was usually the most difficult part of flight with low power engines they tended to be optimized for that. The second and third components increased in importance as speeds rose.

Power

WWI was the pioneering war in the air, and in some ways it played out as a battle over engine power. The arm of the German government that oversaw the air war, the *Idflieg*, generally assumed that Daimler-Benz's excellent water-cooled, in-line, six cylinder, 160 hp engines would be adequate for the short war they envisioned. Some Daimler-Benz engineers realized that their engine could be improved, but the company refused to take them seriously and the engineers left to form the *Bayerische Motorenwerke*—the Bavarian Engine Company, ancestor of today's BMW automobile company. At 185 hp BMW's engines were more powerful than Daimler-Benz's but not radically so, although they performed significantly better at altitude. By 1918 the allies had much more powerful engines and better fuels than did the Germans, routinely getting 250 hp from the French water-cooled Hispano-Suiza V-8 that powered their best fighter airplanes, the British S.E.5a and the French SPAD XIII. Although no one yet knew why, petroleum from the Dutch East Indies that was naturally highly volatile and that was marketed by the Anglo-Dutch Shell Company to the allies also allowed higher power outputs.

A major consequence of what was, over four years of war, an increasing shortfall in engine performance was that by 1917 German airplane designers were focused on aerodynamic efficiency rather than brute power. Further, *Idflieg* began to use competitions to weed out inferior designs instead of insisting, as the British tended to, that the market should do it. Such competitions were marred by bribery of pilots to prefer the products of certain manufacturers, but pilots such as Manfred von Richthofen were incorruptible and reported honestly on the best airplanes (Weyl 1965, 263). By 1918 a new generation of much more aerodynamic airframes was beginning to appear, such as the Fokker D.VII biplane that won *Idflieg's* first fighter competition in January 1918, which was to replace the increasingly outdated Albatross D.V. The D.VII had markedly less drag than contemporary British or French designs, not least because "the combined drag of thin airfoils with their bracing wires was significantly more than the drag of the thick [cantilever] Fokker wing which needed no bracing wires" (Herris 2013, 36). Wrongly, however, Fokker thought that it was his experience as a pilot that made him a great designer. He thus paid no attention to the accumulation of aerodynamic knowledge in Imperial Germany and refused to pass it on to his workers. Fokker's real designer was Reinhold Platz, who had no formal training in aerodynamics, but was obsessed with efficiency and instinctively knew what would work (Hassell 1997, 141). Among German airplane companies Fokker's was, however, the exception.

By the time of the second *Idflieg* competition in June 1918 two German fighter designs stood out. One, Hugo Junkers' D.1, was a low wing aluminum monoplane that would not have looked out of place on an airfield twenty years later and it was ordered into limited production, though many pilots disliked the way it flew. Another, the Zeppelin D.1 designed by Claudius Dornier, was an even more structurally advanced stressed skin, aluminum, cantilever-winged biplane without any interplane struts. Both these airplanes would have a major impact on the postwar development of civil aviation. At the third *Idflieg* fighter competition of late October 1918 the best-regarded designs were sophisticated cantilever-winged monoplanes, their structures grounded in theoretical aerodynamics (Herris 2013, 103, 104, 144–145). Poorer aerodynamic understanding meant that allied airplanes had no better performance than less powerful German airplanes, and all parties regarded the Fokker D.VII with a BMW engine as the best fighter to see service in the war (Stinton 1997, 129).

The main solution to the engine problem, the development of much better air-cooled engines, was in sight at the very end of WWI. As in the rotaries, the cylinders in these new radial engines were arranged in a circle, but unlike the rotaries the new engines stayed still and only the propeller rotated. The best early radials emerged in Britain, although The All British Company's radial engine, the ABC Dragonfly, was very nearly a disaster prevented only by the war's sudden end: "a technically illiterate Government purchasing machine" had foolishly ordered it in considerable numbers before it had proven itself. Even more foolishly they had canceled contracts for most other engines (Gunston 1999, 122). Success came from a company called Cosmos Engineering, whose designs were taken over by the Bristol Aeroplane Company which then developed their *Mercury* and *Jupiter* into the most effective and reliable power units of the 1920s (Barnes 1988, 30). By 1930 Bristol was the world's main aeroengine company with "17 foreign licensees . . . Jupiters had flown in 262 different types of aircraft" (Gunston 1999, 126).

Initial success in radial engines also eluded America, although the Lawrance J-1 emerged in the early 1920s as a basically sound design. By that time the USN had decided to order only air-cooled radials since water-cooled engines were more vulnerable in combat and cooling systems were prone to such problems as leaks even in normal use. The USN did not believe Lawrance could produce enough engines for its needs and insisted that Lawrance be taken over by the Wright Company in 1923, which developed the J-1 as the *Whirlwind.* Developed steadily, by the later 1920s the *Whirlwind* was light, powerful, and extremely reliable, rated for a maximum operating time of 9,000 hours. A big problem with aeroengines at the time was the need for regular external lubrication, which on most engines meant before every flight and even in-flight. The *Whirlwind* "self-lubricated" and was therefore rated for forty continuous hours of operation—Charles Lindbergh used one to power his *Spirit of St. Louis* on its 33½ hour flight from New York to Paris in 1927. In the middle 1920s several of Wright's engineers left to form what would become Wright's main competitor, Pratt & Whitney, also making air-cooled radials. Continuous development combined with the development of a robust domestic market ensured that Wright and Pratt & Whitney would overtake Bristol in the early 1930s to become the world's main suppliers of high power, air-cooled aeroengines.

The other route to increasing the power output of the internal combustion engine without much increase in weight was to force the air and gasoline mixture into the cylinders under pressure. Supercharging was initially seen in WWI as a way of producing better power output at altitude, where the air was thinner. In fact, the conventional mechanically driven supercharger allowed much higher take off power, which was important given the poor power-to-weight ratios of early engines. But a mechanically driven supercharger bled off a reasonable amount of power to drive it and was less effective at altitude. It was obviously more efficient to use the exhaust flow from the engine to drive the supercharger, which version became known as the turbosupercharger, or just turbocharger, the practical development of which came from the work of Sanford Moss at America's General Electric Company. American aeroengine companies such as Wright and Pratt & Whitney began to specialize in turbocharging and, as a result, developed engines that performed very much better at altitude as well as on takeoff. Flying at altitude in thinner air reduced drag and this allowed higher speeds and reduced overall fuel consumption, which was important alike for both long-range airliners and strategic bombers.

Engine Aerodynamics

In-line engines seemed initially to have an aerodynamic advantage over bulky radials and early aerodynamic understanding outside Germany tended to focus on frontal area. The radiators required for water cooling imposed a substantial drag penalty, however, and much thought and work went into how to reduce this. Although radial engines performed very well in many areas they suffered from high frontal area. Like rotaries they were low in weight relative to their power but unlike rotaries they could be scaled up, could much more easily use more than one row of cylinders, and could thus be made very large and powerful. Designers initially believed that the cylinders of the radial engines needed to be completely exposed to the air to be properly cooled, which looked sensible but in fact imposed a huge drag penalty. Close cowling of radial air-cooled engines, developed by NACA in 1928, massively reduced aerodynamic drag and actually improved cooling substantially. It even contributed some thrust, all of which together made long transoceanic flights far more practicable. Early tests showed that the reduced drag of the NACA cowl gave an equivalent power increase of 40 percent (Hassell 1997, 145). In 1929, without any knowledge of the work at NACA, Hubert Townend at Britain's National Physical Laboratory developed a more limited engine cowl, the Townend ring, which demonstrated aerodynamic improvement over uncowled engines, though only at speeds below 250 miles per hour (mph). Designing the NACA cowling to ensure proper cooling was always rather a black art and, as long as speeds stayed relatively low, companies such as Boeing preferred the simpler Townend ring.

Variable Pitch Propellers

The third crucial component of the propulsion system was the variable pitch propeller, akin to the transmission in an automobile. The Wright brothers had realized that the propeller was, in effect, an airfoil that, when generating thrust, behaved much as a wing did when it generated lift. Early wooden propellers were fixed pitch and optimized for takeoff, the most dangerous part of the flight with high fuel loads, not high-speed. Metal propeller blades, the angle of which could be adjusted relative to the air they cut through, allowed the engine to perform at its best throughout the airplane's entire speed range, delivering maximum thrust at takeoff yet allowing efficient air movement at high speeds. The first such propellers, developed by the American company Hamilton-Standard in the early 1930s, had only two settings, "fine" for takeoff and "coarse" for cruise. By the mid-1930s fully variable pitch propellers had been developed (Munson 1997, 18).

Fuel Quality

Despite the empirical understanding that some oilfields produced better petroleum than others, early designers of internal combustion engines had very little sense of how fuel burned in a combustion chamber and why some fuels burned better than others. They had certainly recognized before WWI that getting the fuel in and the exhaust out was critical, and racing car engine technology had shown that higher outputs were possible when valves were properly designed and positioned in the cylinder head. Daimler-Benz's

160 hp aeroengine was a logical development of the engines used in their successful Mercedes racing cars just before WWI. But what went on in the combustion process was still a mystery until the British engineer Harry Ricardo started to build a series of experimental engines. His first, in 1919, had a variable compression ratio. Later he added quartz inserts to allow visual analysis of how fuel entered, was burned, and left the combustion chamber as exhaust gases (Ricardo 1931, 30–37, 105–106). Ricardo was trying to solve a problem that bedeviled all early seekers after power. Increasing the compression ratio allowed higher power outputs, but at high compression ratios the highly volatile fuel/air mix ignited on the compression stroke before it was supposed to, which was when the spark plug fired. Ricardo mentions that, since weight of fuel matters greatly for aircraft, what matters most is the heat value per pound of fuel, not per gallon. As a result "a very light and highly volatile fuel . . . was prepared by the Asiatic Petroleum Co. for, and used in, the cross-Atlantic flight" of 1919 by the British fliers Alcock and Brown (Ricardo 1931, 7).

Ricardo was seeking a mechanical solution to preignition. He theorized that if you could design the combustion chamber and its valves more scientifically you would be able to get higher power outputs from higher compression ratios without preignition. Although he was correct, Charles Kettering and Thomas Midgeley Jr. found an easier solution in 1921 in America—change the chemical composition of the fuel to retard preignition. Thanks to their work on adding tetraethyl lead to fuels octane ratings started to climb in the 1920s, from around forty to around sixty-five. How to measure octane ratings more accurately was an issue throughout the 1920s and 30s and subject to much work by the Co-operative Fuel Research Committee (Hamilton & Falkiner 2003, 62–63). Chemical treatment with lead allowed fuels to reach eighty-five octane in the late 1930s. Fuels with a rating of 100 octane became available at the onset of WWII for aviation use (Strauss 2003, 89). Fuel quality thus became directly linked to engine power and indirectly to reliability, since a given engine could be made a little heavier and stronger when better fuels allowed it to generate more power from a given capacity.

Aerodynamic Theory

The most besetting problem through the late 1920s was, despite the diffusion of German technology after WWI, the generally poor understanding of airframe aerodynamics in Britain and America compared with Germany, although the improvement thereafter was extremely rapid. This had implications for speed, endurance, and take off performance. In the early 1900s two competing theories had emerged out of fluid dynamics, developed strongly in Britain by such as Froude and Reynolds, "the discontinuity theory and the circulatory or vortex theory. . . . The names derive from the particular character of the postulated flow of air around the wing" (Bloor 2011, 3). Lord Rayleigh, a highly respected member of the influential Cambridge School of mathematical physicists, well published in the area of fluid dynamics that was important to ship hull designers, developed the discontinuity theory: unfortunately, the respect afforded Rayleigh ensured its continued influence in Britain despite plentiful evidence that it was wrong. Rayleigh's academic influence was finally ended in Britain by the publication of Herman Glauert's work by Cambridge University Press (Glauert 1926).

In the first decade of the 1900s the British engineer, Frederick Lanchester, published the first major work on the vortex theory in two books on aerodynamics (Lanchester 1907; 1908), but the respect afforded Lord Rayleigh ensured that Lanchester's work was generally ignored in Britain. Lanchester was well known as a highly educated engineer and as a pioneering manufacturer of sophisticated automobiles, but not as a theoretician, a mathematician, or a fluid dynamicist. The Physical Society rejected for publication Lanchester's earliest work on the vortex theory, read to the Birmingham Natural History and Philosophical Society in 1894. Disheartened, Lanchester never resubmitted it. In fairness his writing style was difficult, but the Cambridge School attacked his work for not being quantitative and even recent authors have tended to continue this line of criticism: "he made no substantive aerodynamic calculations of lift and drag" (Anderson 1997, 246). However, his books were quickly translated into both French and German (Lanchester 1909; 1911) and it was clear that "the coming-of-age of theoretical aerodynamics began with Frederick Lanchester" (Anderson 1997, 247).

What should have been a bright spot in the development of British aeronautical theory was the founding of the Advisory Committee for Aeronautics in 1909 by the British secretary of state for war, Richard Haldane. Although chaired by Rayleigh it included Lanchester. However, during the run-up to WWI and throughout much of the war itself the committee was mercilessly attacked for not including practical men who actually produced airplanes and condemned as nothing but "a committee of professors" (Bloor 2011, 21). By the time the war broke out the committee had effectively abandoned Rayleigh's discontinuity theory but, despite Lanchester's presence on the committee, it failed to shift to the vortex theory. As Lanchester himself noted, he was treated on the committee not as an aerodynamicist but as an expert on the production of light, reliable, and powerful engines, at which he was certainly very skilled. Many of the attacks on the committee, on Haldane, who was hounded out of office by false charges he was pro-German, and on the British government owned and run Royal Aircraft Factory emanated from Cecil Grey, an "accomplished polemicist" and an apostle of free market solutions only (Bloor 2011, 27) who hated government and was the influential editor of the important weekly magazine, *Aeroplane*. Lanchester was one of the few people willing to stand up to Grey and Grey repaid him by describing his *Aerodynamics* as "some science of symbols invented by Professor Lanchester" (quoted in Bloor 2011, 137). The combination of Rayleigh's continued academic influence and Grey's vicious, unprincipled, and very public attacks placed Lanchester's theories at a disadvantage in Britain.

But Lanchester was at no such disadvantage in Germany. By the early 1900s German science and engineering was becoming the most theoretically grounded in the world and its technical universities excellent. German theoretical and practical work in areas such as chemistry and physics was moving well ahead of the rest of the world (Hugill & Bachmann 2005). Lanchester's theories were massively improved upon before WWI by the work of Ludwig Prandtl and his doctoral students at the University of Göttingen's Institute for Technical Physics. Lanchester visited Göttingen in 1908 at the invitation of Carl David Runge, who taught a seminar course on hydro- and aeromechanics with Ludwig Prandtl. Together with his wife it was Runge, whose English was excellent, who translated Lanchester's work (Bloor 2011, 133). Prandtl then moved Lanchester's theories forward, quantifying them, using his wind tunnel at Göttingen to demonstrate their correctness, and diffusing them to much of the German aircraft industry via technical memoranda and his large number of well-trained doctoral students. When Graf Zeppelin consulted Prandtl in 1913 about the possibilities of using one of his dirigible airships to win the *Daily Mail* prize Prandtl dissuaded him on aerodynamic grounds and persuaded the Graf to turn his attention to large, multiengine airplanes.

By the armistice at the end of WWI the Germans had developed two categories of very large four and five engine airplanes, called *Reisenflugzeug* or R-planes. Both owed their origins to Graf Zeppelin. The first category emerged directly out of the *Daily Mail* competition for the first Atlantic crossing. In 1913 Hellmuth Hirth, a skilled pioneer flier, proposed a six-engine seaplane in which he intended to "cross the Atlantic in 1915 and then fly on to the San Francisco World's Fair." Hirth joined forces with Graf Zeppelin, but WWI intruded and Hirth proposed instead a large bomber derived from his proposed transatlantic airplane. The Hirth/Zeppelin collaboration developed initially into the VGO Company, which became Zeppelin-Staaken when it moved to Staaken airfield outside Berlin. This company focused on rather empirical designs: huge, thin-winged, strut-braced, four and five engine, wood and canvas biplanes with a wingspan bigger than a modern Boeing 747. Although many companies built prototype R-planes to this formula, Zeppelin-Staaken eventually produced the only ones that saw combat, bombing London in particular. Graf Zeppelin also supported a second company, Zeppelin Lindau, set up at Lindau on Lake Constance for Claudius Dornier to produce what would become the second category of large airplanes. These were very much more advanced designs, initially giant all-metal flying boats (Haddow & Grosz 1969, 209–211, quote from 209). Any of the successful R-planes could have taken the *Daily Mail* prize for the Atlantic crossing, but when the prize was reopened after the war the Germans were banned.

Only a few R-planes were successful. The empirical approach to their design resulted in some that were truly bizarre and often disastrous, the Linke-Hoffman R.I being memorably described by one of its pilots as "not an aircraft but a sickness" (Haddow & Grosz 1969, 144). The most ludicrous extreme of this approach to large aircraft design was reached by an Italian flying boat of 1921, the Caproni Ca 60. Caproni had designed several successful large bombers in WWI for the British, American, and Italian air forces. To get enough lift for its supposed 100 passenger payload the Ca 60 had nine wings in three sets of three and eight 400 horsepower American *Liberty* engines. It staggered briefly into the air over Lake Maggiore, its maiden flight ending when its wings collapsed under the aerodynamic load. Mercifully it crashed without killing anyone, which is more than can be said for too many other aircraft of the period.

One of Dornier's coworkers, Adolph Rohrbach, took Dornier's sophisticated construction techniques to Zeppelin-Staaken to build large land planes by 1919. By the time of the second *Idflieg* fighter competition in May/June 1918 Dornier was showing his Zeppelin D.I, an all-metal, stressed skin, cantilever-winged biplane, an extremely advanced structure for the time. Its lack of interplane struts and bracing wires combined with the naturally smooth surfaces of its stressed skin structure to give it excellent aerodynamics. Rohrbach would adapt this stressed skin construction for large landplanes. Many pilots who flew the Zeppelin D.I regarded it as the best airplane at the second *Idflieg* trials, but it had not been properly tested and failed structurally, killing its pilot. The third *Idflieg* completion in November 1918 saw a strengthened version and an order for fifty. Both the USAAC and the USN purchased examples for evaluation when the war was over (Herris 2013, 104–105, 124).

The research material diffused to America by NACA included voluminous information on German aeronautical theory and on all-metal, cantilever-winged airplanes such as Dornier's Zeppelin D.I and Junkers' J-9 fighters, the Junkers F-13 airliner, and the hugely innovative all-metal, stressed skin construction of Rohrbach's Zeppelin-Staaken E 4/20 four-engine airliner. Any of these planes would not have looked out-of-date in the mid-1930s (Kosin 1988, 38; Kay 2004, 28; Hugill 1993, 255). The low wing monoplane Junkers J-7 was shown at the first *Idflieg* fighter competition of January/February, 1918, and its more sophisticated successor, the J-9, at the second in May/June 28 (Herris 2013, 58, 103). The J-9 was considered so promising it was awarded a development contract although its "all-metal construction . . . represented an insuperable obstacle to mass production at the time" (Kosin 1988, 39). The J-9, slightly enlarged as the D-I, saw very limited service in the last weeks of the war (Herris 2013, 128). The single-engine Junkers F-13 was derived from the J-9 and was the world's first successful commercial airplane, 322 examples being built between 1919 and 1932. With a passenger capacity of four it was more like an air taxi than an airliner, but it was robust and reliable. Some F-13s were built in the clandestine factory Junkers set up at Fili, a Moscow suburb, to train Soviet aircraft manufacturers in the latest German technologies and turned over to the Soviet manufacturer, Andrei Tupolev, in 1925 (Kay 2004, 47). Twenty-eight F-13s were sold in America but plans to build the F-13 there collapsed after a series of fiery crashes caused by ill-considered local modifications to the fuel supply. The F-13 also sired a series of descendants at Junkers, most significantly the W-33. This was intended as a cargo only version of the F-13 and in 1928 would become the first airplane to cross the Atlantic nonstop from east to west (Kay 2004, 67). Almost unknown is the Junkers J-14, a large four-engine airliner along the same lines as the Zeppelin-Staaken E.4/20. Construction began in 1920, but it was destroyed unfinished on the orders of the Inter-Allied Control Commission immediately it was discovered (Kay 2004, 39).

Pretty much all the practical work on aerodynamics in the late teens and early twenties came from German companies such as Junkers and the two branches of Zeppelin. At the practical level the most important figure was Rohrbach at Zeppelin-Staaken. Although Rohrbach does not figure in the academic history of aerodynamics—his name is absent from the index of Anderson's otherwise seminal book (1997)—his work was crucial in the ability of designers to move beyond the limitations of the Lanchester/Prandtl style design, limitations that Lanchester himself had identified. Rohrbach correctly recognized that the limits on load carrying ability identified by Lanchester could only be transcended by increasing wing loadings. Higher wing loadings would make airplanes "cheaper in initial cost and running work, . . . less sensitive in squalls and winds . . . , superior in general robustness . . . and . . . [able to] carry considerably higher load with a noticeably higher speed, than "old" type large planes of the same weight" (Rohrbach NASM/CO523900). The E.4/20's wing loading was 80 kilograms per square meter, twice that of some of its most advanced contemporaries (Haddow & Grosz 1969, 290). At the theoretical level most of the work came from Prandtl, Prandtl's acolytes, such as Glauert in Britain, and Prandtl's Ph.D.s, such as Max Munk, who was at NACA from 1920 to 1926, and Theodore von Kármán, who went to the California Institute of Technology (Caltech) in 1930 (Hanle 1982). Munk developed NACA's wind tunnel and a new and important family of airfoils, but was an arrogant and difficult character. He alienated most of his coworkers and failed to push American aerodynamics as far forward as he should have (History of NASA Online). It was left to von Kármán to complete the diffusion of German technology to the American aircraft industry.

Despite this initial German focus, one of the main breakthroughs in aerodynamic understanding in the late 1920s came from a British scientist in an influential article entitled "The Streamline Aeroplane" (Jones 1929). This proposed "a new rule of thumb for aircraft designers" and showed that "most contemporary designs were using at least three times as much power at cruising speed as his ideal aircraft required" (Hassell 1997, 143). Jones offered no new theories, but he both validated Prandtl's theories very publicly for the English speaking world—Glauert had previously validated them to the academy—and persuaded many British and American designers to begin to think differently about the advantages of streamlining via such things as stressed skin construction, retractable undercarriages, and the like as the 1930s began.

Retractable undercarriages had been used since the very early 1920s in a very limited way, but they seemed to offer no real advantages as long as speeds were low. One of the first was on the almost unknown American Dayton-Wright RTB that competed in the 1920 Gordon-Bennett races in France and was designed for up to 200 mph. Such devices offered an obvious disadvantage if the pilot forgot to extend the undercarriage during landing—a regrettably common occurrence—and early designs such as the Boeing 247 and the Douglas D.C.2 and 3 only partially retracted wheels so that they would reduce the costs of such a mistake! But they were an obvious way of reducing drag as speeds rose and by the early 1930s they had become *de rigueur* on advanced military and commercial designs.

Airframe and Wing Design

Airframe Design

The key to better airframe structures was clearly the switch to metal, either for part or all of the airplane's primary structure or for the entire airframe. In Britain Vickers had used fabric covered steel fuselage structures from its inception, unsuccessfully proposing such structures to the admiralty as early as 1911 (Andrews & Morgan 1969, 34). In Germany Reinhold Platz created elegantly simple welded mild steel fuselage structures for Anthony Fokker. Platz's Fokker Dr-I Triplane is famous, but his Fokker D-VII was exceptional (Gray & Thetford 1962, 106). Also in Germany Junkers experimented with corrugated iron structures. Although his first iron airplane crashed in 1913 he went on to produce the world's first airworthy all-metal airplane in 1915 with his J1. In his J3 of 1916 he switched to the much lighter duralumin (Kay 2004, 15–18), moving quickly through the J-7 to the J-9 and D-I.

In their use of aluminum Dornier and Rohrbach followed suit although they innovated further in embracing stressed skin structures.

Three forces drove the early adoption of metal structures in Britain that were unique to that country. One was simply the problem of securing a large enough supply of high-grade lumber to build the huge number of airplanes that were needed in the last two years of WWI as the air war ramped up. Britain had long been short of lumber, and had, for example, pioneered the shift to iron-hulled ships in the 1840s as a consequence (Hugill 1993). Western Canada was a major source of the high-grade spruce and other lumber needed for British airframe construction but it was the entry of America into the war in 1917 that ensured British access to a huge, reliable supply of high-grade lumber. By 1918 British production was running at 3,000 airframes per month, but even more would have been needed had "Plan 1919" gone into effect. American entry into the war allowed Britain to just avoid a serious shortage. The Bristol Aeroplane Company had recognized this problem early and had produced its experimental, steel-framed, duralumin clad M.R.1 Metal Biplane in 1917, as well as successful sample steel wings for other aircraft (Barnes 1988, 126–127). Such airframes were simply metal versions of the conventional British wooden airframes of the period, not the sophisticated structures being pioneered in Germany.

The potential lumber shortage avoided by the entry of America into the war "was not lost on the [British] Air Ministry and, as a matter of policy, it was decided that, from 1924 onwards, all aircraft ordered into production for the RAF [Royal Air Force] should be of metal construction." The question was, steel or aluminum, and the Air Ministry initially preferred steel as a domestic product that was a known quantity in terms of its ease of working and corrosion protection, but sheet steel would have been much too heavy as a skin. Aluminum, at least to begin with, was both difficult to work with and expensive in production terms. The logic in 1920s Britain thus pointed to a simple substitution of tube steel for wood in the airframe and the retention of fabric covering (Tapper 1973, 22–23). The first aircraft to incorporate this shift was the Armstrong Whitworth *Siskin*. This emerged in 1919 as a classic all-wood, fabric-covered structure and showed great promise. The Air Ministry encouraged its rapid development but also a shift to metal construction. The Mark II *Siskin* of 1922 thus had a fabric covered, steel tube fuselage and wooden wings and the Mark III of 1924 made the final transition to an all steel structure with fabric covering (Tapper 1973, 117–123). At Shorts Brothers, Oswald Short far preferred duralumin, using it from 1916 on in airship construction and proposing an all-duralumin airplane to the Air Ministry in 1919. When they refused to fund it he built his *Silver Streak* as a private venture. It was a sensation at the Olympia Aero Show in 1920. Compared to the far more radical German airplanes it was, however, a very conventional thin-winged biplane. The Air Ministry finally bought it in August, then proceeded to pretty much ignore it despite its exceptional performance in tests in 1921 (Barnes 1989, 162–167).

The German switch to metal structures before and during WWI does not seem to have been resource driven to the same extent as Britain. Reinhold Platz at Fokker had long used steel structures with fabric covering for engineering reasons. In any case Southern Germany, Bavaria in particular, had plentiful supplies of high-grade lumber and the Pfalz airplane company continued to produce excellent wooden airframes for the Bavarian air force through the end of the war.

A second force pushing the British to metal construction was that the wooden airframes used in WWI proved to have extremely short life expectancies, even when not lost in combat. When the war was over austerity set in, and airplanes were expected to last much longer. Although the British cut their air force severely it was given major new responsibilities in "air policing." Since this was an entirely colonial venture much of the area to be policed had extreme desert or tropical climates. Problems with wooden airframes showed up quickly. The austerity minded Geddes Committee of 1921 noted that the RAF had more than enough war surplus Bristol F2.B Fighters and de Havilland D.H.9As for pretty much any predictable future use, and argued that there was thus no need to procure any new airplanes. After the decision to "police Iraq and the North-West Frontier of India mainly with the RAF instead of the Army" reality set in. Wooden airframes were totally unsuited to areas where they would be "subjected to extremes of temperature and humidity, aggravated by termites and tropical fungi. There was thus the strongest possible case for early replacement of these aging veterans by all-metal aircraft," the first of which was a militarized version of Shorts' *Silver Streak*, the *Springbok*. But while "there were plenty of skilled carpenter-riggers available to minister to orthodox wood-and-fabric airplanes, however fragile, trained sheet-metal tradesmen capable of repairing aluminum-alloy stressed skin structures were as rare as icebergs in the Sahara" (Barnes 1989, 168–169). This helped ensure a switch to steel and fabric structures and airplanes such as the *Siskin* rather than any early move to more advanced all-duralumin airplanes such as the *Springbok*.

The third force pushing British companies to metal construction was that British maritime geostrategy needed a considerable number of large flying boats. During the war these had proved their worth as naval patrol vessels, especially against U-boats, and Britain had a large maritime empire to patrol. In addition, commercial flying boats were the obvious way to connect a maritime empire politically, since they incurred no cost for airfields. Wooden structured flying boats suffered similar problems to wooden landplanes, perhaps absent the termites except when they were beached. But wooden hulled boats could easily soak up several hundred pounds of water, which ruined their payload. Here, of course, steel could not be made to work, being simply too heavy.

Shorts' breakthrough in getting the British to accept metal construction was the Shorts S.2 of 1924, which added a duralumin hull to the strut braced wooden wings of a much older design, the Felixstowe F.5, itself a design originating with the Curtiss America designed in 1913 to attempt to win the *Daily Mail* prize for the first transatlantic flight (Barnes 1989, 197). Shorts were concerned about the poor life expectancy of wooden structures in the tropics, about the substantial weight gain as wooden hulls became waterlogged, and about leakage after heavy landings. For example, the metal-hulled version of the Supermarine *Southampton* weighed 500 pounds less than the wooden-hulled version, which also absorbed up to 400 pounds of water: the 900 pound weight saving allowed 200 miles more range or greater load carrying (Andrews & Morgan 1981, 100). On one trial a Shorts S.2 was deliberately stalled in from 30 feet when landing without its hull springing any leaks (Barnes 1989, 197). The S.2 evolved into a series of duralumin-hulled biplane boats that kept Shorts ahead of their competitors, but the Air Ministry ordered other British companies to outfit similar flying boats with metal hulls, since such boats were considered crucial to both ocean patrolling and imperial communications. These were the Blackburn *Iris*, English Electric *Kingston* Mk. II, an up-dated version of the Saunders *Valkyrie* called the A.14 "metal hull," and the Supermarine *Southampton* Mk. II (Jackson 1989, 196; Ransom & Fairclough 1987, 142; London 1988, 71; Andrews & Morgan 1981, 100). Shorts' technological lead was not great, and there was little to choose between the S.2 and the four other airplanes in performance terms.

Wing Design

The initial key to modern wing design was the adoption of the cantilever wing, pioneered in Germany during WWI. Such wings needed no complex drag inducing struts and bracing wires. Beyond that the issue was how a wing could be designed to be (a) safely resistant to stalling and (b) capable of high lift at low speeds and low drag at high speeds.

It was a British designer, Frederick Handley Page, who pioneered high-lift devices on wings with his leading edge slot. He began thinking about this in 1911, did considerable wind tunnel work on it from 1917 on in great secrecy because of its obvious commercial value, and finally acquired British and American patents in 1919 (Barnes 1976, 210–211). Orville Wright patented a split flap at the rear of the wing in 1920, and later that year Handley Page patented a combination of the two ideas for the rear of the wing as the slotted flap. The leading edge slot alone increased the wing's lift coefficient from the value of 1 for the simple wings of the airplanes that fought WWI to 2.3, but its big advantage was that it delayed the stall, a safety problem that greatly worried all early designers. By itself, the split flap at the rear of the wing gave a lift coefficient of 2.6: the slotted flap pushed that figure up to 2.9. For comparison, the complex wing designs of modern airplanes have lift coefficients of 3.9 (Anderson 1997, 362–367). Rear edge flaps proved far more useful and effective on cantilever-winged monoplane designs than on strut and wire braced biplanes, which benefited more from the leading edge slot.

Handley Page's work on leading edge slots was paralleled by that of a German designer, the former WWI pilot Gustav Lachmann. Recovering from an accident in which his airplane stalled and spun in, Lachmann pondered its cause, experimented, and drafted a German patent specification that was refused in early 1918, just before Handley Page's first patent application. In 1920 Lachmann, finally hearing of the work of Handley Page, persuaded Ludwig Prandtl to test his idea in the wind tunnel at Göttingen, and was then able to acquire a German patent backdated to February 1918. Lachmann and Handley Page met in Berlin in 1921, came to an amicable agreement, and began collaborating. Both really had different goals: "Lachmann was chiefly concerned with improving safety, while Handley Page hoped to use much higher wing loadings." Lachmann went on to obtain his doctorate under Prandtl, moved permanently to Handley Page's in 1929, and became the company's chief designer in 1931 (Barnes 1976, 33–34, 215–216, quote from 216).

Take Off Performance

The problem of the Zeppelin-Staaken E.4/20 was that to get low drag and exceptional speed and range Rohrbach went for a very high wing loading, more than twice that of his contemporaries. The E.4/20 thus needed a long run to accelerate to a high enough speed to get off the ground, whereas biplanes with low wing loadings lifted off quickly. In the postwar environment, with his E.4/20 destroyed, Rohrbach turned his attention back to the flying boats he knew so well from his days with Dornier. In 1922 Dornier produced his *Wal* flying boat to the same formula as the E4/20 and in his Ro II flying boat of 1923 Rohrbach followed suit, both boats using a very high wing loading since they had no real limitations on the length of their take off runs. The Ro II was built in tiny numbers in Denmark in the early 1920s to circumvent the Inter-Allied War Commission's restrictions on the production of large aircraft in Germany. The *Wal* was built in Italy and Spain.

The first major American airplane to be built along the lines developed by Rohrbach in his Zeppelin-Staaken E4.20 was the Boeing Model 214 that became the B-9 bomber and gave its wings to the Model 247 commercial airliner of 1933. The 214 and 247 had retractable undercarriages and used

Townend rings to smooth airflow over their radial engines, but their take off performance was relatively poor. As speeds increased in the early 1930s it became obvious that retractable flaps at the rear of the wing had become a necessity to improve take off performance. As a result, the most crucial design in the development of commercial airplanes was the Douglas DC-1 of 1933 and the designs it spawned. Douglas used flaps to give the DC-1 and its successors excellent take off performance, and Douglas was the first American aircraft company to pay full attention to German aerodynamics. The DC-1 effectively combined Jones' arguments for the streamline airplane with a thorough understanding of German aerodynamic theory and the work of NACA in improving engine aerodynamics. Douglas quickly stretched the DC-1 from ten passengers to twelve as the DC-2, then produced the radically larger DC-3, initially as the Douglas Sleeper Transport for the fledgling American airlines, which used it to replace their superannuated Curtiss *Condor* biplanes on overnight transcontinental routes. For the design of the DC-3 Douglas consulted with one of Prandtl's greatest students, Theodore von Kármán. In Britain Shorts used the same suite of technologies to come up with their S.23 *Empire* C Class flying boat and its descendants. Imperial Airways used the *Empire* boats to replace airplanes such as the Shorts S.17 *Kent* boats on imperial routes. Both the *Condors*, which were a civil version of the Curtiss B-2 bomber (Bowers 1979, 215), and the *Kents* would have looked at home in 1919.

Navigational Equipment

As early as February 1919 the forerunner of the USAAC, the Army Air Service, began planning for a flight to Hawai'i, a far more difficult navigational problem than crossing the Atlantic. The instrument branch of the air service engineering division was established under the leadership of Second Lieutenant Albert F. Hegenberger, an aeronautical engineer trained at MIT, to study techniques of air navigation, and by February 1920 had developed a program for a transpacific flight from California to Hawaii. While the primary goal of the instrument branch was effective instrumentation, development of all-weather and night navigation capability was espoused by the USAAC's assistant director, Brigadier General Billy Mitchell. One of Mitchell's protégés was First Lieutenant Lester J. Maitland, who was transferred to Hawai'i in May 1919. Maitland tried to organize a flight to Hawai'i using Martin's new NBS-1 twin-engine bomber, but was refused permission. Meanwhile Hegenberger was working to test and adapt for military purposes the earth inductor compass first patented in 1912 but heavily developed by 1924 by the Pioneer Instrument Company in conjunction with the National Bureau of Standards.

Magnetic compasses did not work on aircraft, there being far too much metal in engines and the like that affected them. The first key to accurate navigation on long over water flights was therefore the earth inductor compass.

> The outstanding feature of the earth inductor compass is the disassociation of the magnetic element from the indicating element. Instead of using magnetic needles, the direction responsive element of the earth indicator compass is an electric generator the same in principle as any electric dynamo except that no artificial field is used, the earth's flux serving for a field. The output of such a generator is dependent upon the angular relation between its brushes and the earth's flux. (Goldsborough 1927, 544)

At the USAAC Hegenberger worked on radio compasses as well as the earth inductor compass. The latter was first successfully used in the round the world flight of the USAAC's Douglas World Cruisers of 1924. The successful USAAC flight to Hawai'i in 1927 by Maitland and Hegenberger used both a radio compass—that failed—and an earth inductor compass. Charles Lindbergh used an earth inductor compass in his flight from New York to Paris in 1927. Such compasses fell out of favor in the 1930s and were replaced by more accurate gyroscopic heading indicators (Smithsonian Time and Navigation Online).

Airport Infrastructure

The apparent advantage of the flying boat in the 1920s and 1930s was that, on a planet predominantly covered with water and with plenty of rivers and inland lakes it did not seem to require much in the way of infrastructure, other than docking facilities. Whichever direction the wind was blowing, the boat could take off into it. Flying boats also offered a psychological advantage. Given the relative unreliability of early engines, engine failure on long over-ocean flights merely meant landing on the water and taxying along or waiting for rescue. Of the three USN Curtiss NC boats that set out to cross the Atlantic in 1919 NC-1 alighted at sea short of the Azores: her crew were rescued by a ship. NC-3 came down 200 miles from the Azores and could not take off again, so taxied in. Of the three boats that set out, only NC-4 made it (Bowers 1979, 118–119). In 1925 a USN crew attempting to fly 2,400 miles nonstop from San Francisco to Hawai'i without proper navigational aids came down with engine failure 559 miles short of their destination. The crew rigged sails using fabric from the lower wing to reach the islands (Swanborough & Bowers 1976, 335).

Low wing loadings made sense for landplanes when such were expected to take off from airfields that were no more than flat grassy areas. Such fields were often quite small and high take off speeds were hard to reach on grass, especially when the field was wet. Other than flaps, a critical key to the development of aircraft with high wing loadings was thus the use of long, hard-surfaced runways. Such runways were a by-product of the road making technology that developed in America in the late 1920s.

The development of hard-surfaced roads was a long and tedious process, although important pioneering efforts had appeared in France and Britain during the early 1800s. As automobile ownership exploded in America the Federal Highway Act of 1916 began to move the country forward, and by 1922 every state had a state highway department. During WWI many water bound macadam roads built under the state "Good Roads" programs of the late 1890s through 1916 failed under truck traffic, but a new highway construction technique using concrete surfaces seemed to be the answer (Hugill 1982). Unfortunately concrete highways often failed structurally for no apparent reason. The problem, as it turned out, was not so much in the structure of the road itself as in understanding what could be described as the road foundations. This required a sophisticated understanding of soil mechanics, and the beginnings of such was not available until the publication in 1925 of Karl Terzhagi's pioneering work (Terzhagi 1925).

However important Terzhagi's theoretical work, a new generation of reliable concrete highways could not be built in America until two other things happened: a cadre of road engineers had to be trained with the proper knowledge of soil mechanics; and sophisticated road making machinery had to be developed to get beyond the labor intensive use of shovels and wheelbarrows by which the early state "Good Roads" had been built. First, after the publication of his work on soil mechanics the Austrian born Terzhagi spent the years 1925 to 1929 at MIT and as a consultant to the US Bureau of Public Roads, developing a cadre of graduate students who would both build roads and populate highway and university engineering departments, both at the state level and in many Morrill Act universities. His work was also central to the construction of earth dams, subways, and building foundations. In 1929 he returned to Vienna and eventually consulted on autobahn development for the Nazis, but fell out with them over his having also consulted for the Soviet Union and the fact that his second wife was American, all of which made him politically suspect. In 1938 the *Anschluss* pushed him back to America, and he eventually settled at Harvard University, although he was heavily recruited by the Texas Agricultural and Mechanical College, the ancestor of Texas A&M University and the Morrill Act University for Texas, as the construction of dams to tame Texas' wild rivers got under way. Second, the 1920s also saw a massive upswing in the development of earth-moving machinery, most notably motor graders, the first of which appeared in 1919, motor scrapers, developed by the Texas company, LeTourneau, from 1923 on, and Caterpillar bulldozers, the first of which appeared in 1925 (Hugill 2011).

Terzhagi's work and that of his students and acolytes from M.I.T. and Harvard was as applicable to airfield runway construction as it was to roads. During WWI Terzhagi had been in charge of airfield construction for the Austro-Hungarian Empire, although at that time no hard-surfaced runways were in use. What the work of Terzhagi, his students, and companies such as LeTourneau and Caterpillar made possible for the allies in WWII was rapid construction of long concrete runways for the large, heavy, four-engine bombers needed by the strategic bombing campaign. Whereas DC-3's could take off and land on grass fields, the late 1930s generation of four-engine airliners needed hard-surfaced runways as much as the strategic bombers they were related to. Such hard-surfaced runways were a direct consequence of better soil science, soil mechanics, and road engineering. The ability to build such runways cheaply and quickly was the death knell of the long-range flying boats, which suffered much more of a weight penalty from the size of their hulls than land planes. By the time WWII was over the long-range land plane using a new generation of airports equipped with long, reliable, concrete runways would thus reign supreme.

Real World Results of the Nine Technological Developments in America and Britain

The Case of America

In many ways the real beneficiary of all the German and British work on aerodynamics and structures in the late war and early postwar period was an American trading state that reaped the benefit of being a "fast second" and not an early adopter of these complex new technologies. America was also an adoption environment that was particularly conducive to both the concept of precision strategic bombing as put forward by Billy Mitchell and to commercial air transport as it began to mature in the early 1930s. Large, long-range bombing airplanes offered the possibility of extending America's "moat defensive" by making it possible to attack enemy fleets far out to sea. Commercial air transport made sense for the rapid movement of highly paid executives around a large, fully integrated economy spread over a considerable but relatively well settled continental area over which income was relatively evenly spread.

By the late 1800s America had developed an excellent national transportation system, its railroads. These did an excellent job of connecting its main cities and showed there was considerable demand for long-distance travel. By the early 1900s this system was in a constant state of acceleration, with intercity trains getting faster, more frequent, heavier, and more comfortable

(Stilgoe 1983). Two national routes stood out based on the rail system, both of which quickly became important for air transport as it developed. Three over-ocean routes, two of which were technologically very challenging, would emerge from air transport as America became as a global rather than merely a continental power.

1. The first national route connected the two main cities of the country, New York and Chicago, one the interface with the Atlantic world and a major player in the global economy, the other the powerhouse center of the national economy and the nation's main rail hub. As with route (2) the airplanes that flew this route had to be commercially viable.
2. The second national route connected New York to the rapidly growing and highly innovative west coast, especially Los Angeles. Los Angeles, in part because of its reliably sunny weather, was the home to two new industries that would generate a great deal of wealth and an aura of glamour in the 1920s and 1930s: movie making, focused on Hollywood, and airplane manufacturing (Hugill 1993; Meinig 2004).
3. The third route was really a complex set of routes by which America began to cement its geopolitical control over the Caribbean and Latin American states. It had long claimed something akin to this via the Monroe Doctrine, but aviation made it far easier to move financial instruments and high-ranking political and business figures around the region. The airplanes that flew this route did not need to be commercially viable, nor have very long-range.
4. The fourth route, San Francisco to the Hawaiian Islands, developed out of America's increasing concern with its Pacific geostrategy, the need to contain an increasingly aggressive Imperial Japan, and the central role of the Hawaiian Islands in that emerging geostrategy. The airplanes that flew this route did not need to be commercially viable, but they did need very long-range, as did route (5).
5. The fifth route connected the world's most important trading cities. By the early 1900s it was becoming clear that New York was beginning to challenge London as the world's premier trading city, and in 1913 the British tabloid newspaper, the *Daily Mail*, offered a prize of £10,000—$1.16 million in 2017—for the first flight between the two cities in less than seventy-two hours. As with routes (1) and (2) the airplanes needed on this route had to be commercially viable. As a consequence route (5) took much longer to become viable than routes (1) through (4).

The New York to Chicago Route

Once the Confederacy had been broken by the Civil War the city of Chicago exploded as the principal rail hub of the interior of the North American continent as well as a major port on the Great Lakes. The area between New York and Chicago, especially the cities along the rail corridor across New York State and immediately surrounding the Great Lakes—Albany, Syracuse, Rochester, Buffalo, Cleveland, Toledo, and Detroit—emerged as a great industrial region, the country's manufacturing core. Railroads did well promoting the industrial growth of this region.

The problem with the railroads was that they were slow in transmitting the mail Chicago depended upon in its role as the country's major futures' market. In 1902 the New York Central introduced its crack train, the Twentieth Century Limited, timed at twenty hours from New York to Chicago (Hugill 1993, 182). But such timing required a twenty-four hour "float" so that financial transactions between the two cities essentially "lost" twenty-four hours interest in either direction. This represented only 0.1 percent of the transfers, but it was 0.1 percent of a great deal of money. The development in 1919 of overnight airmail service between the two cities and, by 1934, daylight flights of five hours using Douglas DC-2s removed this loss (Hugill 1993, 275–279).

The New York to the West Coast Route

San Francisco, whose stock exchange opened in 1882 to serve America's growing Pacific interests, had a similar, if much lesser case of the financial problems of Chicago since far less money was involved. Even the fastest trains took between four and five days. By 1934 Los Angeles was displacing San Francisco in importance, and Transcontinental and Western Air was scheduling its DC-2s for eighteen-hour flights from New York with no change of airplane and only three refueling stops.

As it developed in the early 1900s it became clear that the Los Angeles area would generate two main products for the national economy: movies from Hollywood and airplanes from Southern California more generally. In turn, Hollywood itself generated two main products, film stars and the films they starred in. Both were, in prosaic economic terms, compact, easily transported, in demand everywhere, and perishable. Both needed rapid transportation anywhere in the country, although the major cities were the logical centers in which they would generate the most income. By the early 1930s the very fastest transcontinental trains took some three days, although five was common. Early air travel between the coasts required several refueling stops, changes of airplane, and even partial use of the railroad. This was faster but inconvenient compared to the increasingly luxurious trains, the best of which were luxury hotels on wheels (Stilgoe 1983). Douglas's new generation of airplanes radically changed the connections between New York and Los Angeles. Although no faster than the DC-2, American airlines began service between Los Angeles and New York in 1936 using the new and more luxurious Douglas DC-3 in 17 hours 30 minutes (Francillon 1988, 223–224). Airplane travel was expensive, the DC-3 carried only twenty-one passengers and airmail and had to land several times to refuel, but for wealthy movie stars and their movies air transport was ideal.

The second industry, airplane manufacturing, was epitomized by the Douglas Aircraft Company of Santa Monica, founded in 1921. Airplane manufacturing began to concentrate on the Los Angeles area in the twenties for the same reason as movies, the area's constantly clear, sunny weather. Southern California's climate made test flying easy, and in the early history of airplane development test flying was extremely necessary. Theoretical understanding of how a particular structure, control system, or engine might perform in the real world was almost nonexistent, and there were almost no wind tunnels for even the simplest empirical ground testing. Flight-testing was central to the fledgling industry.

Flight-testing new airplanes in the 1920s and 1930s was dangerous, which made it rather glamorous, and a logical early link grew up between Hollywood and airplanes, epitomized by one individual, Howard Hughes. Hughes inherited and then developed a huge fortune from his father's rotary drill bit for oil wells, and his own decision to lease rather than sell these. Hughes had a penchant for attractive women that drew him to Hollywood, where he developed a dual career as film producer and airplane builder. In one of his early films, *Hell's Angels* of 1930, he epitomized the attitude of many to the fighter pilots who had fought on the Western Front in WWI, that the best of them were "knights of the air" who lived a glamorous if highly dangerous life. Hughes showcased Jean Harlow, "the blonde bombshell," in Hell's Angels. In 1935 Hughes designed and built his H-1, and set a world landplane speed record of 352 mph with it that same year. In early 1937 he flew the H-1 nonstop from Los Angeles to New York in just over seven hours. The H-1 used a NACA cowled Pratt & Whitney radial engine but Hughes' design did not seem to have had much impact on American designers. The air force stayed with in-line engines for its fighters, believing they allowed the best streamlining and speed, and the main supplier of fighters to the navy, Grumman, stayed with bulky radials that they did not cowl carefully.

Markets and weather were not the only reason for the success of the airplane industry in Southern California. The hiring of German educated Theodore von Kármán and the development of proper wind tunnels by him at the California Institute of Technology, Caltech, were also crucial. The measure of the success of Southern California in the development of this first generation of modern commercial airplanes is that whereas Seattle's Boeing produced 75 Model 247s Douglas had delivered 193 DC-2s and 430 DC-3s by the time America entered WWII in 1941, and held orders for 149 more of the latter. Douglas would go on to build over 10,000 more such airplanes as C-47 military transports for WWII. What Douglas pioneered in the DC series with the help of von Kármán and Caltech was essentially the process of design and testing of models that then became the industry norm. Whereas the DC-2 was designed empirically von Kármán's wind tunnel showed that the initial design for the DC-3, with a simply scaled up DC-2 wing, would not fly.

Whereas the Boeing 247 and the Douglas DC-2 and DC-3 were ideal for the New York to Chicago service longer-range airplanes were needed for the New York to Los Angeles service and the proposed services to London and Hawai'i. From the mid-1930s to 1947 a family of five commercial airplanes were developed by America's rapidly improving industry that owed their origins to the success of the DC-3. Four were land planes: the Boeing Model 307 *Stratoliner*; the Douglas DC-4; the Lockheed L-049 *Constellation*; and Boeing's Model 377 *Stratocruiser*. One was a flying boat: the Boeing Model 314. All three Boeing airplanes owed their origins to bomber designs. In each of the five cases the designers chose four engines to get enough fuel and a large enough payload of passengers into the air to be profitable. Another way to increase range was to fly higher, in thinner air that caused less drag. The DC-3 flew in the relatively thick and turbulent air found at altitudes below 10,000 feet, which lowered speed, raised fuel consumption, and could make passengers airsick. The goal of the *Stratoliner*, the first pressurized commercial airplane to enter service, as it was advertised to passengers, was to fly "above the weather," but higher altitudes meant less drag and thus higher speeds and longer range. The *Stratoliner* used the wings, engines, and tail surfaces of Boeing's Model 299 B-17 bomber married to an entirely new fuselage. Pan-American World Airways initially bought four in 1939 for Latin American routes, but its 2,390 mile range attracted Transcontinental and Western Airways, which ordered six to begin coast-to-coast nonstop service in 1940. Although only ten *Stratoliners* were built and it seems to have routinely needed a fuel stop in transcontinental service it showed the rapid improvement in technology occurring in American airliners of the late 1930s.

Douglas initially proposed that its DC-4 would be a rather larger airplane than the *Stratoliner*, with forty-two seats to the *Stratoliner's* 33, but also with transcontinental range. Even expanded to fifty-two seats Douglas could not demonstrate effective operating economics, and the DC-4, which first flew in 1938, also turned out to be expensive to maintain. The single prototype was renumbered as the DC-4E—for experimental—and disposed of to Japan, where it was used as the basis for an equally unsuccessful long-range strategic bomber. The second DC-4, the *Skymaster*, emerged in 1942 as a smaller airplane with excellent operating economics, Douglas having clearly learned their lesson, but they did not pressurize it. The DC-4 entered production just as America entered WWII, and it served as the C-54 transport. In its L-049 *Constellation* Lockheed proposed an airliner that would not only be pressurized but also able to cruise much faster than the *Stratoliner*, at 300 rather than 200 mph, and with the much longer range of 3,500 miles. Howard Hughes, by now the major stockholder of Transcontinental and Western Airways, had a substantial influence in the airplane's design. The *Constellation* first flew in 1943 and was initially used as a military transport, the C-69. As the war wound down Hughes demonstrated its potential by flying nonstop from Los Angeles to New York in just under seven hours in 1944, beating the transcontinental record he himself had set in 1937 with his H-1, an airplane designed purely for racing. The DC-4 and the *Constellation* were the land planes that convincingly closed the Atlantic gap when WWII was over. A DC-4 flew the first commercial landplane service across the Atlantic on October 23, 1945, refueling in Newfoundland and Ireland, a route made possible by the hard-surfaced runways built for the strategic bombing campaign and to ferry strategic bombers such as the B-17 and B-24 across the Atlantic gap. Only with much later versions of the DC-4 and *Constellation*, the DC-7C and the L-1049 *Super Constellation* that entered service in the mid-1950s, were London, Paris, and Berlin connected nonstop to New York. In June of 1947 Pan-American World Airways emphasized the "World" that it had added to its name in 1943 by using the *Constellation* to open the first regularly scheduled commercial flight around the world. The last of the five airplanes to appear was Boeing's Model 377 *Stratocruiser*, essentially the wings and engines from their Model 345 B-29 *Superfortress* bomber grafted onto a new fuselage. This first flew as a military transport, the C-97, in 1944 and as a civilian airliner in 1947.

The Caribbean and Latin American Routes

Outside continental America, in the 1930s America's geopolitical interests in the Caribbean and Latin America resulted in its encouragement of Pan-American Airways as its "chosen instrument" in order to both cement American geostrategic control and resist attempts by the Nazis to use *Lufthansa* to increase German influence there. The logical air routes from continental America followed the coasts, since all the important cities of the Caribbean and Latin America, with the exception of Mexico City, were established as ports in the Atlantic trade by the Europeans. Such routes were also long enough that the large, multiengine flying boats developed in WWI were the logical technological solution.

As with British flying boats, the genesis of the large American flying boat of the early 1920s lay in the Curtiss *America* and its descendants that had served Britain and America on anti-submarine duty in WWI. The most important Curtiss boat was the F-5, an Americanized version of the British Felixstowe F.5 that had itself evolved from the earlier Curtiss H boat, also known as the *Large America*. The F-5s were built in considerable numbers and served through the mid-1920s. They were replaced by a set of descendants produced by the Naval Aircraft Factory, the PN series, PN-7 through 12, which were in turn replaced by a further redesign as the Hall PH series, the PH-1 through 3, "some of which served into WWII, thus resulting in the same basic flying boat design spanning two world wars" (Swanborough & Bowers 1976, 254–255, 334–336, quote from 336).

From Pan Am's point of view the descendants of the Curtiss *America* were serious anachronisms when compared to what companies like Consolidated, Martin, and Sikorsky were beginning to build in the late 1920s and early 1930s. Sikorsky built a series of effective boats for Pan-American's commercial service in the Caribbean, starting with the S-36 and culminating in the S-42. Such boats were not commercially viable.

The Route to Hawai'i, The Philippines, and China

The real technology forcing that produced such relatively successful flying boats as the Sikorsky S-42 occurred, however, because the American navy began to be concerned in the late 1920s with acquiring a boat with the range to reach Hawai'i from the west coast, a distance further than that across the Atlantic from Newfoundland to Ireland. Hawai'i had, of course, long been central to American commercial interests because of its connections to the trade such east coast cities as New York and Boston had developed with Asia following the revolution. But by the late 1800s and early 1900s, with the beginnings of an American political empire in the Philippines and increasing worries about the expansion of Imperial Japan, Hawai'i had also become crucial to American geostrategic designs in the Pacific.

Although the USN initially used flying boats from the Naval Aircraft Factory, the PN-7 through 12 to (just) reach Hawai'i, the company that eventually best met the USN's needs was Consolidated, which built a series of important, mostly twin-engined, Rohrbach-style boats in the late 1920s and early 1930s. The first of these, which the USN ultimately rejected after forcing the ill-considered design change of adding a third engine it believed was needed for reliability, was the Consolidated Model 9 *Admiral*. When completed in late 1928 it was "the largest flying-boat in the USA" (Wegg 1990, 54). The *Admiral* lost out to the Martin XP2M, which though designed with three engines took Martin much too long to develop. Consolidated's *Admiral* was followed in 1929 by its Model 16 *Commodore*, which used the twin-engined *Admiral* wing over a passenger carrying hull. The *Commodore* was sold as a commercial airplane to the New York, Rio, and Buenos Aires Airways, which was forced into a merger with Pan Am by the American postmaster general. The last direct descendant of the *Admiral* was the Model

22 *Ranger* of 1932, which used the *Admiral's* wing in a sesquiplane layout and was bought by the navy in moderate numbers. In January 1934 the USN showed it had in the *Ranger* a flying boat capable of making the non-stop flight to Hawai'i with a useful military payload. Consolidated did not sit on their laurels and evolved the design one more time into the much more sophisticated Model 28 *Catalina* that first flew in 1935. The *Catalina* would become the most numerous flying boat ever, over 3,400 being built, about 150 of them under license in the Soviet Union (Wegg 1990, 66, 69–74).

Pan Am initially attempted to use Sikorsky's S-42 boats on the San Francisco-Honolulu run, but Hawai'i was too far for them to carry a useful load. Pan Am then turned to Martin's M-130, which Martin built specifically for the Hawai'i, Philippine, and China routes that America needed to cement its geostrategic control over the Pacific basin, although Pan Am hoped they could also use it to close the Atlantic gap. The M-130 was underpowered, lacked high-lift devices on its wings, and could carry only eight passengers on the Hawai'i run, although it was designed to carry up to forty-six passengers on shorter flights. It could be used only for proving flights on the Atlantic.

Douglas built its first modern boat, the XP3D, to compete with the *Catalina*, but by then Consolidated had accumulated a wealth of experience with flying boats, and the Douglas design failed to get any orders (Francillon 1988, 176–179). Consolidated developed two more naval flying boats before WWII, the Model 29 *Coronado* and the Model 31 *Corregidor*. The *Coronado* was a large, four-engine boat but was underpowered and never markedly successful. The large twin-engined *Corregidor* is important because it pioneered the efficient Davis wing, which was then converted in four-engine form for use on Consolidated's very successful B-24 landplane bomber, built in larger numbers than any other strategic bomber of WWII (Wegg 1990, 78, 81). None of the Consolidated boats had much presence on the commercial market since they were optimized for the USN's use.

The New York to London Route

Although in the late 1930s Pan Am used the S-42 and M-130 experimentally on the transatlantic route, neither airplane was able to combine the necessary range flying west against the prevailing planetary winds with the carrying capacity needed to be commercially viable. The real breakthrough on the Atlantic route was thus Boeing's Model 314 *Clipper*, which married the wing from their massive Model 294 XB-15 bomber to a flying boat fuselage designed to carry passengers. The XB-15, which first flew in 1935, was the largest and heaviest airplane yet built in America. With an 8,000 lb bomb load its range was 5,030 miles cruising at 152 mph. The XB-15 was the first airplane Boeing built to the DC-2 standard, with carefully cowled engines, variable pitch propellers, and split flaps to ensure good take off performance. The Model 314 first flew in 1938. It was the first commercial airplane with genuine transatlantic range to enter fare-paying passenger service and it did so in late June of 1939 (Bowers 1968, 237–238). Pan Am also used it extensively on its Hawai'i services since, unlike the M-130, it could carry a real payload on that route.

The Case of Britain

Britain may have been a trading state through much of the 1800s but by the 1920s the New Imperialism was taking its toll, and one of Britain's main concerns had become how to hold onto an increasingly complex and geographically spread out Empire in which dissidence was clearly on the rise. "Air policing" offered an attractive, low-cost solution to controlling such dissidence without putting too many British soldiers' lives at risk. A second problem was how to extend the navalist geostrategies that had served Britain well from the 1500s on into the aerial age. Mahan had identified blockade and *guerre de course* as two of the most important of these. Blockade, especially of needed food supplies, was clearly an attack upon civilian populations, and strategic air war was a logical extension of such tactics into the aerial age. A population such as that of Imperial Germany depended on food imports. In WWI Germany was starved by sea blockade and its population driven to revolution and mutiny (Offer 1989; Sondhaus 2014). The aim of terror bombing was no different: terrorize the population so that it would force its government to sue for peace. *Guerre de course*, the war on an opponent's commercial shipping, could easily be extended into the aerial age if one abandoned the Cruiser Rules put in place by Henry VIII. The Cruiser Rules required that merchant ships be captured, not sunk, but the German practice of unrestricted submarine warfare in WWI effectively threw them out. If the goal was simply to sink merchant ships this could be also done from the air using long-range bombers. Even Mahan's ultimate goal of a climactic battle between opposing battlefleets could be reimagined using a variety of aerial vehicles launched from shore bases or floating platforms to attack surface ships. In that sense the Battle of Midway (1942) was no more than the aerial age version of the Battles of Trafalgar (1804), Tsushima Strait (1905), or Jutland (1916).

The strength in British air policy in the period between WWI and WWII thus did not come from civil aviation but from the obsession with strategic bombing that the RAF inherited from the Royal Naval Air Service (RNAS) and that originated in the Mahanian geostrategy of blockade. Between 1919 and the beginnings of German rearmament at the end of

1933 the RAF procured thirty-two bomber prototypes and put six into mass production. Between 1934 and 1945 it procured ten more, seven of which it put into mass production. The American focus was a little different, since it emphasized extending America's defensive perimeter well out to sea and, thanks to Billy Mitchell, focused more on destroying the enemy fleet in a climactic battle. This still required bombers. Between 1919 and 1933 America procured 15 and mass-produced 2, then between 1934 and 1945 America procured 18 and mass-produced 8.

And these were merely the landplanes. As did America, Britain procured a plethora of large, multiengine flying boats as long-range patrol bombers, all of which traced their lineage back to the Curtiss *America*. Between 1919 and 1933 this amounted to a total of 18, several of which were produced in reasonable numbers. After 1934 the old Curtiss design was, just as in America, left behind and from then until 1945 the number of designs fell drastically to three, although one, Shorts' S.25 *Sunderland* was produced in very large numbers, 749 in total. Although the British developed aircraft carriers at the end of WWI postwar austerity measures forced the Royal Navy to accept unsuitable RAF designs for its aircraft. Its development of long-range single seat fighters, dive and torpedo bombers that could cope with the rigors of naval service thus languished compared to USN efforts. American procurement of long-range, multiengine, naval patrol bombers, most of them flying boats, is just as hard to disentangle. In the period 1919 to 1933 these were, again, modified versions of the Felixstowe F.5 that derived from the Curtiss *America* and subsequent Curtiss boats. These modifications, generally less heavy than those the British indulged in, emanated mostly from the Naval Aircraft Factory as the PN series, –7 through –12. Since the Naval Aircraft Factory had no real production facilities these were then farmed out for production: Douglas built 25 as the PD-1, Martin 55 as the PM-1, and Keystone 18 as the PK-1. Hall Aluminum built an even more modified version as the PH. If one counts all these as one mass-produced design the USN procured another six patrol/attack boats between 1919 and 1933, two as prototypes, two of which saw small scale production, and two which entered mass production. 1933 saw the initial design of the most successful patrol flying boat in history, the Consolidated PBY *Catalina*, with over 3,400 built. From 1934 on the USN began to procure specialized landplanes as patrol bombers as well as flying boats. Between then and 1945 this amounted to four flying boat and three landplane designs. Of the flying boats two were prototypes, one entered small scale and one mass production. Of the landplanes there was one in each category. The lesser activity in the later period is because attack aircraft had moved to the decks of aircraft carriers and matured as single-engine torpedo and dive-bombers. The USN procured some ten of these before 1933 and thirteen between 1934 and 1945. In WWII these aircraft along with naval fighter aircraft were made freely available to the British, first for purchase, then under Lend-Lease.

The link between airliners, naval patrol bombers, and strategic bombers became increasingly clear. Shorts' S.23 C class *Empire* boat evolved into the S.25 *Sunderland* patrol bomber, and then gave its wing to the S.29 *Stirling*, which first flew in 1941, the first British four-engine strategic bomber since 1919. Boeing's Model 294 XB-15 bomber gave its wing to the Model 314 *Clipper* commercial flying boat and its Model 299 B-17 to the Model 307 *Stratoliner* airliner. When the war ended Boeing married the wing from its Model 345 B-29 *Superfortress* to a new fuselage to produce the Model 377 *Stratocruiser* airliner. The British also elaborated wartime bombers into airliners, though with no success: the Handley Page *Halifax* bomber gave its wings to the unsuccessful *Hermes* civil airliner and the Avro *Lincoln* bomber to the disastrous Avro *Tudor*, the prototype of which crashed killing all aboard and two of which vanished utterly in the Atlantic when hastened into airline service they were never to be ready for. In part this was because, when America entered the war the British had agreed to concentrate on fighter and bomber development and leave civil transports to America. In the aftermath of WWII this proved to have been a disastrous decision, as America had developed several prototypes into highly effective civil transports during the war years, the DC-4 and *Constellation* in particular.

Britain simply did not see commercial aviation as, in many ways, early 1930s America did, a technology that would simply speed economic integration and increase efficiency over the substantial but generally compact earth space of the forty-eight states. To Britain, political unification of an increasingly complex and troublesome Empire mattered far more, and aviation promised to draw the centers of a far-flung and geographically discontinuous imperial system much closer to Whitehall. The maritime structure of the Empire had worked well for several hundred years, but demands for more rapid movement of people and goods were driven by the acceleration of communications wrought in the late 1800s by the submarine cable system. Aviation mattered relatively little within the British Isles, both geographically compact and in possession of an excellent railroad and ferry system. Connections to Europe were of far less importance to the government than connections to the Empire. But in the immediate aftermath of WWI aviation rapidly became as much a matter of civil as of military imperial policy. Air policing was joined, as it were, by "air politicking."

There were three attempted technological solutions to the problems of "air politicking" to politically unify the Empire following WWI. The first was to use airships because of their immense range and carrying capacity, the second was to use landplanes, the third was to use flying boats: both types of airplane were derived from WWI bombers or maritime patrol airplanes. Neither type of airplane was ever built in anything like reasonable numbers when compared to such properly commercial airplanes as the Boeing 247, DC-2 and DC-3. The fatal attraction of the airship was that, with the emergence of fascism in Europe and the increasing problems of Britain's imperial adventures in the Middle East, notably in Iraq, the crucial range problem for British aircraft was getting across Europe to imperial territory with an effective payload. Short-legged landplanes and flying boats had to fly either through the Balkans or via Italian territory to get to Egypt. Fascism had closed Italian airspace to the British by 1930 and German influence rose rapidly in the Balkans after Hitler took power. The crucial range requirement was thus around 800 miles, enough to fly from Southampton across friendly France to Marseilles, refuel, then cross the Mediterranean to imperial territory in Alexandria, Egypt, and on through the Middle East to India.

Airships

In the immediate aftermath of WWI it was obvious that airships had immense range and carrying capacity, but they were expensive to build and lift with lighter-than-air gases such as hydrogen and helium, needed huge ground crews, and could fly only in relatively calm conditions. Hydrogen is dangerously flammable and helium only became available in commercial amounts in the early 1920s, at which time it came only from America, its export being banned in 1927. Hydrogen lifted airships proved suicidal as bombers when used to attack air space defended by fighters using incendiary bullets. Britain built several dirigible airships based on the captured German airship L33 (LZ76—*Luftschiff Zeppelin* hull #76) during the latter half of WWI, notably R33 and R34. In 1919 R34 made the first east west aerial crossing of the Atlantic, flying direct to New York from England, a performance well beyond that of any airplane at the time. In January 1921 R34 was badly damaged in a gale and scrapped. The British government then stopped all development of airships for financial reasons. R33 was mothballed until 1925. She was then used in experiments with mooring masts, an attempt to cut down on ground crew costs. In late 1925 she was torn from her mast in a gale and blown to Holland. She was scrapped in 1928. A third British airship to a lengthened version of the same design, R36 of 1921, was intended for commercial service from the start but again suffered damage in bad weather, was sidelined by the financial problems, and scrapped in 1926. A fourth, much larger, British airship, R38, built for long anti-submarine patrols, was sold to the USN before completion in 1919. On a proving flight in 1921 R38 suffered structural failure over the Humber Estuary, killing all but five of those on board.

With such a record it is surprising that the British returned to the Imperial Airship Scheme in 1924, but return they did. This proposed a series of airships, R100 through R104, to connect Britain nonstop across the Atlantic to Montreal and via the Mediterranean and Indian Ocean to Karachi. Two experimental prototypes, R100 and R101, were built, the assumption being that the bad experiences with earlier airships would help resolve the main problems. Both ships were overweight and underpowered, and although R100 made a successful flight to Montreal and back R101's career was short and deadly, of *Titanic* proportions. Its maiden voyage ended when it careened into a hillside in France in rain and gusty weather on its proving trip to India in October 1930. Rain adhering to the outer cover dangerously increases the weight of an airship. All but six of the fifty-four passengers and crew died in the resulting fireball, including the British air minister and the director of civil aviation (Davies 2005, 34). R100 was immediately broken up and Britain would build no more airships.

America made no attempt to use rigid airships commercially. The experience of the USN, which proposed to use four airships as scouts for the fleet in the huge expanse of the Pacific, was as bad as the British, despite their exclusive ability to use nonflammable helium as their lift gas. Bad weather was again the problem. *Shenandoah* crashed in 1925 killing fourteen of its crew of twenty-five, *Akron* in 1933 killing seventy-three of the seventy-six on board, and *Macon* in 1935, killing merely two of the seventy-six on board. Only *Los Angeles*, built as LZ-126 and delivered to America in 1924, failed to crash.

The Germans would, of course persist with airships much later, their ships leading an apparently charmed live until LZ-129 *Hindenburg* exploded into flames in full view of the newsreel cameras when landing in New Jersey after its first transatlantic flight of 1937. Of the ninety-seven on board thirty-five passengers and crew died.

No other power experimented with airships on other than the most minor level.

Landplanes

With the formation of the RAF as an independent air arm in 1918 the focus of the RNAS on strategic bombing became securely lodged in RAF doctrine. In 1914 Frederick Handley Page had persuaded the RNAS to procure from his company a "bloody paralyser" of an aircraft, a twin-engine landplane that would be able to "bomb the High Seas Fleet before it ever left the safety of its anchorage at Kiel" (Barnes 1987, 74). Although it would be 1917 before the resulting Handley Page 0/100 flew its first combat mission, this led to a series of increasingly large and powerful strategic bombers. In developed form the 0/100 became the 0/400, and this led to the much larger, four-engine, V/1500, designed to bomb Berlin from bases in Britain and equivalent to the best of the German R-planes. It was logical in the aftermath of WWI to see such airplanes as the possible basis for airliners, but their complex, wire braced fuselages were totally unsuited to human habitation. In the case of the 0/400 a new fuselage in turn necessitated a more general redesign, out of which emerged the W/400 and the fully civilian W.8 and W.10 series (Barnes 1987, 121, 147, 169, 184). The V/1500 was simply too expensive to modify.

Other British companies followed Handley Page's example. In 1916 de Havilland designed his D.H.3 as a strategic bomber for the Royal Flying Corps (RFC). The War Office canceled it on the grounds that "strategic bombing of Germany was unnecessary and that the twin engine bomber was impracticable" (Jackson 1987, 51). This was a strange conclusion given what the RNAS was up to, but the Royal Navy was always a law unto itself and the RFC was legitimately focused on aerial observation for the artillery on the Western Front, not strategic bombing. The D.H.3 was eventually resuscitated in 1918 and redesigned as the D.H.10. The new RAF placed large orders, which were never filled since the war ended. After the war de Havilland moved into the airliner market and solved the fuselage problem by making the fuselage of its D.H.18 a plywood box. De Havilland resisted the temptation to convert his D.H.10 and the D.H.18 was designed from the outset as a commercial proposition. It proved very much cheaper to operate than any converted bomber. It was, however, a single-engine design usable only on European routes because of its short range (Jackson 1987, 140, 158).

In 1925 the Air Ministry decided to phase out the RAF's airmail service from Cairo to Baghdad and subsidize Imperial Airways to carry it instead. Imperial required three engines for safety and to give an ample power reserve under tropical conditions or at higher altitudes, such as in East Africa. Engines that performed well in temperate Britain lost a substantial amount of power under such conditions. Two airplanes emerged from this decision. The Armstrong Whitworth *Argosy* entered service in 1926 but was used mostly in Europe and Africa, and only seven were built (Tapper 1973, 209). De Havilland's D.H.66 *Hercules* was more successful. The experience accumulated by the RAF with the damage wrought by high temperatures, humidity, and termites caused de Havilland to use a tubular steel fuselage with replaceable plywood boxes inside it rather than the plywood structure pioneered on the D.H.18 and its many successors. Eleven were built, and from 1927 on Imperial's D.H.66s built "a reputation for trouble free service under trying conditions" (Jackson 1987, 270).

Both the *Argosy* and the *Hercules* epitomized what was problematic about Britain's "air politicking." Such airplanes were built in tiny numbers because they were designed for very specific routes. They were over-powered to allow them to take off quickly from short airfields with only grass "runways" as well as to compensate for power loss in tropical conditions and at altitude.

This meant they used a lot of fuel. Because they emphasized take off performance and did not follow the design principles pioneered by Rohrbach they lacked speed and range. The last airplane designed in this fashion for Imperial Airways was a stately, four-engine, strut braced biplane, the Handley Page H.P.42, designed to meet the airline's 1928 request for a replacement for its *Argosies* and *Hercules*. Although an exceptionally safe, all-metal biplane with Handley Page's usual high-lift devices on its wings, and with a luxurious interior, the H.P.42 was slow, cumbersome, and lacked the range needed for critical sections of Imperial Airways' routes to India and Africa. Only eight were built, four for European routes and four for the Empire services. Even so they entered service in 1931 and remained in service through the beginning of WWII.

The H.P.42 was the first Imperial Airways airplane to have four engines to ensure safety on long flights, but it was an old-fashioned design. In 1929 Armstrong Whitworth proposed the first British commercial airplane to follow at least some of Rohrbach's design principles This was the *Atalanta*, a cantilever high winged monoplane of mixed wood and metal construction, its four engines faired into the wing leading edge precisely as those of Rohrbach's E.4/20 had been, and which it resembled. As soon as it entered service in 1932 it was obvious that it was too small for the duties it was being called on to perform, although it performed well in tropical conditions. It took until 1934 for Imperial to call for a larger replacement. The cause of this was the Empire Air Mail scheme, the decision by the government to carry all first class mail within the Empire by air at a low rate. Armstrong Whitworth responded with its all-metal, cantilever-winged, four-engine *Ensign*. This was a large and handsome airplane, and their last commercial design before WWII. It was one of the first British commercial airliners fitted with split flaps in the style pioneered by the DC-2 to improve take off performance. Eleven were built, but when the *Ensign* entered service in 1938 it was obvious that it was seriously underpowered (Tapper 1973, 219, 235).

Flying Boats

As indicated earlier, British "air politicking" in the aftermath of WWI began to favor flying boats on long imperial routes. Such boats had proved their worth on anti-submarine patrol in WWI, they could land close to almost any major city in the maritime based Empire, they could be built very large without much concern over the length of their take off run, and they offered survivability in the event of a forced landing on over-the-ocean routes. The first round of such boats of the early 1920s, British and American alike, were all descendants of the Curtiss Model H *America* of 1913 that had been designed to be flown by a joint American and British crew to take the *Daily Mail* prize.

When WWI broke out the British member of the crew intended to cross the Atlantic in *America*, Lieutenant Porte, returned to active duty with the RNAS and helped redesign the Model H as a maritime patrol boat, usually known as the *Small America*. This was then enlarged as the Curtiss Model H-12, or *Large America*. Although generally good designs the *Americas* had two problems: their Curtiss engines lacked power, so they had to be re-engined with British engines; and their hulls "demonstrated notable structural and design deficiencies in British wartime service." Porte, now working at the RNAS station at Felixstowe, redesigned them accordingly, at which point they became known as the Felixstowe F.5, which was in turn built by Curtiss in America as the Curtiss F.5. Confusingly, Curtiss continued to use the original hull design for the H-12 on their H-16, which otherwise was an F.5. The F.5 was built in large numbers by Curtiss (184 examples, 60 of which went to Britain) and by the Naval Aircraft Factory in Philadelphia (150 examples). Thereafter it was redesigned as the PN series boat and built by several American manufacturers (Bowers 1979, 57–60, 83–86, 90–96, quote from 90).

During the 1920s five British companies developed more sophisticated boats either based on or influenced by the Felixstowe F.5, several of them to Air Ministry specification 22/24 for a large general duty/patrol flying boat for imperial duties. Such boats were built by Blackburn (the *Iris*—Jackson 1989, 194–208); English Electric (the *Kingston*, based on the Phoenix *Cork*—Ransom & Fairclough 1987, 110–125, 134–146); Saunders Roe (the *Valkyrie*—London 1988, 63–70); Shorts (the *Cromarty*—Barnes 1989, 151–158); and Supermarine (the *Southampton*—Andrews & Morgan 1987, 96–112). No company did more than Shorts, which pioneered the duralumin hull with its S.2 version of the F.5. Shorts also licensed the S.2 to the Japanese navy's Hiro Naval Arsenal, a relationship with Britain that continued into the 1930s with the licensing of later versions of the *Southampton* as well as the *Calcutta* and their construction by Aichi, Hiro, and Kawanishi (Mikesh & Abe 1990, 70, 93, 95, 135). Shorts also licensed the *Calcutta* to the French company, Breguet in 1931 (Barnes 1989, 230).

Just as the Empire Air Mail scheme had persuaded Armstrong Whitworth to design the *Ensign*, it also persuaded Shorts to design a new airplane, but a much more radical and successful one, the S.23 *Empire* boat. Oswald Short had seen the Douglas DC-2 that came in second in the MacRobertson race from London to Melbourne in 1934. Although the race was won by a pure racer, de Havilland's elegant D.H.88 *Comet*, the remarkable performance of the two American commercial transports in the transport section was what properly impressed Short. The Dutch airline K.L.M. flew a DC-2 to victory in the transport section and was only just behind the Comet overall. For part of the race it was on its commercial route to Batavia complete with mail and even three fare-paying passengers. Second in the transport section and third overall was a Boeing 247, ahead of a second de Havilland *Comet.* This was enough to convince Short that American designers had it right, and he designed his S.23 *Empire* C Class boat accordingly: cantilever wing, stressed skin, four powerful Bristol *Pegasus* air-cooled radial engines closely cowled in NACA style, variable pitch propellers, and an even more sophisticated system of flaps than those used by Douglas that increased the lift coefficient by 30 percent (Barnes 1989, 312–314). The *Empire* boats, which first flew in 1936, represented a total break from the biplane descendants of the Curtiss *America* that had dominated British flying boat design to that point and that Shorts had done much to perpetuate. An advantage of this new generation of flying boats was that, rid of its struts and wire braced wings, it had good aerodynamics and it did not need the weight and complexity of a retractable undercarriage. Third was the fact that most of the Empire offered as landing places relatively closely spaced colonial era port cities or, from north to south in East Africa, a network of large lakes. Long-range continued to seem unnecessary, although Shorts paid a great deal of attention to ensuring excellent take off performance by using their very advanced flap design. Imperial Airways, impressed by Shorts' proposal, ordered first fourteen and then twenty-eight *Empire* boats "off the drawing board," and Shorts delivered. The S.23, which first flew in 1936, had a range of some 800 miles with twenty-eight passengers and cruised at 180 mph. A range of 800 miles was enough for most imperial routes. Shorts would eventually build forty-seven *Empire* boats, the most successful British commercial airplane to date (Davies 2005, 92).

A strengthened version, the S.30, had its range increased to 1,500 miles so that it could fly the 1,348 mile leg from Sydney to Auckland, the longest route in the Empire other than the Atlantic. What neither the S.23 nor the S.30 could do was close the Atlantic gap with a payload without in-flight refueling, which was seriously considered. One *Empire* boat, the S.20 *Maia*, was produced as part of the Short-Mayo Composite airplane to carry a small mailplane, *Mercury*, aloft with enough fuel to complete the trip. The S.23 was extensively used in Atlantic proving flights, for which its fuel supply was raised from 650 to 2,320 imperial gallons. Closing the Atlantic gap with British equipment should have fallen to Shorts' enlarged *Empire* style boat, the S.26 or G Class, which had a range of 2,500 miles. WWII intervened, only three were ever built and they never entered transatlantic service. Had WWII not intruded Shorts were planning on using the S.26 wing on their S.32 landplane, intended to carry up to twenty-four passengers 3,000 miles and designed to be pressurized. There were concerns that such a large landplane would need a longer than usual take off run but "provision was already being made for it when war broke out." At least on paper the performance of the S.32 would have been better than Boeing's *Stratoliner* but worse than Lockheed's *Constellation* (Barnes 1989, 301, 312–345).

Whereas in America the pattern of symbiosis between civilian and military airplanes was from military design to civilian, that pattern was reversed in Britain, and the *Empire* boats spawned two military offspring, the Shorts S.25 *Sunderland* flying boat, intended for mostly anti-submarine and convoy protection work, and the S.29 *Stirling* long-range bomber. The *Sunderland* was a fairly straightforward modification of the *Empire* boats to bring them up to military specifications and it first flew in October 1937 (Barnes 1989, 345–346). The *Sunderland* wing was to have been used on the *Stirling*, but the Air Ministry demanded a reduction in wingspan from 112 to 100 feet to fit existing hangars, which short-sighted move severely reduced the *Stirling's* altitude performance when it first flew in late 1939 (Barnes 1989, 370–371).

The Second String: Germany, Russia and Japan

Although this book focuses on America's struggle to supplant Britain as hegemon there were other global players. Several played important roles in the development of aviation and provide important contrasts to the two trading states. Germany and Russia need to be combined not only because both acted as classic territorial states in a geopolitical sense but also because of the substantial diffusion that occurred between their aircraft industries in the early 1920s. Neither would have succeeded at the level they did without the other. Japan, too, was the recipient of a significant amount of German technology, although it also learned from Britain and America. The German-Russian diffusion resulted in Tupolev designs for long-range strategic bombers that were a better expression of the design principles of Junkers than Junkers' designs themselves. As indicated earlier Junkers was constrained by an Inter-Allied Control Commission that refused to allow him to complete his J-14 airliner on the grounds that it might be used as a strategic bomber. Germany emerged from WWI with three extremely capable designers of very long-range airplanes: Junkers, Dornier, and Rohrbach. Of the three, Junkers was the most conservative, although he pioneered cantilever wings and all-duralumin airplanes. In his G-38 of 1929 he produced the largest landplane yet, albeit a vast lumbering beast (Kay 2004, 80). After WWI Dornier focused almost entirely on very large, all-metal, cantilever-wing flying boats that he produced in Italy and Spain to avoid the attention of the Inter-Allied War Commission. Rohrbach was the most radical of the three, pioneering true stressed skin designs to keep structure weight as low as possible. When the Commission destroyed his brilliant E.4/20 Rohrbach turned his attention back to flying boats and produced, in the *Roland*, another innovative airplane: an all-metal, long-range, twin engine, cantilever wing monoplane that Rohrbach built in small numbers in Denmark where the Commission could not reach him.

Both Rohrbach and Dornier licensed their designs to Japan, thus diffusing to that country German best practice of the early 1920s. The Japanese Hiro Naval Arsenal and Mitsubishi both became involved with Rohrbach, a Mitsubishi built *Roland* featuring in American anti-Japanese propaganda as a threat to America's shores in the mid-1920s, though only prototypes were ever built! A landplane version of Dornier's *Wal* was designed in secret at Dornier's plant in Switzerland, developed by Kawasaki as the Army Type 87 Heavy Bomber, and entered service in 1927. Twenty-eight were built. Kawasaki also imported a Dornier *Wal*, derived an experimental flying boat from it, licensed it direct in 1930, and completed three. To complete the German triumvirate, Mitsubishi licensed the lumbering Junkers G-38 in 1928 as the Army Type 92 Heavy Bomber and built six (Mikesh & Abe 1990, 94, 144, 150, 151, 165, 182).

As indicated earlier, Japan also licensed flying boats from Britain. But Japan's involvement was entirely military, concerned with naval patrol boats and bombers, and with no extension into commercial airplanes,

Although German designs were the most advanced in the immediately post-WWI world Germany was never much of a trading state, especially after Hitler came to power, and even the early attempts at commercial aviation can be read as having politics at their center. In the light of such war aims from WWI as the creation of *Mitteleuropa* (Fischer 1967), the attempts of DELAG to set up a network of passenger services around Europe by Zeppelin before the war, and the success of Junkers after the war in using his F-13 airplanes to set up first the *Europa Union* system of air routes, then the Eurasia wide network combined from three airlines, *Lufthansa*, *Deruluft*, and *Eurasia* (Hugill 1993, 259, 270, 272), politics looked more important than commerce. To be fair to Junkers, although he produced warplanes he focused on commercial airplanes, an emphasis that did not change until the Nazis forced him out of control of his company and turned its efforts to warplanes in the mid-1930s. Although a sophisticated plane for the early 1920s the Junkers F-13 carried only four passengers and was more of a mail plane and air taxi than an airliner. Junkers elaborated a series of larger airplanes from the F-13, initially the three-engined, fourteen-seat G-24 that became the mainstay of German airlines in the late 1920s and opened *Lufthansa's* Berlin-Peking service in 1926. The ultimate expression of Junkers' commercial designs, and one of the last planes he himself designed, was the three-engined Ju 52, the wing of which embraced high-lift devices. The Ju 52 was extremely reliable and used mostly BMW radial engines, but it was not much faster than the G-24 nor did it carry many more passengers. Some 400 were built before WWII, but 230 of these went to one customer, *Lufthansa*. Like the Douglas DC-3 the

Ju 52 was built in large numbers as a military transport for WWII. Perhaps 5,000 were completed, but attrition was extremely high (Kay 2004, 47–49, 100–102, 127).

As Germany moved toward rejecting the Versailles Treaty and rearmament in the 1920s it began developing the *Luftwaffe* from a clandestine organization training in secret in Russia to one that would once again become a central part of German military thinking. Although, in general, German thinking on air power focused on tactical air power, by 1934 the *Luftwaffe* had proponents of strategic air power as well. These thought in terms of a long-range bomber able to bomb deep into Russia from bases in Germany, as far as the Ural Mountains. The four-engine bombers they procured in 1935, the Dornier Do 19 and Junkers Ju 89, were called "Ural bombers" in consequence: both were rejected in 1937. In part this was because the engines of the period lacked power, in part because the *Luftwaffe's* main proponent of the Ural bomber program, *Generalleutnant* Wever, died in an air crash in 1936, and in part because the head of the *Luftwaffe*, Herman Goering, was committed both to tactical air power and a quick victory. Large, twin-engined, shorter range bombers such as the Heinkel He 111 had proven effective in the Spanish Civil War and a large number of such airplanes could be built much more quickly than the more complex Ural bombers. For *blitzkrieg* fought over land or such short sea stretches as the English Channel there was no reason to think them inadequate. Once France had fallen the *Luftwaffe* had bases on the edge of the Channel and all the main British cities seemed to be within easy reach. While hindsight makes it reasonable to see the cancellation of the Dornier Do 19 and Junkers Ju 90 Ural bombers as a mistake it is unlikely they would have helped the Luftwaffe much in the Battle of Britain in 1940, or even in the early phases of the invasion of Russia, where such bombers as the He 111 continued to perform very well. And in any case their cancellation did not end the perceived need for a long-range bomber. A month after cancellation of the first generation Ural bombers the *Luftwaffe* ordered the Heinkel He 177 *Greif*, a much more powerful long-range bomber that should have been ready to enter service in 1941. In this case *Greif—Griffon*—might better translate as "grief," and the He 177 was plagued with short-sighted policy decisions to turn it into a dive-bomber that delayed it and serious technical problems with its development, including a remarkable and frightening tendency to engine fires. As a consequence, it wrote "the most dismal chapter in the wartime record of the German aircraft industry" (Green 1970, 336).

If there was a German breakthrough airplane in the late 1930s it was Kurt Tank's design for Focke-Wulf of the first landplane with genuine transatlantic capabilities, the Fw 200 *Condor*. In August 1938 it flew nonstop from Berlin to New York in just over 24 hours. It was powered initially by four

Pratt & Whitney radials, later replaced by BMW's license-built version of the same engine, and was in advance of both the *Stratoliner* and DC 4, not least because Tank was one of the world's best aerodynamicists. Not until the *Constellation* of 1943 did an American commercial transport emerge that could outperform the *Condor*, and it was the aerodynamic capabilities of such American designers as Howard Hughes that gave the *Constellation* its edge. The *Condor* was pressed into service in WWII as a long-range commerce raider, bombing Allied shipping so effectively that Winston Churchill called it the "Scourge of the Atlantic." Together with the U-boats it brought Germany near to victory in the Battle of the Atlantic. The Fw 200C military version was heavily armed and it defended itself well against the Grumman *Wildcats* and Hawker *Hurricanes* flying from British escort carriers that took it on, though they would eventually defeat it. Fortunately it also had its flaws and the Germans built very few of them. It could not maintain height on three engines, and the heavier military version had a distressing tendency to break its back if landed heavily. Tank was mainly a fighter designer who lacked experience with large metal structures. More *Condors* were lost to landing accidents than combat. Even so, the twenty-six Fw 200Cs built, only eight of which were usually serviceable, sank eighty-five Allied ships between August 1940 and February 1941 from their bases in Normandy (Green 1970, 223; Bowyer 1997, 110).

Germany's great failing here was, however, that it could not break free of the geostrategy of the territorial state to which its suite of leaders had committed it during the Nazi period. Before WWII the navy had lobbied for "a strong naval air arm for anti-shipping strikes and to mine ports in case of war with Britain." The navy had developed a new air-dropped magnetic mine that would have been devastating if it had been used in large numbers at the opening of the war but Goering's commitment both to a single *Luftwaffe* under his exclusive control and to the almost exclusive use of "his" bombers in the land war denied the needed bombers to the navy. This gave the British time to devise countermeasures when they began encountering small numbers of the new magnetic mines. Goering also refused the navy the long-range strike aircraft that would have been able to use bombs and torpedoes far out into the Atlantic. Such airplanes as the Heinkel He 111 and the Junkers Ju 88 were well suited to such roles and would, too late, eventually be employed in them (Corum 2006, 206–207, quote from 206). The battle that Germany most needed to win against Britain in the early part of WWII was not the air Battle of Britain but the air and sea Battle of the Atlantic. Neither battle fitted well with the likely geostrategies of a territorial state and the logical consequence was that the Nazi state failed to procure and deploy the correct suite of weapons to win either, although they were technologically quite capable of developing them.

It is certainly the case, as Corum notes, that the Nazi State made many other fundamental mistakes: betting on a short war; inadequate aircraft production capacity in the early war years; badly organized mass production; inadequate training programs for aircrew; badly flawed strategic leadership under Goering and design leadership under Ernst Udet; and amazingly inept logistics both in supply and repair. These failures were as nothing compared to their commitment to the geostrategy of the territorial state in a two-front war with both trading and territorial states. And the failure of their territorial geostrategy against another territorial state was almost as stunning as the failure to embrace a trading state geostrategy against highly capable trading state opponents. The Nazis attacked the Soviet Union with less aircraft than they committed to the Battle of France and the *Wehrmacht* committed itself and the *Luftwaffe* to fighting "more than a thousand miles from their home bases backed by a tenuous supply line." As Corum rightly concludes, "professional soldiers should have known better" (Corum 2006, 207–221, quote from 209).

Soviet Russia was even more obviously a territorial state than Nazi Germany, which had some roots in the trading state of the Hamburg merchants. But Russia covered vastly more earth space than Germany, and in the aftermath of WWI its aircraft designers had access to the latest German aerodynamic theory, German understanding of metal structures, and German airplane engines. No Inter-Allied Control Commission could refuse them permission to develop large metal aircraft that might be used as bombers. The first of these was Tupolev's large, Junkers' style, all-metal, twin-engined ANT-4 bomber, which first flew in 1928. In August 1930 an ANT-4 flew across the Soviet Union and on to New York via Seattle. Tupolev followed his ANT-4 with the ANT-6, the world's first all-metal, cantilever-wing, four-engined bomber, which first flew in 1930 and was still in front line service when the Soviet Union entered WWII. Some ANT-6s were converted for the Soviet airline, Aeroflot (Andersson 1994, 247–255), though Russia paid no real attention to commercial air transport. Even more than was the case in the British Empire its main use of "commercial" aviation was to move high-level personnel around to ensure political unity.

AMERICAN VICTORY AND BRITISH DEFEAT

Britain's obsession with air policing and "air politicking" within the Empire essentially ensured that its development of commercial aviation, despite starting in the early 1920s, could never catch that of America once America had begun to realize the immense advantages that commercial aviation would confer on the American economy. Part of America's advantage was its geography. It was geographically "large" enough for the airplane to be commercially very useful but also "small" enough to allow the efficient use of limited range landplanes as they began to mature technologically in the early 1930s. Although America needed long-range flying boats for fundamentally the same reasons as Britain, to unify its unstated "empire" in the Pacific and Caribbean and then to expand its political influence throughout Latin America, it did not make them the main focus of its airplane development. At the same time, American defensive geostrategy caused an emphasis on long-range coastal defense airplanes that helped push the development of effective strategic bombers. As American universities began to improve in science and engineering in the late 1920s and 1930s they brought German knowledge and skill to the table and American designs began to surpass their European counterparts. What emerged was the remarkable florescence of the late 1930s, epitomized by the Douglas Commercial series but also by the Hughes H-1 racer, the Boeing B-17 *Flying Fortress* coastal defense bomber, and the Boeing 314 *Clipper* flying boat. All these benefited from each other and led by the early 1940s to such outstanding designs as Boeing's B-29 *Superfortress* and Lockheed's *Constellation* that far out-performed German and British designs.

Britain's Empire was not geographically contiguous but it was spatially extensive. Reaching the outskirts of the Empire from the home islands required an aircraft with range capabilities well beyond those of any airplane that was commercially viable until the late 1930s. In effect this initially pushed Britain up the blind alley of the rigid airship and delayed Britain from adopting a trading state approach to commercial aviation until the late 1930s, although it began to make a remarkable recovery at that point. In the immediate aftermath of WWI efforts to establish commercial air transport without high government subsidies of the sort used by the French at the time foundered very quickly. The policy Britain adopted in the early 1920s was thus avowedly political and imperial. In the words of Samuel Hoare, who became secretary of state for air in 1922, "as a Conservative brought up in the days of Rudyard Kipling, Joseph Chamberlain and Milner, I saw in the creation of air routes the chance of uniting the scattered countries of the Empire and the Commonwealth" (Templewood 1957, 90). In 1923 Hoare was the first secretary of state for air to be given cabinet rank, putting his office equal to that of the secretary of state for war and the first lord of the admiralty. Hoare's strength was as much the role he saw for civil aviation as a central part of imperial policy as his management of the air force. In late December 1926 Hoare was on the first D.H.66 *Hercules* to fly the route to India that opened in 1927, a route that indicates Britain's problems relative to those of America. To reach Delhi from London was a journey of over 6,000 miles with twelve days in the air, nineteen refueling stops, and a maximum stage length of 410 miles between Naples and Malta (Templewood 1957, 127). And India was only halfway to the ultimate goal, Australia. It was the MacRobertson race to Australia in 1934 that at once completed this imperial goal, demonstrated the remarkable progress that had been made in American commercial aviation in the early 1930s, and finally pushed the British into the development of commercially viable long-range airplanes.

The complicating factor in the period before the outbreak of war in Europe in 1939 was the agreement reached in the early 1930s between Pan Am's Juan Trippe and Imperial Airways' Woods Humphrey that they would act as equals, open an Atlantic route only when both companies were ready with the proper equipment, and effectively create a dual monopoly. Juan Trippe particularly wanted to keep the American government out of the agreement, but by the late 1930s the British government had become more and more involved with Imperial Airways. The interests of the British government, Imperial, and Pan Am in restricting access to the potentially highly profitable Atlantic route were aligned, but the American government wanted a much more liberal solution that would let other American airlines than Pan Am access the Atlantic routes (Dobson 2011, 85–86). In the late 1930s proving flights were made with the Sikorsky S.42 and the Shorts C Class *Empire* boats, but neither had the range to carry a payload, especially flying west. The Boeing 314 *Clipper* changed the calculation in Pan Am's favor but Pan Am held to its agreement with Imperial and was waiting until the Shorts G Class boat was ready. Given the discrepancy between American and British technology in civil aviation, as late as the mid-1930s, the speed with which Britain caught up in the late 1930s was impressive, though clearly related to the long British obsession with the development of strategic bombers.

The successful development of strategic bombers in both America and Britain is worth a brief aside. The obsession of both trading states with the strategic offensive from WWI on gave them a huge advantage in WWII's air war. Simply put, the development of large, highly complex aircraft took time: five years was normal, and the likelihood at the time was that many designs would fail in practice. The two most successful strategic bombers of WWII originated as specifications at least five years before they entered service. Boeing's B-17 *Flying Fortress* began life in 1934 and Avro's *Lancaster*—as the *Manchester*—in 1936. The development process radically changed their final versions. The third great Allied strategic bomber and the most technically advanced, Boeing's Model 345 B-29 *Superfortress*, followed suit. Its specification was finalized in 1939 well before America entered WWII and it entered service in 1944 only after prodigious efforts by Boeing. One of the *Superfortress*'s advantages was that it was pressurized to fly very high, and Boeing was the one company in the world with serious experience in pressurization, pioneered on its *Stratoliner* (Bowers 1989; Jackson 1990).

German failure here is interesting, since German technology was more advanced than American or British until well into the 1930s. Despite the cancellation of the two Ural bombers the Germans started the development of the Heinkel He 177 in 1937, so that in the normal five-year scheme of things it should have been in service by 1942. But German air policy kept changing, and Ernst Udet's obsession with dive-bombing resulted in the forced strengthening of the He 177 so it could perform such a role, a ludicrous one for such a large and already heavy airplane. Adding weight always adds problems. Constantly changing a design to suit different aerial strategies is a surefire way of ensuring failure. In addition, the coupled Daimler-Benz DB 610 engines used on the He 177, essentially two of its excellent DB 605s bolted together, never worked properly and frequently caught fire in mid-air. Avro had similar problems with the Rolls-Royce *Vulture* engines in its *Manchester* bomber, essentially two *Peregrines* bolted together. The *Peregrine* was a good engine, but Rolls-Royce preferred to focus on its more powerful *Merlin*. A redesigned wing with four Rolls-Royce *Merlin* engines solved Avro's problem in 1941 and the *Manchester* became the *Lancaster*, one of the war's best bombers, with almost 7,400 built. When problems emerged with the DB 610 Heinkel proposed a version of the He 177 powered by four DB 605s as early as 1940—it emerged as the He 277 only in 1944, just as all Germany's industrial capacity was being diverted into defensive fighters. The Germans completed eight He 277s (Green 1970, 336, 359). The He 177 is a classic example of putting all one's eggs in one basket once the first two Ural bombers were canceled. Both the British and Americans developed numerous failed prototypes to get the strategic bombers they were ultimately able to deploy operationally in large numbers.

After America entered the war in late 1941 a second force emerged that would ensure American victory in the struggle over international transportation by air. This was the deal struck between Britain and America in late 1942 by which the British aviation industry would concentrate on producing fighters and night-bombers and the American industry would produce transports as well (Dobson 1991, 129). Given that the Douglas DC-3 was pouring off the assembly lines in 1939, that the Boeing 307 *Stratoliner* had entered service that same year, that Douglas's DC-4 would fly in 1942 and Lockheed's *Constellation* in 1943, the Americans were well set to dominate the postwar market for airliners. Even so, British companies made a valiant effort to keep up. Avro used the *Lancaster* wing on its *York* transport, developed without official sanction in 1942. Toward the end of the war Avro developed what was basically an enlarged *Lancaster* bomber, the *Lincoln*, and immediately the war ended it married the *Lincoln* wing to a civilian airliner, the *Tudor*. Handley Page did the same, using the wing of its *Halifax* bomber on a civilian aircraft called the *Hermes*. The *York* was a modest success, the *Tudor* a disaster, and the *Hermes* a failure in competition with such airplanes as the DC-4 and the *Constellation*. The merger of Imperial Airways with British Airways formed the successor airline to Imperial in 1940 as the British Overseas Airways Corporation (BOAC). After the war BOAC was pressured by the British government to "buy British": *Yorks*, *Tudors*, and *Hermes*, along with *Sandringham* flying boats that were converted *Sunderlands*. The *Tudors* and *Hermes* were intended for the Atlantic routes, but the *Tudor* lacked the range for the North Atlantic and two vanished in unexplained circumstances on South Atlantic service in 1948 and 1949. The *Hermes* never had the range for the Atlantic at all. BOAC eventually canceled all its orders for *Tudors* and *Hermes* and bought American *Constellations*, *Stratocruisers*, and DC-4s for the Atlantic run, albeit the last were Canadian built and engined with Rolls-Royce *Merlins* (Barnes 1987; Jackson 1990).

This is not to say that the British did not attempt to retrieve the position that the 1942 agreement to let America focus on transports had put them in. After three years of war the British government decided in 1942 that it could turn some attention to the problem of civil aviation in peacetime, and it set up what became known as the Brabazon Committee, chaired by Lord Brabazon, who had a long background in aviation. In its initial work the Brabazon Committee defined the *Tudor* and *Hermes* as interim types to be developed as soon as the war was over along with the *Sandringham*. The *York* was an interim type that was already available. The committee then went on to define a series of new, much more radical airliners to make use of such wartime technological advances as the gas turbine. Although their final proposals were never fully realized, and some that were realized were unsuccessful, the Brabazon Committee report certainly set the tone for the immediately postwar British civil aircraft industry and it did so with a keen eye on responding to American commercial competition (Phipp 2007).

The proposals of the Brabazon Committee produced one great "might-have-been," the de Havilland D.H.106 *Comet*, which first flew in 1949 and in 1952 entered revenue service with BOAC as the world's first commercial jetliner. Although the *Comet* 1 lacked transatlantic range it proved ideal on the routes pioneered by Imperial Airways' *Empire* boats in the late 1930s, and slashed travel times in half on such routes compared to BOAC's DC-4s. The prototype for the Boeing 707, America's first commercial jetliner, did not fly until 1954 and did not enter revenue service until 1957. Unfortunately the *Comet* 1 had a terrifying and fatal design flaw. Without experience in pressurization de Havilland had produced a fuselage that could not withstand the numerous cycles of pressurization and depressurization high altitude flight entailed, and two *Comet* 1s exploded at altitude with the loss of all aboard in 1954. It took two years of painstaking work to show that this was the problem. Boeing had considerable experience with pressurization, dating back to the *Stratoliner* but also incuding their B-29 bomber, and designed the 707 to be much stronger. De Havilland's heavily revised, larger, more powerful *Comet* 4, which did have transatlantic range, did not enter service until 1959, by which time the later and superior Boeing 707 and Douglas DC-8 were ready to sweep the market and usher in the jet age (Bowers 1989, 427; Jackson 1987, 453).

But for all the technological advantages held by American civil aviation coming out of WWII the *coup de grace* to British ambitions to compete with America in the postwar global market was delivered on the policy front by the Bermuda Agreement of 1946. From the very onset of the air age the major states had struggled to define air policy through a series of treaties (Butler 2001), there being two key problem areas. The first was "the right of innocent passage" for commercial airliners over the territory of other states. Over time, this "right" has come and gone. For example, after WWI Germany was not allowed to control its own airspace for some time, and in the Soviet period Russia allowed no such "innocent passage," even shooting down one commercial flight, Korean Air 007, that accidentally strayed over Russian territory in 1983.

The second critical right, especially from the perspective of a trading state, is the right of a vessel operating under the flag of any nation to both drop off and pick up goods and passengers in the territory of any other nation. The denial of this right is known as *cabotage*. The British Navigation Acts, first passed in 1651 and repealed in 1849 at the height of British liberalism, were a centerpiece of mercantilist economic theory. They denied non-British ships the right to do other than drop off cargoes in British ports: they could pick nothing up, classic *cabotage*. Two conferences define the success of America in setting international air policy as WWII ended: Chicago in 1944 and Bermuda in 1946. In fact there were only two powers at Bermuda, America and Britain. The British wanted a more restrictive air agreement, which they almost got at Chicago, whereas the American president Roosevelt wanted "free use of airports throughout the world for nations to operate air services" (quoted in Dobson 2011, 167), although he "was clear about reserving cabotage" (Dobson 2011, 170). Although America got a more liberal agreement than Britain wanted at Bermuda, it made sure it retained the right of *cabotage*, which in airline terms means "the reservation of domestic flights for a state's own national airlines" (Dobson 2011, 6). The consequences of this were simple. By the late 1930s a dense, geographically extensive system of domestic airlines had emerged in America and nowhere else in the world. Although the British imperial system had begun to develop once the *Empire* boats entered service, the system was tiny and geographically strung out. Most of the development of air traffic in the postwar period was obviously going to be within America and across the Atlantic as America flexed its newly found hegemony and New York displaced London as the world financial center. It hardly mattered if an American airline disembarking passengers in London could not pick up passengers to convey onward to some other British controlled location. The likelihood of passengers wishing to go to Paris, Rome, or somewhere else in Europe was much higher, *cabotage* did not prevent selling seats to them, and the right to do so was enshrined under Roosevelt's concept of the "Fifth Freedom" (Dobson 2011, 175). The Chicago Conference did not embrace the "Fifth Freedom," but it was enshrined in the Bermuda Agreement. "Fifth Freedom" rights did not help the British much, which is why they opposed them at Chicago. More to the point, a British airline disembarking passengers in New York and flying on to, say, St. Louis, would not under *cabotage* be able to sell those now empty seats whereas an American airline would. This had two results. First, it discouraged British airlines from offering service past the American coast, which is all the first generation of transatlantic airplanes could really manage in range terms. Westbound most had to land in Ireland to refuel before leaving Europe and in Newfoundland to refuel again before flying on to New York. Second, it froze British airline service across the Atlantic into basically one port of entry, New York, where passengers flying on to other American destinations had to change to American flag airlines.

After Bermuda, and given the success of the first generation of American transatlantic range landplanes and the failure of the British to develop such, it is thus fair to say that Wilson was wrong at Versailles. Since the principal struggle turned out to be in long-distance air, not sea travel, it was America and not Britain that would win the struggle to dominate international transportation in the post-WWI world.

Chapter 4

The Struggle over Telecommunications

As Wilson noted at Versailles, international communications would be one of the three pivotal arenas of Anglo–American struggle. "Economically, telecommunications has been much more important to the world economy than broadcasting or film" (Hills 2007, 2). As indicated in chapter 5, however, broadcasting had significant implications for the hegemonic struggle between Britain and America as well as for a resurgent German challenge. Dealing with problems of technology development is never simple, and the development of broadcasting was in any case technologically totally dependent upon the development of telecommunications.

The tendency in the pioneering technological histories of submarine cables was to focus on the joint attempts to lay the Atlantic cables in 1857 and 1858 and then the successful British cables of 1865 and 1866, not to see those efforts as part of a complex geopolitical and cultural picture involving America, Europe, India, and then the rest of the planet. The more general technical history of both the Atlantic cables and submarine cables in general was well covered by one of their greatest architects, Charles Tilson Bright (Bright 1974 [1898]), as well as by more recent authors (Dibner 1959; Kieve 1973; Coates & Finn 1979; Burns 2004). The development of the American domestic network has also been also well covered (Reid 1879). The remarkable early attempt to compete with Britain in global communications represented by Western Union's Bering Straits and Eurasian project has received far less attention (though see Hugill 1999a). In general, the geopolitical aspects of this competition have been ignored (though see Hugill 2009b). The social and cultural aspects of the struggle, especially between Britain and America, are well treated by Müller (2016).

Technologies themselves are fairly easy to write about at a surface level, but they do not emerge in a vacuum. They are embedded in economic, social, cultural, and political structures that are usually regarded and studied as national in character and are rarely linked to the technologies in question. Many scholars, trained in specific disciplines, fail to see the complexities caused by the interaction of such structures, although Marshall McLuhan memorably described Innis' pioneering work, *Empire and Communications* (1950), as "handing us the keys to understanding technologies in their psychic and social operation at any time or place" (quoted in Innis 1972, xii). There is a vast amount of literature on telecommunications, far too vast to summarize here or perhaps anywhere, which seems highly compartmentalized along disciplinary lines: communications, engineering, history, politics, and the like. There are also histories of specific technologies in their national contexts. Electrical engineers have, of late, become interested in the history of their profession and the British Institution of Electrical Engineers (IEE) has convened a series of conferences on the history of telecommunications, published proceedings from those conferences, and has embarked on a series of very useful and well researched scholarly monographs.

This chapter examines the development of wired and wireless telecommunications in a geopolitical, global context. It contextualizes the struggle over international telecommunications seen by Wilson into three main areas between the 1850s and the 1940s: terrestrial telegraphy, submarine telegraphy, and wireless telegraphy.

TERRESTRIAL TELEGRAPHY

Both America and Europe began to install land-based telegraph lines in the 1840s. In America and Britain the primary role of the telegraph was commercial from its very onset. In Europe, however, it was almost always seen as an agency of state power, thus in geopolitical terms since it allowed the political as well as the commercial control of space. British telegraphy developed slightly earlier than American, but initially in the form of lines tied to specific railway companies to facilitate signaling. Britain's Great Western Railway began to install electric signaling telegraphs as early as 1838, but directors of railways were generally unwilling to spend money on signaling devices. Not until the late 1840s did the railways bring the electric telegraph into general use. Even as late as 1850 Britain had completed 7,231 miles of railway, but only 2,215 miles of telegraphs along them. The railways were the logical users of telegraphs, since they markedly reduced capital costs and accidents by allowing efficient and safe single line working. The main British railways had, however, easy access to capital and could afford to construct their tracks to allow two-way working, which minimized the need for signaling. Even so, the first telegraph company in Britain set up primarily for public use, the Electric Telegraph Company, was incorporated in 1846 and had links to the railways.

In America the telegraph companies were separate from the railroads. In contrast to Britain single track railroad lines were normal, so that as early as 1850 twenty American companies had installed some 12,000 miles of telegraphs, most of it along railroad rights-of-way (Kieve 1973, 43, 51). Western Union, founded in 1856, offered the railroads along which it built "free transmission of railroad messages, all having priority over non-railroad messages" (Beauchamp 2001, 65). Companies like Western Union were also common carriers of information, selling their services to everyone, but it was their relationship to American railroads that mattered most in giving them customers and rights-of-way so they could establish their networks. The early adoption of the railroad in America occurred in a capital poor environment, and American distances and topography were challenging. American railroads innovated substantially in locomotives, track construction, and the way they approached topography, but they were still too short of capital to install double tracking in the British way. Before telegraphy the risks of working trains in both directions on a single line were substantial, and head-on collisions regrettably common. Western Union provided the railroads with near real-time information about where trains were in space and time, so that they were no longer "lost in the system" and the risk of collision was therefore very substantially reduced (Hugill 1993). In many ways the enthusiastic adoption of telegraphic signaling was the single most important innovation of early American railroading.

By the late 1860s significant discussion had arisen in Britain about the proper ownership of the telegraphs. Although Britain was in many ways the classic laissez-faire liberal state, private enterprise had done a poor job of serving rural areas and small towns with telecommunications. Traffic density—and profits—were low in such areas, but the members of parliament for such poorly served districts began to complain. By the late 1860s the notion of "natural monopolies" was both becoming accepted and, to some extent, supplanting laissez-faire reasoning, and the government moved to buy out the telegraph companies and place them under the control of the British General Post Office (GPO) in order to better serve underserved communities. "In the United Kingdom, unlike on the Continent of Europe, there was never a suggestion that government control of the telegraphs was a political-strategic necessity; the state monopoly was set up as a business proposition" (Kieve 1973, 230).

The government buyout of British inland telegraphs was first discussed in 1865, legislated in 1868, but not fully completed until 1870 (Kieve 1973, 128, 152, 177). The consequences of this were substantial. As the government dithered and the terms of the buyout were negotiated and publicized, share prices rose. Ultimately, shareowners were handsomely compensated. Since the government had ceded to the GPO control only of telecommunications within the British Isles, and since the installation of successful submarine telegraph cables was just getting under way in the late 1860s, a significant number of the investors who had been shareowners in the inland systems reinvested their windfall profits in the new submarine cable companies being formed in Britain in the period. Their experience with and profits from inland telegraphs had been good and, correctly as it turned out, they saw no reason to assume that such returns would not continue in the arena of international communications.

Following the American Civil War there was considerable discussion in Congress about whether American telegraphs should be nationalized and an Act of Congress of 1866 allowed for such nationalization. As the decade wore on the rationale of the British government in nationalizing British telegraphs became cited as a primary reason for American nationalization. Senator C. C. Washburn made what seemed a compelling case in a speech of December 22, 1869, and seemed to have considerable support from both much of Congress and the Press. But opposition by Western Union was simply too great. "Not surprisingly, a monopoly gradually emerged during the next decade in the form of the powerful Western Union Company, which co-operated with the US post office to provide a public service paralleling the well-established letter post in America" (Beauchamp 2001, 65).

SUBMARINE TELEGRAPHY

The idea of the submarine telegraph had been suggested as early as 1795, but the most critical technological advance came in 1843 when Michael Faraday identified gutta-percha as an effective insulator that, unlike rubber, would maintain its elasticity at the low temperatures and high pressures of the ocean depths. In 1847 Werner Siemens successfully demonstrated the use of gutta-percha to insulate a cable laid across the bed of the River Rhine. British entrepreneurs rapidly brought under their control the world supply of gutta-percha, which came from Southeast Asia but was mostly marketed through Singapore. The Gutta Percha Company in London became the world's first supplier of submarine cable and produced the pioneering cable laid, initially unsuccessfully, under the English Channel in 1850 (Kieve 1973, 101–104). It was not the cable itself that proved to be problematic but the carelessness of fishermen and others who simply broke the cable by dragging it up near shore.

Telecommunications via submarine cables would confer on the states that controlled the new technology massive structural power in the arena of geopolitics. To begin with the state that most benefited was Britain, which installed after its hesitant start in 1850 a network of submarine cables that quickly became global. Britain did this in the almost totally laissez-faire environment of a British dominated liberal world-economy that was moving from manufacturing to transportation and financial services. By the 1840s shipping earnings were second only to cotton manufactures as a source of foreign exchange for Britain—cotton thread and textiles then accounted for 30 percent of exports and some 7 percent of total national income—and insurance earnings were already "more than double the value of machinery exports" (Foreman-Peck 1995, 18–19). Also by the 1840s the British were well aware of the "financial reward available through the international linkage of stock and commodity exchanges" that better telecommunications were likely to bring about. As early as 1845 John Watkins Brett and his brother Jacob proposed to British prime minister Sir Robert Peel "a general system of land and ocean telegraphy" (Dibner 1959, 10). That same year "the Bretts registered the General Oceanic Telegraph Company with the specific aim of connecting Britain with Canada" (Hills 2002, 23–24). In 1847 they obtained from the French government a concession to link Britain and France, and in 1851, after initial failure in 1850, they successfully connected the two countries (Kieve 1973, 103). Ultimately, the system of submarine cables "made the world market a possibility (Its most efficient use was between the Cotton and Corn [wheat] Markets of Liverpool and the New York Cotton and Chicago Corn [wheat] Markets)" (Kieve 1973, 237). The first truly globalized commodity was cotton, with cotton grown in the American South, spun and woven into textiles in Lancashire, and exported worldwide. For most of the 1800s, cotton was by far the most important and valuable of America's exports, with wheat being the second (Hugill forthcoming 2020).

Connecting Europe with North America would prove vastly more expensive and difficult technologically than connecting Britain with France across a gap of some twenty miles in fairly shallow water where the single biggest problem was fishing trawlers dragging up the cable. The first deepwater cable was the real technology prover for the Atlantic project, and this was laid between England and Ireland under the direction of Charles Tilson Bright in 1853 (Dibner 1959, 14 fn). Although some early experiments with submarine cables had been costly failures, improvement from then on was rapid. Between 1854 and 1861 the London company of Glass, Elliott and Co. "which was in the forefront of the great expansion in submarine telegraphy, laid twenty-four cables with a total length of 3,677 miles." In the early 1850s, however, doubts persisted over whether a signal could be propagated across the near 2,000 miles of the Atlantic gap "at such a speed that messages could be passed quickly enough in succession to be remunerative." In October 1856, a series of experiments by Charles Tilson Bright and Edward O. W.

Whitehouse linked up some 2,000 miles of cable and proved such distances workable (Kieve 1973, 105–106). Samuel F. B. Morse commented in a letter of October 3, 1856 that "the experiments under the direction of Dr. Whitehouse and Mr. Bright, which I witnessed this morning . . . have most satisfactorily resolved all doubts of the practicability . . . of operating the telegraph from Newfoundland to Ireland" (quoted in Dibner 1959, 14). That same month the Bretts, Bright, and the American telegraph pioneer, Cyrus Field, registered the Atlantic Telegraph Company (ATC) in London.

By the middle 1800s telegraphy was also proving its worth in war. When the Crimean War broke out in 1854 the British laid a 340 mile-submarine cable across the Black Sea from Varna to Balaclava. The burgeoning European network and the cable across the English Channel allowed much of the war to be directed from London and Paris. The telegraph system within the Indian subcontinent, which was designed as a tool of the imperial state and not commerce, confirmed the utility of the telegraph to the British in the Indian Mutiny of 1857 when troops were moved rapidly to the trouble spots identified by telegraphic reports (Choudhury 2010). The French made substantial use of telegraphs in their Italian campaign of 1859. But the main confirmation of the utility of telegraphy in war came during the American Civil War, "when over 15,000 miles of wire were used and more than a thousand operators employed." Thereafter the telegraph was obviously indispensable to the military (Kieve 1973, 239–240). Lincoln was able to direct much of the Union war effort from the telegraph room at the White House. The Confederacy had no such advantage.

The 1850s also saw critical pioneering work in oceanography with the mapping of the floor of the Atlantic basin. Virtually all the distance between Ireland and Newfoundland was found to be occupied by "a gently undulating plateau of great breadth, at a depth varying from 1,700 to 2,400 fathoms . . . compared . . . with soundings of 6,000–7,000 fathoms further south" (Kieve 1973, 107). Lt. Mauro of the US Navy, who did much of the mapping, named this "telegraph plateau" and described it as ideal for laying a submarine cable. In November 1856 the ATC raised £350,000 (equivalent to some $33 million today) in 350 ordinary shares of £1,000 each, with the Bretts and Field each taking 25 shares and Bright being appointed chief engineer. Construction, at Field's insistence, was rushed, Bright arguing unsuccessfully for a much sturdier cable. Although British experience had already shown that submarine cable was best stored coiled in water tanks as it was manufactured, Field's insistence on speed caused the ATC cable to be laid dry after manufacture and exposed to the sun, a ruinous mistake. Field, of course, was experienced in building terrestrial systems, not submarine ones. The British government guaranteed an annual payment of £14,000 during the working of the cable and provided one of the two cable laying ships, the steam battleship

HMS *Agamemnon*, the other being the USN's *Niagara*, then the world's largest and fastest steam frigate. The first attempt, in August 1857, failed when the cable broke at 274 nautical miles out of Ireland because the paying out machinery was too rough on it. The cable laid could not be recovered, and the unlaid cable was brought back to Britain and, again, laid dry.

The American effort was not without its detractors. When the scheme was first mooted in 1856 both President Franklin Pierce and his Secretary of State William H. Seward were strongly supportive of an American subsidy and the use of American naval ships to match British contributions. But when legislation came before Congress the next year there was violent opposition, there being "groups in Congress which saw only their own immediate regional interests instead of the potential for national benefit that the transatlantic cable offered" (Dibner 1959, 18). Eventually, with strong support from Seward, whose constituents in New York State saw the advantages of the cable clearly, the bill passed, but it passed the House by only nineteen votes and the Senate by only one. By the mid-1850s America was careening toward Civil War and most of the opposition to the cable came from the South, which saw it as of benefit primarily to the increasingly industrial economy of the northern states.

After considerable practice in the Mediterranean for the second attempt, in the spring of 1858, the *Agamemnon* and the *Niagara* again loaded their cable and met in mid-Atlantic. The two cables were spliced together and reeled out to their respective shores, *Agamemnon* in particular having to fight her way through very bad weather. This second attempt succeeded, at least briefly. The first messages were passed in early August, but the poor treatment the cable had received on land caused partial failure in early September (Kieve 1973, 106–109). An attempt by the electrical engineer for the project, Dr. Edward O. W. Whitehouse, the W standing regrettably for "Wildmon," to "clean out" the cable with some 2,000 volts of electricity, radically reduced the cable's effectiveness and it failed completely on October 20, 1858. Whitehouse had, in fact, been dismissed in September, the ATC Board describing him as "stubborn, uncooperative and insubordinate." Bright argued that it was Whitehouse's "high voltage tests which ultimately ruined the life of the cable, for as the voltage was stepped up the communication results became poorer" (Dibner 1959, 44). Even so, and for all its failures and the huge financial loss, the 1858 cable laid by ATC led British investors to conclude that, with a sturdier cable and more attention to electrical engineering along lines suggested by such as Bright, the project was feasible. The American conclusion, perhaps based too much on Field's lack of experience with submarine cables, was not so sanguine.

British attention then moved to the links with India where the tension between a terrestrial strategy and a submarine one played out even more clearly. In 1859 a submarine cable was laid from Suez to Aden, a distance of 1,358 nautical miles. This line failed frequently and had to be abandoned in 1861. In 1860 another submarine line was laid between Aden and Karachi, a distance of 1,685 nautical miles. This failed in a few weeks. There were two obvious problems, the first technological, the second political and economic. First, the line was laid taut, so that it easily broke when the unsurveyed topography of the sea floor undulated and the line had no slack to descend to that floor. Mauro's survey of the "telegraph plateau" allowed the Atlantic cable to avoid such problems. The line was also built too lightly to have survived for long in any case, although at this time what made for a successful cable was almost entirely in the realm of the empirical. Second, the British government, concerned about Indian security after the Mutiny of 1857, had agreed to a guaranteed return on capital to the Red Sea & India (RSI) cable of 4.5 percent for fifty years, whether it succeeded or not. Between them the ATC and RSI caused financial losses to the British economy of over £1,000,000, nearly $100 million in modern terms, as well as the huge fifty-year drain on the Treasury to pay the interest guaranteed to the RSI cable company. A British government-committee was thus charged to examine such problems in late 1859, producing in 1861 a vitally important report that concluded that the main problems of construction, laying, and operating submarine cables were technological and eminently surmountable, but that the financial risks should be left in future to private enterprise. One consequence of this was the successful laying in 1864 of a cable 1,450 miles along the Persian Gulf from south of Basra in modern Iraq and across the Indian Ocean to Karachi in what is now Pakistan. A second consequence was the development and laying of the much sturdier and successful Atlantic cables of 1865 and 1866 (Kieve 1973, 110–115).

Even with the Persian Gulf cable the route to India was only partially submarine, and it ran overland the rest of the way. There were three competing terrestrial routes, all of which suffered from the same fundamental cultural, geopolitical, and linguistic problems: wayleaves had to be given by foreign governments, and such governments insisted on their sovereign rights to control what happened on their territory and that their communications and languages be given primacy. The need for constant retransmission at state borders caused massive delays and garbling of messages. The first line ran through France and Italy, thence across the Adriatic into the Ottoman Empire. The second line ran through Belgium, Germany, and Austria into the Ottoman Empire where it joined the first and ran to the Persian Gulf. The third line ran through Germany and Russia, thence around the Black Sea and through Persia to Bushire on the Persian Gulf. The first two lines were seriously problematic: "telegrams took days rather than hours . . . and frequently arrived mutilated. . . . The problems were attributed almost entirely to the

Turkish section of the route, where the variety of scripts used by the Turkish operators and their unfamiliarity with European languages led to retransmission errors." A select committee of the House of Commons argued at the time that the line was worthless since it "was at the mercy in Asia of a set of incapable and unimprovable Arabs, Turks, and barbarians" (Beauchamp 2001, 165). The Black Sea route through Germany, Russia, and Persia was completed in January 1870 and suffered none of these problems, but it was clear it could be subjected to geopolitical interference, especially in Russia. The only geostrategically secure solution was an all-submarine route. Three separate companies, all of which were controlled by John Pender, laid this cable. The steamship *Great Eastern* that had successfully laid the 1865 and 1866 Atlantic cables completed the work by June of 1870. Pender's success led him to consolidate his companies and found in 1872 two more submarine cable companies, the Eastern and Eastern Extension Companies "to link all of Britain's Asiatic and Australasian possessions" (Beauchamp 2001, 167–169).

THE FIRST TELECOMMUNICATIONS STRUGGLE BETWEEN BRITAIN AND AMERICA: A TERRESTRIAL SOLUTION OR A SUBMARINE ONE?

Although it is correct to point to an emerging global economy by the mid-1800s the truth is that there were really two rather separate such economies, though both were centered on London: a highly commercial one centered on the Atlantic basin and a commercially important but much more political one focused on the British Empire in India. The two most vibrant cities in the Atlantic basin were London and New York, and linking them electronically was an obvious priority as submarine telegraphy emerged, hence the pioneering attempts of 1857 and 1858. Six things pushed Americans toward a terrestrial solution between the joint failure of 1858 and 1866, when the British finally succeeded without American help.

1. The serial failures of the first Transatlantic, Red Sea, and Indian Ocean submarine cables between 1858 and 1860 at high cost to the British state.
2. The success of the England–Ireland cable of 1853 that showed that short, deepsea cables were practical.
3. The sectional tensions that were leading to the American Civil War and that would render Congress increasingly dysfunctional in the late 1850s ensured that no more subsidies would be forthcoming to a technology that favored northern business interests.
4. The fact that one powerful American company, Western Union, was constrained by agreements with competing American companies from

connecting in New England to the proposed British transatlantic cable, so favored a terrestrial route.
5. The fact that the Civil War had both trained a large cadre of telegraph engineers and demonstrated that large-scale terrestrial systems were perfectly feasible technologically.
6. And that by 1865 the massive cultural, political, and linguistic problems the British would have with their terrestrial links to India had not yet surfaced.

By 1865 Western Union had thus been able to raise a large amount of money, in northern cities and at the height of the Civil War, for its Eurasian project to link America with Europe, New York with London, across two great land masses with the short sea link across the Bering Straits. It may have been technologically conservative and culturally and geopolitically naive but to American investors it made a great deal of sense.

The Civil War had proved the worth of telegraphy to the North, and the Union Army had trained a huge cadre of telegraph engineers easily recruited by Western Union for its Eurasian project as the war wound down. A further consequence of the war had been to guarantee the construction of the first transcontinental railroad, a single line in the usual American style, and thus the increasing importance of San Francisco and California to the reunited Republic. Once California was acquired from Mexico by the Treaty of Guadelupe-Hidalgo in 1848 San Francisco, driven to even greater importance by the gold rush, quickly became central to American commercial and thus geopolitical designs on the Pacific Basin (Delgado 2009). This importance was such that Western Union had pushed the telegraph across the continent to San Francisco in 1861, well ahead of the railroad. From San Francisco, a telegraphic link up the northwest coast toward the Bering Straits was a simple proposition.

The American secretary of state at the time, William H. Seward, had geopolitical designs on the whole of the west coast of North America based in part on the process of settling the *Alabama* claims, which can better be seen as three claims. The first, and the one eventually paid, was simply for the value of the ships lost. The second, by far the most radical and pushed by Charles Sumner, the chair of the Senate Foreign Relations Committee, was based on the argument that the *Alabama* and her sister ships had prolonged the Civil War by two years and that the proper claim was for the whole of Canada. The third was Seward's more modest proposal, that the *Alabama* claims should be settled by Britain ceding to America at least what is now the province of British Columbia (BC). When this failed his fallback position was to attempt to manipulate BC public opinion into preferring to join the American Union rather than the Dominion of Canada, newly minted in

1867. This ultimately failed when BC joined the Dominion in 1871 with the promise from the Canadian government of supporting what would become the Canadian Pacific Railroad to connect BC to Eastern Canada. The other part of Seward's plan was to purchase Alaska from Russia, which he accomplished in 1867. Alaska had long been a drain on Russia, and Russia had consistently failed to establish secure communications with or properly supply its far-flung outpost. Attempts to establish a Russian presence on the island of Kaua'i in the Hawaiian chain began in 1815 but failed in 1817, although three forts were constructed. Attempts to establish an agricultural colony in northern California to feed an Alaskan colony that could not feed itself also came to nothing in the end. Established in 1812, California's Fort Ross was abandoned in 1841 in the light of clear legal claims to the Pacific Coast by Mexico, Britain, and America. These were settled in America's favor by 1848. Britain gave up on the Oregon question in 1846 and America was victorious in the Mexican War of 1846–1848.

To an American interested in global telegraphy in the mid-1860s it would have seemed likely first that all the Pacific Coast of North America would soon be under American political control, and second that there was no obvious technical problem to extending a telegraph line across the Bering Straits and thence to London via Eurasia. In 1865 and 1866 Western Union sent hundreds of telegraph engineers trained by the Union Army in the Civil War to the Amur River region of eastern Russia to begin construction of its Eurasian project. At the same time as the British were raising the capital for the 1865 and 1866 cables Western Union representatives, at the height of the Civil War, were raising capital in northern cities for their alternative. Cyrus Field seems to have been convinced by ATC's failures that submarine telegraphy was just too risky, despite the conclusions of the 1861 British report that the technological problems could be resolved. The initial failure of the 1865 British attempt must have seemed to confirm this, but the remarkable British success of 1866, laying an entirely new cable as well as grappling up and repairing the 1865 line, caused Western Union to abandon its attempt and pull back to being a purely American company.

In many ways control over global telecommunications was crucial to the continuation of British power after 1861, when the Morrill Tariff marked the beginning of the American challenge to British hegemony (Hugill 1999a; 2009c). In this high-Victorian world technology diffusion was becoming increasingly rapid, aided by telecommunications, rapidly improving higher education as universities and academic disciplines began to develop into something like their modern form, and the emergence of recognizably modern academic organizations and scientific journals. The struggle over the telegraphic connection between the Old and New Worlds was a complex one, shifting back and forth technologically, politically, and economically. It was however, marked by American failure and British success, a success underscored by the success of the Atlantic cables of 1865 and 1866 and the all sea route to India in 1870. In *Great Eastern* Britain had a large enough ship to carry huge mileages of cable safely stored in water tanks. Once the first three cables had proven themselves the way was clear for the British Eastern and Eastern Extension Companies to girdle the earth with submarine cables and establish a British telecommunications hegemony that was never effectively challenged in the period covered by this book. British persistence, after Americans cried off, in developing the submarine cable technology that finally proved effective in 1866, and British willingness to invest vast sums of money in what seemed to American investors a huge money pit paid off in a remarkable fashion.

There was one last region of the planet to be conquered by submarine cables after the cable laying frenzy of the 1870s through 1880s. The Pacific was so vast that it required at least one, if not two, repeater stations en route, stations manned by telegraph company personnel whose job was to rekey messages on these very long routes. British companies completed the first transpacific line in 1902 as part of *The All-Red Line* (Johnson 1903), linking the areas on the world map colored red in British school atlases to show they were part of the Empire. In general the focus of the literature on submarine cables has been away from the Pacific, perhaps because the cable there coincided with the rise of wireless telegraphy (though see Hugill 1999a and Starosielski 2015, the latter dealing mostly with the modern fiber-optic cable system).

THE SHIFTING SANDS OF HIGH-VICTORIAN GEOPOLITICS AND THEIR ATTENDANT POLICIES, AND THE BEDROCK BELIEFS OF VICTORIAN CULTURE

Geopolitically the struggle that broke out between Britain and America in 1861 was characterized by America's return to mercantilism in the guise of protection for its "infant industries." As Wallerstein long ago pointed out, free trade is the preferred policy of the core state in the world-economy, which by definition means the hegemonic power, whereas mercantilism and protectionism are the preferred economic policies of the states he defined as semi-peripheral, which by definition will include powers seeking to challenge the hegemon in a period of multi-polarity (Wallerstein 1974). At the level of policy two forces began to operate very strongly in the developing global economy of the late 1800s: an increasing insistence on patent rights, and a building concern with international treaties to control emerging forces and technologies that would impact all members of the global economy.

The Return of "Mercantilism"

The developing constraint for high-Victorian geopolitics was the "new nationalism" that developed out of the American shift away from the British liberal world-economy into a neomercantilist world-economy. This became increasingly defined on the basis of formalized geographic boundaries around what we now know as the nation-state. This neomercantilism was begun by America in 1861 by the Morrill tariff (Hugill 2009c), then continued most notably in Imperial Germany. Appropriately, given this neomercantilist agenda, "the submarine cable license introduced by the U.S. Government under President Grant in the 1870s was the first . . . regulatory control on foreign investment" in America. America used protectionist policies early in its struggle to compete with Britain in the realm of global telecommunications. Such neomercantilism "greatly benefitted those companies protected and promoted by the protectionist state" (Hills 2007, 4, 12).

There was recognition that by no means all of the new technologies developing in the mid to late 1800s could be controlled by protectionist economic policies that impacted only the geographic region of the new nation-state: many such technologies had impacts across such increasingly formalized national boundaries. Control over the impact of these technologies by nation-states developed in three main ways. Neomercantilist protectionism, though important early in states like America and, later, Imperial Germany, was by no means universal. The patent system, which was a way of assigning rights to new technologies at the level of private property, though by no means perfect, was universal at the level of the nation-state. And in this period international conventions emerged as a way in which nation-states attempted to apportion rights to various international resources accessed through new technologies. The International Telegraph Union, founded in 1865, was, in fact, the first such body founded to deal with such issues (Codding 1972, 4; Codding & Rutkowski 1982).

The Problematic Patent System

Although the primary focus of patent rights was to protect the private property interests of entrepreneurs and, increasingly, corporations, it became evident by the late 1800s that radically different national patent laws had serious international ramifications. Embedded as they were in the domestic economies of the rising nation-states, both entrepreneurs and the ever-larger though not yet multinational corporations that characterized the late 1800s struggled to patent new technologies in all national markets they saw as potentially profitable, then to use their patent rights to exclude other entrants. A mark of the new nationalism was the development of an increasingly bureaucratized state apparatus. National elites sought to control this apparatus to ensure that states favored elite interests when imposing such control on areas previously lacking controls (Mann 1993). Absent international agreement on the appropriate rights of patent holders, rights varied markedly from state to state. Patents are clearly a serious state control of technologies, albeit for the private profit of patent holders. The most serious problems arose when states and governments failed to recognize how critical some new technologies could be to national security interests and how carelessly written or cleverly manipulated patent law could fail to protect such interests. This problem emerged very clearly in the arena of organic chemistry in the late 1800s and early 1900s, with German patent law being markedly better at protecting national interests than American and British (Hugill & Bachman 2005).

The Trouble with Treaties

International telecommunications was an obvious and fertile arena for the economic and geopolitical struggles that grew out of the "new nationalism." Initially, international communication between states was mediated through individual treaties, the first being in 1849 when Prussia signed a treaty with Austria–Hungary regarding the telegraph line being built alongside the railroad from Berlin to Vienna, which passed through multiple polities (Codding 1972, 13). The result was a messy patchwork of discordant treaties negotiated polity by polity. The International Telegraph Union (ITU) was founded in 1865 in part to impose some order on this dysfunctional mess. The order imposed by the various international conferences on wired telegraphy that followed was mostly limited to "landing rights," cross-border operations, the technical conditions for interoperability, and the rate structure for international messages, but it was better than nothing. Since the telegraph systems of all the states that subscribed to the ITU were state-controlled, such treaties were as much about politics as technology and economics.

America held back from the ITU on the grounds that telecommunications within its borders were the domain of private enterprise, although it subscribed to the technical aspects of the agreements reached in ITU conferences. In the first two decades of the 1900s America took the same position with regard to wireless telegraphy, but in 1920, in the immediate aftermath of WWI, America proposed a new "Universal Electrical Communications Union . . . with the object of simplifying communications by bringing all methods of electrical transmissions as far as practicable under the same rules." The process that began in Washington with the victorious allies from WWI in 1920 finally bore fruit in the Washington Conference of 1927 at which wired and wireless telecommunications were brought under a single organization, some order was imposed on broadcasting, and America finally became a signatory. From this point on America took the lead in shaping the international regulatory environment, culminating in the Atlantic City Conference of 1947, which established the framework for regulation in the period of American hegemony (Codding & Rutkowski 1982, 3–26, quote from 14).

The Bedrock Beliefs of Victorian Culture

And yet . . . the Victorian world that stretched across the North Atlantic was much more a homogenous culture in its bedrock beliefs than a heterogeneous set of economies divided by geopolitical strife. Both American citizens and British subjects held hard to their cultural certainties. On the American side religion was of even greater importance than it was in Victorian Britain, and Müller notes that many of the American investors and promoters that she characterizes as "The Class of 1866" were promoting a *Societas Christiana* across the Atlantic world intended to eventuate in a global peace movement (Müller 2016). The most central of those certainties was, however, a belief in the absolute certainty of progress, a belief not really challenged until the loss of RMS *Titanic* in 1912 and certainly not shattered until the horrors of WWI bore it home upon an increasingly destabilized and demoralized western world that global peace was an unattainable goal. Even then, such horrors were not fully understood in an America that was insulated geographically from the worst of them. Even today such beliefs in the progress of the era persist, witness the bizarre description of the Victorian cable system as a "Victorian internet" by one recent author (Standage 1998).

To Britons and Americans much of the Victorian era's progress was a result of its burgeoning technology, increasingly informed by science, in which such things as steamships, railroads, and telegraphs loomed largest. Along with other innovations, such as radically overhauled postal systems, these technologies shrank the Victorian world in a radical fashion. When the young Queen Victoria ascended the throne in 1837 "a letter from London would have taken nine months to reach San Francisco: by Victoria's death in 1901 mail took nine days and telegrams nine minutes. In 1840 the British introduced the "penny post" anywhere in Britain: in 1898 they extended it throughout the British Empire" (Hugill 2011).

The investors in the 1858 Atlantic cable were drawn pretty equally from America and Britain and they would become Müller's "Class of 1866." The prospectus of 1857 said "The Anglo-Saxon Race which is in the van of all social progress, dwells on both sides of the Atlantic" (quoted in Coates & Finn 1979, 131). Such investors "hoped, it is true, for excellent returns on their investment. But the historical records show that they also expected great returns for their nations and mankind in general. If they had any doubts on the latter point, the necessity of appealing to the investors' Victorian ideal of Progress and Civilization as well as the Victorian love of profits kept them from revealing such reservations" (Coates & Finn 1979, 120). To the Victorian cultural mind-set progress increasingly called forth an imperialist impulse to spread civilization. Imperialism, whether of the American variety as the American state spread itself across North America and on into the Pacific, or the British variety as the Empire spread itself around the globe required a massive geographic expansion of the Anglo-Saxon peoples. Better, faster forms of transportation and telecommunications were the central tools enabling that civilizing imperative to take imperial form.

At the level of international politics a crucial but often under-remarked event took place in diplomacy in the mid-Victorian period. Until the telegraph, diplomats abroad, especially ones separated by considerable geographic distance from their home country, had remarkable latitude in dealing with the country to which they had been appointed. The telegraph reduced the independence of diplomats and required they keep in constant communication with their own foreign offices. In general diplomats did not like such a reduction in their influence, and complained that the telegraph tended to exacerbate crises. Lord Lyons, the British representative in Washington at the time of the Trent Affair in 1861, argued that he successfully prevented that crisis from escalating into war because he was not being pressured telegraphically to "do something." Cyrus Field argued the exact opposite, that the crisis would have faded away with a brief telegraphic exchange (Nickles 2003, 73). Lyons hated Washington's weather and politics and he distrusted Seward, Lincoln's secretary of state. Such distrust was hardly justified. Seward had long supported Britain, and had argued strongly and successfully before a resistant Senate in 1857 in support of American involvement in the transatlantic cable. The Foreign Office certainly had a more nuanced and positive view of Seward than Lyons, as did Lyons successor of 1865, Sir Frederick Bruce, whose career, which ended abruptly with his unexpected death in 1867, was heavily shaped by the telegraph, the crisis over the *Alabama* claims, and perceived American naval power in the form of the *Miantonomoh's* visit to Britain in 1866 (Hugill 2009b). The telegraph might have been constraining to a few diplomats of Lyons' massive experience and abilities, but it seems overall that it allowed better management of international relations.

The crucial impact of the telegraph was speeding and greatly expanding the flow of information, especially in the developed world. This impacted the economy and culture of the North Atlantic basin very much more than its politics. By the late 1800s the telegraph had radically closed an Atlantic gap that had seemed to be narrowing rapidly in 1858 but had then widened because of the events surrounding and immediately following the Civil War.

But the Treaty of Washington (1871) clearly stabilized things and as the century wore on steamships became faster, better, and much more comfortable, and there were far more submarine cable lines. The American industrial boom that followed the Civil War saw vast amounts of British capital invested in a rapidly growing and increasingly wealthy America and a developing pattern of cultural contact at the level at least of the moneyed elites of Britain and America that resulted in a rapid rise in intermarriage between the two countries. The telegraph greatly facilitated all this. By the time Julian Pauncefote became ambassador to the United States in the late 1880s the scene was fully set for economic and elite cultural rapprochement as well as the mutual realization on both sides of the Atlantic that Germany was a building geopolitical threat.

WIRELESS TELEGRAPHY: A THREE-WAY STRUGGLE DEVELOPS

When wireless telegraphy was proved practical in the late 1800s it quickly acquired two main customers: the world's navies, and the world's merchant ships. The Italian born engineer, Guglielmo Marconi, moved to Britain in 1896 to try to sell his wireless technology to the British GPO. When that body demurred Marconi's mother, the Irish born Annie Jameson, persuaded her relatives in the whisky distilling industry to invest in her son's new venture. Marconi switched his focus to the Royal Navy and in annual maneuvers in the summer of 1899 his wireless technology showed its worth. By the end of 1900 the navy had commissioned Marconi equipment for forty-two ships and eight shore stations (Austin 1995, 44, 47; Hugill 1999a, 93).

In 1900 there was still only one globally dominant navy, that of Britain, but there were now three growing powers intent on building their own naval forces. Of those three, Japan would become allied with Britain in 1902 and its real geographic interests were, in any case, far from the North Atlantic basin. Two of the three challengers, America and Germany, were focused on the North Atlantic, and one, Germany, was even more heavily focused on the seas around Britain, the North Sea in particular. At the level of merchant fleets, Britain in 1900 was clearly globally dominant because of the remarkable explosion of shipbuilding that had occurred in northern Britain from the 1880s on, especially along the River Clyde and focused on Glasgow (see chapter 2). In 1900 there was simply no obvious challenge to the cheap, mass-produced, efficient, British built, steel hulled, triple expansion engine, coal powered, general cargo ship.

Navies saw wireless as massively improving communications, command and control, thus allowing centralized control of fleets at sea. By 1913 every ship in the Royal Navy could communicate with the Admiralty in London: "wireless telegraphy had become the eyes and ears of the fleet at sea and the days of independent action had definitely passed" (Vyvyan 1933, 125). But wireless technology was just as easily embraced by other technologically advanced economies, such as America and Germany. The British company, Marconi, had a slight advantage in that it was the first to develop a successful wireless technology, one that used electrical sparks to generate radio frequency waves. This had, however, been quickly copied by the German company *Telefunken*, which translates as "sparking at a distance." The USN had pioneered technologically much better wireless through its clandestine acquisition of the California company, Federal Telegraph. Federal had licensed and then improved on the electric arc technology of a Danish inventor, Valdemar Poulsen. Arcs generated much more stable, continuous radio waves that could be tuned to a fixed frequency, where spark transmissions emitted radio signals across a wide range of radio frequency. Although it took a lot of electrical energy and skill to run the arc, energy was not a problem on a large warship.

But one of the consequences of geography was that wireless conferred some advantages that were not evenly spread. Britain thus gained a crucial geostrategic advantage over Germany during WWI. German admiral Scheer noted in his memoirs that Britain's long east coast allowed Britain to install many direction-finding stations that were rapidly able to locate enemy vessels as soon as they used their wireless, which they had to use in order to keep a large fleet operating in unison. Germany had only a very short coastline on the North Sea, could not depend on its ships being able to exit the Baltic through Danish and Swedish controlled waters, and was thus highly dependent on the Kiel Canal to link the Baltic with the North Sea. If the German fleet showed by increased radio traffic that it was sortieing it had to be sortieing in the general area of the Jutland peninsula. Vyvyan suggests that had neither side had wireless in 1914 the advantage would have accrued to Germany, since the British fleet would have to have been constantly at sea instead of doing what it did, coming out only when increasing radio traffic indicated when and where the German fleet was operating, as was the case at the Battle of Jutland in 1916 (Vyvyan 1933, 128–131). The only question the British had at Jutland was how much of the German High Seas Fleet was on the move. Since the British had sortied almost their entire Grand Fleet, the result was a Mahanian clash of battlefleets the Germans only just avoided losing disastrously. The British lost more ships, but their damaged ships were ready for battle in two weeks. Many German ships were so badly mauled that they were still being repaired two years later, and the High Seas Fleet never sortied again for the rest of the war (Hough 1983).

The problems raised by wireless telegraphy for merchant shipping quickly became a fertile field for international conferences, problems that went far beyond the simple "landing-rights" type issues of telegraph lines crossing the land-based political boundaries of territorial states. The first three such conferences were held in 1903, 1906, and 1912 and dealt mostly with maritime issues, although the Berlin Conference of 1903 is usually described as "preliminary" to the First International Radiotelegraph Conference held in Berlin in 1906. These conferences arose out of two crises, the first, in 1902, was political and economic in nature and took two conferences to resolve, the second was the stark tragedy of the loss in 1912 of RMS *Titanic* on its maiden voyage.

The first crisis arose out of the events of 1902. From its inception the Marconi Company had refused to carry signals from non-Marconi sets because of the provisions of the British telegraph acts of 1868 and 1869. Essentially, "a private company could send signals for its own use, but not for gain as a public telegraph service" (Burns 2004, 352). Once Marconi's technology was proven, Lloyd's only covered ships at the highest level of insurance if they carried wireless, and Marconi's insisted that their wireless sets only be worked by Marconi employees and that they communicate only with other Marconi equipped stations as part of a private company (Hugill 1999a). In 1902 Marconi shore operators refused a message from the brother of the German Kaiser who was returning from America aboard the *Deutschland* since it emanated from the ship's Slaby-Arco transmitter. "The incident had immediate political repercussions" including the 1903 conference, which focused on interoperability between Marconi and non-Marconi users, and at which America refused to support German complaints (Headrick 1991, 120).

In 1902 the American navy began to experiment with wireless sourced from six companies, not including Marconi's. In 1903 it concluded that the German Slaby-Arco sets from the company that became known as *Telefunken* were the best in technological terms (Burns 2004, 356). Although Lloyds had initially accepted Marconi's preeminence they became increasingly uncomfortable with the idea of interoperability. In 1904 British law was changed to effectively bring wireless telegraphy under the control of the GPO. By the time of the 1906 Conference in Berlin, America had come to side with Germany on the issue of interoperability. The Conference put forward a protocol that would force interoperability and open up international airwaves to non-Marconi equipment users. The British government ratified this protocol and the International Radiotelegraph Union emerged from this 1906 Conference.

The second crisis resulted in the second International Radiotelegraph Conference, held in 1912 in London. This was essentially a response to the loss of the RMS *Titanic* on its maiden voyage. *Titanic's* Marconi operators should have been able to get help from the American ship, *Californian*, hove too only a few miles away because of concerns about ice. But *Californian's* radio operators had shut down their wireless early and the watch on board that ship thought *Titanic's* distress rockets were a firework display. *Titanic's* operators were able to contact the Marconi operators on a British ship, RMS *Carpathia*, which arrived after *Titanic* had gone down, but was able to rescue those who had been able to take to the limited number of lifeboats. It was a harrowing tale but it made very clear the role wireless could play in saving lives in such an emergency. The 1912 conference led in turn to the International Conference on the Safety of Life at Sea held in London in 1914 which required that all ships carrying more than fifty passengers and engaged in international trade be equipped with wireless (Codding 1972, 107).

Although the period 1815 to 1861, the height of the "*pax Britannica*," was a period of geopolitical stability with an unchallenged hegemon, Britain was showing signs of decline by the 1880s. Britain was slow to embrace the Second Industrial Revolution and, their economies increasingly protected by tariff walls, first America and then Imperial Germany were challenging Britain's dominance in manufacturing by the 1880s. Geopolitical instability returned as these two powers intensified their economic challenges and began to see themselves as global powers that could match Britain. What weakened their challenge was Britain's global network of submarine cables, focused on London, which in turn focused the world's financial services on London. By the late 1800s the financial services sector had replaced manufacturing as Britain's main profit center, and its profitability and London's wealth and power increased rapidly. By the turn of the century, however, it was becoming clear that Britain's competitors recognized the advantages control over global telecommunications gave Britain, and this became for the challenger powers a, if not *the*, crucial geopolitical as well as economic issue. Germany and America both began their own networks of submarine cables for reasons of national security, and both sought to exploit the technological end run around British cable dominance that wireless seemed to offer, hence their interest in the early radio conferences. This increasing economic and geopolitical competition created a technological ferment in telecommunications by the early 1900s.

The Problems with Marconi's Spark Technology

The problem with long-range wireless telegraphy was that, as initially developed by Marconi, it used mainly very low frequencies. Very low frequency "long waves" hugged the curvature of the earth, but long-range came at the expense of huge inputs of electrical energy and the need to construct massive antennae covering many acres. The main problem with Marconi's technology was that it generated such low frequency waves using electrical sparks. The spark generated a range of radio frequency waves, not a single, easily tuned frequency. Spark transmitters were thus fundamentally untunable. Marconi certainly understood the importance of tuning and was awarded a patent to that effect in 1901, British patent 7777 (usually known as the four sevens patent). But spark transmission made it impossible to have many stations operating without them substantially interfering with each other, hence the demand for the highest possible power for a given transmitter to override the signals from all weaker stations.

Early Alternatives to the Spark

Two alternatives to Marconi's spark transmitter developed before WWI broke out, the arc and the alternator. The Danish engineer, Valdemar Poulsen, recognized by 1902 that the electric arc, effectively a continuous spark, generated a stable radio frequency. Poulsen initially proposed that his technology would allow hand-held radio voice transmitters and receivers akin to a modern cell phone! But the arc was hard to keep in tune, though less so than the spark: generating it still needed a great deal of electrical energy, and maintaining it needed skilled operators. Cyril F. Elwell, a student at Stanford University, persuaded Poulsen to license his technology to him in 1909 and founded Federal Telegraph in San Francisco in 1911 to exploit it. By 1912 Federal was able to demonstrate a reliable wireless connection to Hawai'i, at least after dark, and the USN took note. Absent a submarine cable, the only effective telecommunications link to Hawai'i until then was via the United Wireless Marconi-style spark transmitter on the California coast north of San Francisco. But in 1911 the Supreme Court found United in violation of Marconi's patents and its property was transferred to American Marconi. The USN had no trust in Marconi because it was British (Hugill 1999a). Federal was American, its continuous wave arc technology was more easily tuned than Marconi's spark system, and the USN could reach its geostrategically crucial base at Pearl Harbor from Arlington, Virginia, with only one repeat at San Francisco. From 1913 on the USN invested heavily in Federal's technology, which was well suited to use on board ships (Mann 1946).

The second alternative source of radio frequency waves was the alternator, which produced relatively easily tuned and stable radio waves but was bulky, required mechanical engineering of a very high order, and still had a high demand for electrical energy to drive it. It was not at all suited to use aboard ships although it was highly effective in shore stations. Alternator technology developed both in America and Germany. Ernst Alexanderson at America's General Electric Company, which already made water powered electricity generating turbines that were able to spin up to very high speeds, developed the most effective alternators. The Alexanderson alternator span at over 2,000 rpm, fast enough that it was able to generate the needed radio frequency directly. Although *Telefunken* used Goldschmidt and Arco-Telefunken alternators before WWI, and installed a Goldschmidt alternator in their station at Tuckerton, New Jersey, just before the war broke out in 1914, these designs span at lower speeds than the Alexanderson design and thus needed a frequency multiplier, which made their output less stable. Goldschmidt used a gearing of four at the Tuckerton installation (Freris 1995, 71, 75; Hugill 1999a, 114).

The Real Solution: The Thermionic Triode

The development of the vacuum tube (British "valve") made for cheap, reliable, simple generation of radio frequencies that were stable enough for wireless telephony and that could be generated without excessive amounts of electrical or mechanical energy. As R. N. Vyvyan, a senior engineer with the Marconi Company noted, the spark, arc, and alternator

> were used in commercial wireless telegraph stations with success, but the wave form they emitted, while continuous, was not sinusoidal, and not suitable for distortionless telephony. The master invention of the thermionic valve by Sir Ambrose Fleming and the subsequent invention by Dr. Lee de Forest in the U.S.A. of the three-electrode valve, and the later discovery that the valve could be utilized as a generator of oscillations, as well as an amplifier or detector, rendered all other methods of creating continuous oscillations obsolete and made telephony by wireless and broadcasting possible. (Vyvyan 1933, 200)

Ernst Alexanderson, the developer of General Electric's very high quality alternator, believed by 1918 that, even with the alternator displacing spark and arc transmitters, "there would be room" on the world stage in the long wavelengths generated by his alternator for only "a total of twelve first-class transmitting stations" (Alexanderson 1984, 627). In fact, only twelve such, each capable of an output of 200 kilowatts, were ever built and installed (Weedon 1995, 70). Such a calculation certainly informed the USN at the end of WWI. The USN thus successfully froze out Marconi's attempt before and during WWI to create a global network of spark transmitters, which would have ensured no one else's signals could have been heard, in the process creating during the war its own global network of Poulsen arc transmitters. To solidify this advantage after the war the USN used its political clout to first prevent American Marconi buying GE alternators in 1919, then to force American Marconi to sell out to the Radio Corporation of America (RCA), a company it believed it would be able to manipulate and which, through inheriting the USN's Poulsen arc network, would be able to monopolize global wireless communications using alternator technology (Hugill 1999a; Winkler 2008).

Forced by the needs of WWI wireless communications technology matured rapidly between 1914 and 1918. In 1905 British engineer Ambrose Fleming patented the electronic diode, a vacuum tube that detected flows of electrons and thus made reception of wireless signals possible. In 1906 American engineer Lee De Forest added a third electrode to make the triode, patented in 1908, and which he called the *audion* tube. In this early period, both tubes used a relatively "soft" vacuum since they were manufactured using machinery developed to evacuate electric light bulbs, which did not require very much in the way of vacuum. These early tubes were hard to make and maintain, and their service life was short, though they had great promise and three huge advantages over sparks, arcs, and alternators. First, they used very much less electrical energy than all other ways of generating the necessary frequencies for wireless transmission. Second, they produced a stable and easily tuned set of frequencies. Third, they could perform functions other than the generation of radio frequencies.

In early 1910 De Forest demonstrated that his *audion* was an effective emitter of radio frequency, broadcasting part of a live performance of Giacomo Puccini's great opera *Tosca* from the Metropolitan Opera in New York City. De Forest then moved to Federal Telegraph in San Francisco, developing his *audion* as an amplifier as well as a radio frequency emitter. Federal's history is of interest. Elwell had initially licensed Poulsen arc technology believing he would be able to use it to create mobile telephony. As that prospect faded, Federal shifted its concentration to long-distance wireless telegraphy using the Poulsen arc and ended up being effectively taken over by the USN (Howeth 1963, 45; Hugill 1999a, 95). The evolution of De Forest's *audion* as an amplifier and oscillator was complex and involved both diffusion and independent invention.

> At least four engineers, and perhaps six, arrived at the amplifying and oscillating audion circuits almost simultaneously. . . . The validity of their US patents was contested in what was known as the "four-party" interference proceedings. . . . The activities of these individual genius engineers were mingled with the strategies of large corporations. [Alexander] Meissner worked for the German firm Telefunken; [Irving] Langmuir was a well-known scientist-engineer at General Electric; [H. J.] Round was the foremost engineer at the British Marconi Company. AT&T [American Telephone and Telegraph] bought [Lee] de Forest's audion patents; Westinghouse later bought [Edwin Howard] Armstrong's; [Fritz] Lowenstein provided much information to Western Electric. (Hong 2001, 156)

This technological ferment and its resulting complex set of diffusion paths and patent rows played out on an evermore competitive geopolitical as well as economic stage. Faced with the problem of poor service life for the early "soft" vacuum tubes in aircraft flown by Britain's Royal Flying Corps, Round was one of the first engineers to see that a much more thorough evacuation of such tubes would improve service life: these became known as "hard" tubes. Faced with severe personal financial trouble and Federal Telegraph's embracing of Poulsen arc technology, De Forest sold his amplifier technology to AT&T and the Bell Telephone System in 1913. De Forest's *audion* amplifier then became the centerpiece amplification technology for AT&T's successful transcontinental telephone line of 1915. American vacuum tube technology also diffused out of America, albeit via industrial espionage. A deserter from the French Army had settled in Germany, trained as an electrical engineer and started work as a commercial representative with *Telefunken.* He learned of the *audion* during a sales trip to America and fed that information back to *Telefunken.* In Vienna Robert von Lieben patented a vacuum tube in 1906 that may have been based on this diffusion path. German companies at the time were notoriously fond of illegal copying and patent rights in such emerging areas of science and technology as organic chemistry and physics were exceedingly difficult to enforce across state lines (Hugill & Bachmann 2005). However it was arrived at, by August 1914 the German Army's central headquarters in Luxembourg were connected by a telephone line amplified by Lieben tubes to the staff headquarters in East Prussia (Hugill 1999a, 76). When war broke out in 1914 the Frenchman acting as an industrial spy gave himself up to the French authorities, to whom he handed over both the latest American and German vacuum tube technologies. These became the basis for the excellent French "hard" vacuum tube, the *Telegraphie Militaire (TM)* that in turn became the basis for much of the early British "hard" vacuum tube industry (Burns 2004, 414).

The initial use of wireless telegraphy was by the military for mobile warfare and because landlines were often quickly cut by the enemy. Following the successful use by the Royal Navy of wireless in its summer maneuvers of 1899 the British Army experimented with wireless in the Boer War (1899–1902). Before the war broke out the Germans offered substantial assistance to the developing Boer insurgency, and that included wireless sets from the Berlin electrical equipment company, *Siemens und Halske*. These were promptly impounded in customs by the British in late 1899. For a variety of reasons, but mostly because the electrical conductivity of the ground in South Africa was very different from that in Britain, as well as from the ocean, the army had little success and the experiment was abandoned (Austin 1995, 44, 47). Even so, it was obvious that mobile warfare on land would benefit as much from wireless as mobile warfare on sea. The next decade saw much experimentation and rapid improvement of the technology.

The danger with wireless was that messages could be easily intercepted. Even so, wireless proved its utility immediately in the war of mobility that characterized the clash between the German and Russian armies on the Eastern Front in WWI. Immediately the war broke out in 1914 two Russian armies invaded East Prussia, and the Russian First Army under General Rennenkampf forced the Germans to retire after the Battle of Gumbinnen on August 20. However, the Germans made excellent use of intercepted Russian wireless messages and observation from airplanes to realize that the Russians were being slow to reunite their forces. The Germans thus moved the troops that had retired from Gumbinnen rapidly by rail to smash General Samsonoff's Second Army at the Battle of Tannenberg on August 23 before the Russian First Army could rejoin it. This was the most crucial German victory of the entire war since in the long run it would effectively remove Russia from the war. Victory at Tannenberg also allowed large numbers of German troops to be released from the Eastern to the Western Front in time for the Marne offensive in September 1914 (Vyvyan 1933, 119).

The initial use of wireless on a Western Front that quickly became a static war of trenches, barbed wire, and machine guns required bulky spark transmitters the British called "trench sets." Each trench set required three pairs of men to carry it and such teams were usually quickly identified and dealt with by enemy machine gunners. From late 1917 on the British deployed much smaller and lighter sets using vacuum tubes. The tube sets proved invaluable as relative mobility returned to the ground war in 1918 after a long period of stalemated trench warfare. A further advantage of the tube sets was that far more could be deployed since they occupied less space in the frequency spectrum. Only about one spark set was operable per division (up to 15,000 men) if mutual interference was to be avoided. Since two tube sets could be deployed per brigade (3,000 to 5,000 men) the tube sets increased the density of radio use at least sixfold (Vyvyan 1933, 112, 115–116).

This effort during WWI to develop vacuum tube wireless focused on the low frequency long and medium waves. There were two applications for this technology: the first was to use the long waves for long-distance speech transmission; the second, more important, was to use the medium waves for short-range telephony on the ground or in the air. In November 1915 AT&T used a complex transmitter with over 300 "soft" tubes to transmit speech from Washington to Honolulu. Next month they repeated that success by briefly connecting Arlington to Paris, but the technology was highly experimental, the tubes were unstable and difficult to maintain, and AT&T was really acting on behalf of the USN rather than developing anything even vaguely resembling a commercially viable service (*The Electrical Experimenter*, November 1915, 321, and December 1915, 393). By 1919 however, the rapid progress made during the war, especially the development of much more stable "hard" tubes, allowed Marconi's chief wireless engineer, H. J. Round, to transmit "speech during daylight hours from Ireland to Canada, using only 2.5 kilowatts with a wavelength of 3800 metres" (Vyvyan 1933, 202).

The more important use of wireless telephony was in the medium waves. As military authorities realized that such "hard" tubes as the French *TM* were excellent radio frequency emitters they pushed them into mass production to meet the evolving military needs of command and control in the new war in the air. "Plan 1919" was the British plan to break out of the trenches in 1919 using advances by air and ground units coordinated by wireless telephony. It called for every multi-place aircraft on the allied side to be equipped with wireless sets capable of voice transmission and reception, and operating in the medium waveband over relatively short-ranges, thirty to fifty miles. Each wireless set needed five to seven "hard" vacuum tubes, which required a massive production effort. Although reliability had improved over the "soft" tubes maintenance was still an issue, and another massive effort went into the training of radio maintenance engineers. By the end of the war the French had manufactured more than a million *TM* tubes. In America General Electric manufactured some 200,000 and Western Electric as many as 500,000. In Germany *Telefunken* was producing 1,000 "hard" tubes per day by the end of the war. By the end of the war the total strength of wireless personnel in the Royal Air Force (RAF) comprised 520 officers and 6,200 operators and mechanics (Burns 2004, 414, 417).

GEOPOLITICAL FAILURE, GEOPOLITICAL SUCCESS

As WWI wound down it seemed clear that the geopolitical sands had shifted away from Britain and toward America because technological innovation had shifted in America's favor. American engineers at Federal Telegraph had perfected Poulsen arc technology, Alexanderson at General Electric had perfected alternator technology, and De Forest working primarily for AT&T had been a prime force behind the rapid improvement of vacuum tube technology. The USN had clandestinely begun construction of the world's first global wireless communications system once it began acquiring the Federal Telegraph Company's Poulsen arc technology in 1911. What President Wilson had come to believe by the time of the Versailles conference was that the stage was set for America to make a technological end run around Britain's dominance of global telecommunications. In 1918 it seemed that the Alexanderson alternator was the best technology available for long-distance wireless communications: British Marconi wanted ten of them and American Marconi another fourteen. If Marconi had acquired them it seemed to Wilson it would cement Britain's control of global telecommunications, adding wireless to its submarine cable network (Lewis 1991, 142). To prevent this what was needed was the right set of policy decisions.

Given the limited technological understanding of politicians such as Wilson and senior USN officers such as Bullard, for all that he was director of naval communication, a state-controlled monopoly seemed a logical necessity. During the war all patent rights had been temporarily suspended in America, which had allowed the USN to come to control wireless telecommunications. In December 1918 the secretary of the navy, Josephus Daniels, was pushing for passage through Congress of a bill sponsored by Cincinnati Representative Joshua Wills Alexander that would "secure for all time to the navy department the control of radio in the United States, and will enable the navy to continue the splendid work it has carried on during the war." The then assistant secretary of the navy, Franklin Delano Roosevelt, took a similar position (Sobel 1986, 22–25, quote from 23). The problem was that Congress gave little support to Alexander's bill or any continuation of the USN's monopoly.

To Wilson and the USN the solution to the postwar resumption of patent rights was, therefore, to fold those various patent rights into a single company, RCA, which would act as if it were under state control. Bullard returned from Versailles to visit General Electric (GE) and reported to GE's vice president, Owen Young, that "Wilson wished to checkmate British domination of worldwide communications" (Lewis 1991, 143). The major problem was that the American Marconi Company controlled many of the patents, and even though this was an almost wholly American company it had, for the USN, the taint of a foreign connection that it did not trust. In the early days of radio before WWI the USN had rejected Marconi equipment since Marconi's were insisting that their equipment could only be used to communicate with other Marconi stations, a by-product of the row over interoperability (Sobel 1986, 17). After tests the USN had preferred Telefunken equipment and then developed its own via its ownership of Federal Telegraph and its Poulsen arc technology. By September 1919 GE acting under Bullard's instructions had pressured British Marconi to sell to them its shares in American Marconi, which they did only after GE agreed that it would not enter the field of wireless communication itself (Sobel 1986, 29). RCA was thus created out of a modified American Marconi with GE holding the largest bloc of shares at 30.1 percent. Westinghouse held 20.6 percent, AT&T 10.3 percent, the United Fruit Company 4.1 percent, and a host of smaller companies held the remaining 34.9 percent (Sobel 1986, 35). Although they could not have known it, nor did RCA to begin with, what Wilson and Bullard had done was to create a company that would soon realize that it controlled all the patents that made radio broadcasting for entertainment purposes profitable (see chapter 5). This started a huge fight between the RCA shareholders, who argued that their willingness to cede their patent rights to RCA was for the purposes of telecommunication, not broadcast radio. Worse, it helped set the stage for America to snatch geopolitical defeat from the jaws of victory in the arena of international communications.

At America's first national radio conference in early 1922 Secretary of Commerce Herbert Hoover compounded the problem when he assigned the low frequencies of the long waves to the USN, the intermediate frequencies—what we would now think of as the medium waves or the AM band on a modern radio—to the broadcasters, and then opened up what we now recognize as the most valuable high frequencies to anyone, without restriction. To make matters worse, Marconi had, by 1919, already demonstrated that such high frequency short waves could carry telegraphic signals over huge distances using very little power. Hoover's advisers were clearly incompetent technologically.

In 1902 two physicists, the American Arthur Kennelly and the British Oliver Heaviside, had independently predicted the existence of a reflecting layer in the upper atmosphere to account for Marconi's claim that he had successfully transmitted a wireless signal across the Atlantic from Poldhu in Cornwall to St. John's, Newfoundland, the previous year. Radio waves are propagated in a straight line and, with the exception of very low frequency waves that are affected by gravity to an extent, should proceed quickly into space as the curvature of the earth took its toll. In this time frame the effect of gravity in "curving" the very low frequency waves so that they followed the earth's curvature for some distance was unknown. Because of his use of a crude form of spark transmission and his use of "the most powerful spark transmitter ever built" Marconi's experiments in 1901 occupied a vast amount of radio frequency spectrum and a modern reconstruction of his experiment showed that parts of his signal may indeed have been at high enough frequencies to be reflected off the Kennelly–Heaviside layer (MacKeand & Cross 1995, 26, 29). Later, he and other early experimenters used better-tuned systems that focused their efforts in the very low frequencies that are affected by gravity.

Kennelly and Heaviside's predictions of 1902 were not, in any case, experimentally verified until the British physicist Edward V. Appleton did so in 1924. Marconi's experiments of the early 1920s, where different frequencies gave different results at different times of day, suggested that the Kennelly–Heaviside layer was by no means a single layer. Its variations were first mapped using radio pulses by American physicists Gregory Breit and Merle Tuve in 1925 (Breit & Tuve 1926). Different reflective layers in the planet's upper atmosphere varied in their ability to reflect different frequencies back to earth at different times of day. The precise nature of this reflective ability took a great deal of time to research and map accurately. Britain, with its far-flung Empire and its dependence on global trade, had a vested interest in cheap, efficient global communications, and Marconi's high frequency experiments promised precisely that. The British thus used their National Physical Laboratory to oversee precisely the mapping effort needed to allow the commercial and political use of Marconi's energy efficient, long-range, high frequency, short wave wireless technology (Hugill 1999a, 87, 128).

It was thus in the arena of imperial telecommunication that Britain wrested back control of the new wireless technology, and here geopolitical concerns were clearly uppermost, although the shift by RCA to a primary focus on broadcasting and profit taking was not immediately obvious in Britain in the mid-1920s. Since it was not, increasing fear of RCA's likely ability to dominate international communications pushed the British to force a merger between the British submarine cable interests, the Eastern and Associated Telegraph companies set up by Pender, and Marconi Wireless in 1928. This created Imperial and International Communications, Ltd., which matured as Cable and Wireless, Ltd., in 1934. Although nominally a private company, Cable and Wireless had no choice but to act as if it were under state control, not least because the BBC had emerged as a state-owned entity in 1927. Once the British state was able to monopolize broadcasting for entertainment purposes no company, however strong its technical position in terms of broadcasting might have been, could have enjoyed in Britain the ability to generate profit through broadcast radio that was enjoyed by RCA and the National Broadcasting Company (NBC) in America. RCA was able to simply evolve as a commercial company and leave behind its geopolitical origins. Marconi and Cable and Wireless had no such options.

As American telecommunications power increased, especially following the 1927 Washington Convention of the International Telecommunications Union (ITU), Britain still had to adjust. From 1927 on the ITU allowed AT&T, which had previously been restricted from international operations, to offer international telephone service.

> By 1936, as well as having a virtual monopoly on domestic telephone transmission and a share of the domestic telegraph market, the company had opened twenty-seven direct international radiotelephone circuits. During the 1920s AT&T established interconnection with telephone systems in Cuba, Canada, and Mexico without any form of U.S. license. High international tariffs, low traffic volume, and the ITU system all reinforced the company's international voice monopoly. (Hills 2007, 32)

This was a direct attack by AT&T on British interests. In 1937 the British argued successfully that, under article 13 of the Madrid Convention of 1932, they were entitled to treat the

> whole conglomeration of Dominions and colonies as one international end-to-end bloc served by one monopoly international carrier—Cable & Wireless Ltd. For the U.S. companies this was an impediment to global expansion, and [only] the European war provided the necessary political leverage to open up the British Empire. (Hills 2007, 32)

During WWII American interests began to press increasingly on Britain's Cable and Wireless. In line with American behavior from 1861 on the Federal Communications Commission (FCC) took a strongly neomercantilist position, refusing in 1943 to allow American companies carrying international messages to negotiate with foreign governments and making it "a condition of international licenses that international settlements be made on a 50/50 basis" (Hills 2007, 37). By early 1943 the American military had "completed an around-the-world wireless pipeline . . . carrying fifty thousand messages a day" (Hills 2007, 43), by 1944 the US Army had "the largest wire and radio system the world has ever seen," and the British ambassador to Washington was urging a united front "before the Americans strengthen their own position by effecting a merger of their international telegraph undertakings" (Hills 2007, 35). The chair of both the FCC and the American Defense Communications Board, James Fly, was seen in London as advancing "American commercial dominance under guise of a democratic philosophy for the free flow of international communications" (quoted in Hills 2007, 43). The Commonwealth Telecommunications Conference of July 1945 recognized that the British had to defend themselves against the American advance, although it stopped short of allowing Cable and Wireless to proceed with its "scheme to bring all the international communications facilities in the Empire under one entity—an Empire corporation" (Hills 2007, 36).

After 1945 the FCC strengthened its position, cutting international rates to liberated Italy without any consultation of the supposed 50/50 partners. At this point "the British ambassador in Washington was instructed to inform the State Department that the British Government 'received the report of United States action in this matter with incredulity'" (Hills 2007, 41). In the first postwar conference of the ITU, in Atlantic City, NJ, in May 1947, "the control exercised by the U.S. government mirrored their control over the 1927 Washington Radiotelegraph Conference that followed World War I" (Hills 2007, 49). Worries over this increasing level of American influence prompted the British to use their political connections with their Dominions and their European allies to prevent the FCC from gaining a much higher level of control over international communications and the manufacturing of radio equipment. They "defeated the U.S. proposal that the ITU should be sited with the U.N. in New York: it would now be located in Geneva.

The U.S. manufacturing companies had not gained overt influence" (Hills 2007, 51).

Of the three arenas of competition between America and Britain the failure to move the ITU to New York ensured that America would not gain outright victory once WWII was over. Even so, in period after 1947 the FCC exerted much energy exempting American companies from ITU regulations. The completion of the first transatlantic telephone cable, TAT-I, in 1958 gave the FCC and AT&T a much higher level of control of international telecommunications, prompting the British and Canadians to construct the Canadian transatlantic telephone cable, CANTAT-1, in 1960 to bypass the FCC's attempts at control (Hills 2007, 63). That part of the continuing Anglo–American struggle past 1947, however, lies outside the scope of this book.

Chapter 5

The Struggle over Broadcasting

At first sight and to modern eyes broadcasting would seem like an almost entirely national enterprise. Modern FM radio and television broadcast services have a geographic range limited to line-of-sight and in the modern world many are also delivered by geographically localized, subscriber only cable systems. Such systems do not seem a likely arena for international struggle. But to begin with radio transmissions had considerable geographic range and radio broadcasting came to fruition, especially in the Europe of the 1920s and 1930s, in a much more geopolitically volatile environment than today. Although broadcasting developed immediately after WWI in Britain and America it developed in very different geopolitical situations: America was geographically isolated, Britain was not; America was in the process of forgetting about the vestigial empire it had started to acquire in 1898, Britain's Empire was reaching its greatest geographic extent.

As with telecommunications there is a vast amount of literature on broadcasting that seems compartmentalized along disciplinary lines, although the three volumes of *The Museum of Broadcast Communications Encyclopedia of Radio* cross many of these, albeit in a format that requires a great deal of mining (Sterling & Keith 2004). The history of broadcasting, for example, is often studied with little or no reference to the history of telecommunications despite the intimate technological connections between the two areas (see, for example, Smith 1973; Street 2006). Historians can be quite capable of crossing such lines but are almost invariably trained in national histories and tend to produce precisely these (Hilmes 1997; Lewis 1991). They can, on occasion, produce transnational histories although Hilmes' later work deals only with America and Britain and covers television as well as radio (Hilmes 2012). Some historians pay lip service to, but do little to unbundle, such tangles, and almost never the technological ones (see, for example, Briggs 1961; 1965; 1985).

There have been attempts to develop more comprehensive approaches (Innis 1950; 1951; Hugill 1999a; Pool 1983; 1990). Histories written by engineers, who are usually not embedded so formally in an academic sense in a national knowledge framework, can be extremely useful in clarifying the emergence of different technologies at different times in different places and their diffusion elsewhere (Vyvyan 1933; IEE 2002; Burns 2004). There are useful histories of the various attempts at international regulation (Codding 1972; Codding & Rutkowski 1982). By examining the regulatory environment from a comparative international perspective Hills attempts "to bring communications within the mainstream of international political economy" (Hills 2002, 7; 2007). Some technologies are, however, embedded in global versions of a variety of these structures that produce a context that can be broadly characterized as geopolitical in nature: broadcasting, like telecommunications, is one of these. This geopolitical context is usually roundly ignored in the communications literature. This chapter examines the link between wireless telecommunications and radio broadcasting in the 1920s and 1930s in such a context.

Three social forces emerged in the mid to late 1800s that would exacerbate the ferment over telecommunications in the 1920s: rapid increases in the popular culture of the developed world in the primacy of the written word, the sung word, and the spoken word.

1. From the 1870s on industrial states developed compulsory educational systems to educate their citizens through age fourteen or so to read and write. This had two impacts: first, it produced populations that would be able to operate more and more complex machinery via written instructions in the emerging industrial economies of the period; second, it was the lesson of such wars as the American Civil War that the broadly literate military of the Union Army, one that could read written orders, performed far better in the field than that of the Confederacy, in which only the officer class was literate. The resulting rapid increase in mass literacy created a massive demand for written news and entertainment just as innovations in printing technology and paper manufacture radically cheapened the price of newspapers, magazines, and books.
2. Second was the realization that cultural power also inhered in such forms as music. Traditionally music in European culture had tended to be oriented toward religious observance or dancing. The rise of grand opera and its increasing linkage with the "new nationalism" of the mid- to late 1800s shifted musical orientation abruptly toward the politics of the nation-state (I am indebted to a colleague, Brett Cooke, from the Department of Modern Languages at Texas A&M University for this insight).
3. Third was the realization that the spoken word carries much more immediate emotional content than the written. The development of the telephone

and its superiority over the telegraph for certain forms of communication made that increasingly evident from the late 1800s on. Broadcast radio allowed orators to reach huge numbers of people, even entire national populations. WWII demonstrated how important the speeches of such national leaders as Winston Churchill and Adolf Hitler could be in creating popular opinion.

The power of radio was intensely cultural. Speaking of American broadcasting, but in a voice applicable to all the other cultures that adopted broadcast radio enthusiastically in the turbulent 1920s and 1930s, Hilmes notes:

> that in speaking to us as a nation during a crucial period of time it helped to shape our cultural consciousness and to define us as a people in ways that were certainly not unitary but cut deeply across individual class, racial, and ethnic experience. Radio was in many ways unique: significantly different from any preceding or subsequent medium in its ability to transcend spatial boundaries, blur the private and public spheres, and escape visual determinations while still retaining the strong element of "realism" that sound—rather than written words—supplies. And its institutionalization as one of our central social structures for transmission and control of information about the world we live in makes its study relevant to almost all aspects of . . . social history—hardly anything that has happened . . . since 1922 has not been shaped by radio . . . in one way or another. (Hilmes 1997, xvi)

In this work Hilmes mainly critiques the "consensus" view of American culture that emerged from the American focus on the commercial voice of radio as "a naturally arising, consensus shaped, and unproblematic reflection of a pluralistic society, rather than the conflicting, tension ridden site of the ruthless exercise of cultural hegemony" (Hilmes 1997, xvii). That cultural hegemony was more submerged in America than in Europe and in some polities radio would rapidly become, openly and clearly, a "ruthless exercise" in political as well as cultural hegemony. The British geopolitician, Halford Mackinder, recognized this when he argued that radio was the ideal medium for exactly such an exercise of ruthlessness by potential dictators because they could reach via radio a national audience. One problem with Hilmes' largely American view is that it pays less attention to political hegemony and almost none, given the vast geographic size of the American market, to the use of radio across political boundaries, though Hilmes does point to the use of Mexican stations to "blast . . . a concentrated violation of cultural standards into US airspace and define daily the limits of the permissible in radio" for American markets in Southern California and Texas in the 1920s and 1930s (Hilmes 1979, xxii).

The very first consumer radios of the early 1920s were usually unpowered crystal receivers that could only receive signals within ten to fifteen miles of a transmitter. Such sets had to be listened to using headphones, but they were in use only a short time. As families embraced radio for household entertainment by the mid-1920s they needed sets with speakers they could listen to in their living rooms. Such sets needed vacuum tubes to amplify the signal they received so that it could power a speaker, and they also used vacuum tubes as receivers. To power these vacuum tubes they needed electricity, initially from batteries that had to be taken into specialist stores to be recharged, later from the supply of mains electricity that was starting to reach most homes in the Western world by the early 1930s. Vacuum tube sets were much more sensitive and could pull in much more distant signals than simple crystal radios (Pegg 1983, 36–39). What we now call Amplitude Modulated (AM) radio, operating on what to modern eyes are low frequencies in what used to be called the long and medium wavelengths, was broadcasting at the only part of the frequency range easily accessible to the vacuum tube technology that emerged from WWI. Between then and WWII broadcast AM radio became a central part of life in the developed world. By day the longer AM waves could travel substantial distances and many European long wave stations could be received over much of Europe. Although the maximum range of medium wave broadcasts by day was pretty much line-of-sight, at night the Kennelly-Heaviside layer reflected these frequencies back to earth and allowed considerable range, upward of 500 miles. Broadcasters reached general agreement in the early 1920s on the distribution of frequencies, especially in the limited number of spaces available for long wave stations, to avoid chaos and interference as demand for radio broadcasts increased. This was especially important where multiple countries in close proximity were concerned, such as in Europe after WWI, where radio broadcasting took off almost as quickly in America, in both places driven by the training of a large number of radio engineers in the last year of WWI for the war in the air and by the release of large numbers of war-surplus vacuum tubes once the war was over (Hugill 1999a).

A second arena of struggle concerns the issues of ownership of the radio frequency spectrum and who was to be licensed to operate in it. The latter would, ultimately, turn out to be central in the struggle between Britain and America. Radio frequencies were only ever leased, never sold absolutely, but their use in broadcasting was controlled in three main ways.

1. America in general pursued an aggressively free market solution, though many Americans admired the British system.
2. The British government created the British Broadcasting Corporation (BBC) as a public utility with some level of independence from government, though many in Britain admired the American free market system.

3. Although other states often initially followed the BBC model the rise of Fascism and Communism pushed many to direct government control in the 1930s and many came to use their stations to broadcast overt political propaganda as well as entertainment.

AMERICA AND BRITAIN AS FIRST ADOPTERS OF RADIO BROADCASTING

America and Britain were the first polities in which radio broadcasting became widespread and it was their experiences that shaped the onset of radio in almost all other polities. And even though America and Britain constructed themselves as in opposition (democratic freedom versus government oversight, chaos versus order and the like) each side of the Atlantic drew heavily on the experiences of the other. Hilmes makes two main points. First, that there was considerable bleed-over between the two polities and that, especially in the later 1930s, that bleed-over increasingly favored the dynamism of the American free market solution.

> The BBC underwent enormous transformations in the four brief years between 1936 and 1940. . . . Even before war loomed on the horizon . . . pressures for change had been building. . . . Each door the BBC opened to relieve the pressures . . . , or to let new practices in, freed a new route for American influence. (Hilmes 2012, 87)

Second, the delay in America's entry into WWI allowed many people there to get involved with radio on what was essentially a small-scale amateur level, whereas in Europe radio was completely taken over by the governments and the military when war broke out in August 1914. In America government did not take over radio until late 1917, and when the war ended in 1918 it proved very difficult for the government to continue to exert much control. Even so, many Americans looked admiringly to Britain's relatively high level of government control and a level of centralization developed after the third American radio conference of 1926. However, this ceded a great deal of control to local stations—the "A" stations broadcasting in the 350 meter wavelengths. Centralization came in the form of the "B" stations in the 400 meter wavelengths, led by RCA's NBC, but the genie was so far out of the bottle that what the British saw as commercial and cultural chaos continued to rule.

On the other hand many Britons admired the commercial and cultural vitality of America's more democratic and free market system, and this saw British companies being set up in London to produce English language commercial broadcasts for long wave AM stations on the European continent to beam back to Britain. By the mid-1930s, and especially in "working-class enclaves of the Midlands and North nestled in industrial valleys where BBC's signal did not reach very clearly" relay services grew up. Subscribers to what was akin to a modern cable television service had their houses wired and received up to four channels of radio, one of which was always a commercial continental station broadcasting in English. By the mid-1930s such exchanges "served more than 230,000 mostly working-class households" (Hilmes 2012, 91). Street notes that from 1934 on London housed what were really two British broadcasting systems, the noncommercial BBC and its commercial competitors producing programs in England but broadcasting to Britain on continental stations and using cabled relay systems (Street 2006).

As indicated in chapter 4, at the end of WWI President Wilson and Admiral Bullard, had maneuvered the various American companies that held patents appropriate to wireless, most notably General Electric (GE) and American Telephone and Telegraph (AT&T), into helping create a new company, the Radio Corporation of America (RCA). GE was its major shareholder, and in late 1919 successfully pressured British Marconi into selling to GE its shares in American Marconi. RCA thus held all the patent rights for wireless, including those of the American Marconi Company. Wilson and Bullard wanted any taint of foreign ownership removed from RCA since they conceived it as a geopolitical construct focused on telecommunications. Its purpose was to ensure that the American navy in particular would be able to continue to control long-distance wireless communication as it had during WWI when it clandestinely created its global network of Poulsen arc stations operating in the very long wavelengths. While RCA certainly had the ability to do that, it also found itself in control of almost all the patents to allow it to broadcast for entertainment purposes in the medium wavelengths, where it became quickly evident that there was a great deal of money to be made. It was missing the regeneration and supersonic heterodyne (superheterodyne or superhet) patents Westinghouse had acquired from Howard Armstrong and that both American Marconi and RCA had initially ignored. These had little value for telecommunications but the superheterodyne technology was important for the success of broadcast entertainment since it made tuning easy. In a complex deal RCA acquired the Westinghouse patents in March 1921 in return for 20.6 percent of RCA's stock (Lewis 1991, 152–154). Once in control of the Westinghouse patents and without the constraints of the telecommunications needs of a far flung Empire, RCA quickly abandoned its geopolitical roots and restructured itself to make money via broadcast entertainment (Hugill 1999a).

One of the main agencies of RCA's shift from telecommunications to broadcast entertainment came in the person of its de facto chief operating officer by the end of WWI, David Sarnoff. Although Wilson and Bullard had placed all the patent rights that allowed broadcasting in the hands of one company, Sarnoff was the person who recognized their value. Sarnoff came to RCA from American Marconi, where he had been chief inspector and contracts manager. In 1913 he had unsuccessfully recommended the purchase of Armstrong's regeneration technology. He had experimented with and advocated the use of radio for broadcast entertainment as early as 1915. Whereas GE followed President Wilson's vision for RCA Sarnoff created his own, convinced as early as 1920, when he heard Armstrong's first demonstration superhet receiver, that the future of RCA would lie in "the creation and marketing of radio music boxes" and perhaps in broadcasting entertainment itself (Lewis 1991, 152). In March 1921 it was Sarnoff who pushed RCA to finally acquire Armstrong's technologies. In May Sarnoff was appointed as RCA's general manager just as American adoption of radio sets as household items began to skyrocket. In July that same year Sarnoff arranged the broadcast of the heavyweight boxing match between Jack Dempsey and Georges Carpentier and, more importantly, the installation of "loudspeakers in about a hundred theaters, Elks, Masonic, and social clubs from Florida to Maine. . . . In Manhattan, an estimated 100,000 . . . gathered around loudspeakers attached to the *New York Times* Building" (Lewis 1991, 158). Many early adopters of radio were male enthusiasts who built their own sets and to whom such a match was important. This Dempsey/Carpentier fight was heard by up to 300,000 people both in social settings and using simple crystal sets as well as radio receivers built from chassis kits and war-surplus vacuum tubes. But the match also greatly increased the awareness of the possibilities of broadcast radio to nontechnically minded adopters of radio. In September of 1922 Sarnoff was appointed as vice president and given what amounted to complete control of the entertainment wing of RCA. On May 16, 1923, RCA began regular broadcast entertainment from its WJZ station in New York, and in its first year focused on talks and lectures (723), music solos, recitals, and concerts (522), bedtime stories (205), church services (67), plays (40), and sporting events (18) (Lewis 1991, 164).

However, the agreement that had been forged to create the company as a geopolitical construct caused RCA problems in the early 1920s. Once it was evident that there was a great deal of money to be made in broadcast radio AT&T defected from its initial agreement to join RCA, arguing both that GE had no rights under it to manufacture radio receivers for sale to consumers and that only AT&T had the right to establish commercial radio stations. This pulled RCA into a struggle that would dominate its history for the next several years and deflect it from international telecommunications as it fought successfully to hold on to its vastly more valuable broadcasting component (Sobel 1986, 47–53). By 1926 RCA had begun purchasing radio stations, and

Sarnoff arranged a secret deal with AT&T by which NBC, the first American broadcasting network, was brought into existence by RCA. AT&T agreed to abandon their attempt to create a broadcast network "in return for RCA's commitment to use AT&T's lines rather than Western Union's for networking" and WEAF, New York, AT&T's pioneer broadcast station, became WNBC (Pool 1983, 35). Sarnoff became RCA's president in 1928.

The situation in Britain was more confusing, more geopolitical, and certainly more driven by imperial than commercial concerns. Although the Marconi Company held all the crucial patents, and although it was able to control broadcasting in the very early 1920s via a cartel, Marconi was never allowed to fully control long-distance telecommunications because of the privileged position afforded the British GPO and the submarine cable companies in that area. Complicating the broadcast arena was that two BBCs came into existence: the first, the British Broadcasting Company, was formed largely by Marconi's cartel in the early 1920s and was intended as a somewhat free market construct; the second was the government licensed BBC of 1927, a public utility. The Company was founded by the "Big Six": Marconi's, Radio Communication, Metropolitan Vickers, Western Electric, British Thomson Houston, and the GE Company, although several hundred more manufacturers joined so they could legally sell sets and parts. Manufacturers were to pay royalties and listeners had to buy a license to listen from the GPO (Pegg 1983, 42). The Company was legally allowed to sell advertising via sponsorship of its programs, although it never did. When control shifted to the Corporation the right to sell sponsorships was removed. The listeners' license monies went to the BBC to fund programming, to the GPO as a collection fee, and to the government.

By founding the British Broadcasting Company, the six largest manufacturers had a clear advantage in the radio market. Marconi's, in particular, had great influence, because of the company's control of a large number of vital wireless patents. Even the largest rivals were obliged to take out expensive licenses for their use. This gave Marconi's a great financial as well as technological advantage (Pegg 1983, 49).

On the telecommunications front, as early as 1919 the Marconi Company had demonstrated that the higher frequency part of the radio spectrum as it was then understood, the short waves, was ideal for very long range, indeed global telecommunications, that transmissions in those wavelengths required relatively little energy compared to the lower frequencies, and that they could be "focused" using carefully designed antennae. The problem was that their use required a thorough understanding of the ways in which the Kennelly-Heaviside layer in the upper atmosphere reflected radio waves. This was forthcoming from Britain's National Physical Laboratory in the 1920s (Hugill 1999a).

As Marconi's developed their shortwave, high frequency, beam antennae, long range telecommunications services they were forced to operate with the GPO because of its privileged position with the British government. As the British government came to fear RCA they then forced a merger in 1929 between the long established submarine cable interests and Marconi's to create the company we now know as Cable and Wireless. This allowed the British much more geopolitical leeway in establishing long range radio communications in the 1920s and 30s, but it also drove Marconi himself into revolt against what he saw as the unjustified takeover of his company and back to Italy where he came to support the Fascist Party (Hugill 1999a, 130–134).

Telecommunications, Broadcasting, and Geopolitics

Two forms of power inform the geopolitical context in the arena of communications, structural and representational:

1. structural power is conferred by telecommunications via the submarine cable and wireless transmitter networks it uses as well as by one-way communication via networks of radio broadcasting
2. representational power is made possible by structural power, which allows control of the networks that disseminate cultural information, including news and entertainment, whether that control be commercial or political.

The "new nationalism" of the late 1800s resulted in increasing demands for representations of national "culture" as an increasingly important part of the power projection of the evolving nation-state. After WWI the ability of radio broadcasts to reach beyond the political boundaries of the nation-state evolved as one of the most central arms of that national projection of cultural power. "If a picture or a map is worth a thousand words, then power in the realm of representation may end up being as important as power over the materiality of spatial organization itself" (Harvey 1990, 233). The crudest form of representational power was the direct political propaganda that emerged from WWI, but much subtler forms of "power in the realm of representation" arose rapidly thereafter.

"Cheap mass products, . . . dazzling technology, and . . . alluring mass culture" were central to the innovative form of American cultural imperialism first on display at Chicago's great Columbian Exposition of 1893 as America stepped out onto the world stage (Rosenberg 1982, 6). Cheap American goods poured onto world markets, many sold in those markets by American stores such as Woolworth's, a far more successful international marketer in its time than Wal-Mart is today, and also America's first multinational company

(Hugill 2006). American magazines, books, jazz, other popular music, and movies all conspired by the 1920s to create an ultimately almost irresistible cultural representation of America, even where banned or restricted (Meinig 2004, 302–303).

As Agnew notes, American hegemony has "been based on a rejection of territorial limits to its influence, as would necessarily come with empire" (Agnew 2005, 52). The price of rejecting "the materiality of spatial organization" that a conventional reading of empire represents has, however, been the need to establish supremacy of "power in the realm of representation." This was first realized in the 1920s and 30s by American movies and American commercial broadcasting. American broadcast radio may have "failed" in the traditional geopolitical arena: it certainly did not establish "the materiality of spatial organization" such an arena required and that to some extent the BBC managed to create in its Empire Service. Where it did not fail was in in establishing its "power in the realm of representation," effectively a new form of cultural "imperialism."

As indicated earlier nation-states took one of three positions with regard to such "power in the realm of representation." Most subtly of all, America consigned broadcasting to the commercial sphere, pretended it was neutral, and let its "alluring mass culture" carry a clear message of American cultural superiority to its audience. Britain sought a level of state regulation through the BBC that attempted to minimize overt commercial or political propaganda while still using radio to further the goals of Empire as these were seen by Britain's ruling elites. The BBC also cultivated a reputation for honesty and accuracy that stood it in good stead in the realm of broadcast news, though in the end it fell short of the attractions of America's "alluring mass culture." Most other states by the mid-1930s came to openly embrace an overtly propagandist "power in the realm of representation" and brought broadcasting completely under state control, most obviously and notoriously under national socialism in Nazi Germany and communism in the Soviet Union.

Some idea of the complexity with which British elites quickly came to regard radio broadcasting to the Empire comes from comments made by Prime Minister Stanley Baldwin at a dinner in 1926 celebrating the BBC's first year of service as a publicly controlled corporation. In his address he compared the BBC's approach to radio favorably with film on implicitly geopolitical grounds:

> It is too early yet to say what the influence on civilisation of the moving picture may be, but I confess that there is one aspect of it upon which I look with the gravest apprehension, and that is the effect of the commoner type of film, as representing the white races, when represented to the colored races of this world. . . . The whole progress of civilisation is bound up with the capacity that

> the white races have, and will have, to help the rest of the world to advance, and if their power to do that be impeded by false ideas of what the white races stand for, it may well be that their efforts will not only fail, but that the conception of the white races generated in the heart of the colored peoples throughout the world may well be an initial step in the downfall of those white races. (Quoted in Hilmes 2012, 53)

Given that Britain's Empire was composed mostly of the "colored peoples," Baldwin was clearly arguing here on geopolitical grounds against the "commoner type" of radio broadcasting associated with American commercialism. Baldwin was indicting American film in this speech, since film had fallen to American commercialism in the form of Hollywood during WWI. Hollywood's success, plus the cessation of European production during four long years of warfare, was such that almost all other national film industries, and certainly that of Britain, had collapsed by the early 1920s. With film a lost cause, Britain had to control radio's "power in the realm of representation" to do two things. First, it had to maintain the civilizing force of an Empire that by the 1920s was at its geographic peak but that British elites clearly recognized must be wound down as more and more imperial possessions gained independence, albeit over time. Second, it had to seek to prevent the downfall of the white races who ran the Empire via a backlash caused by offensive representations of white racism in popular culture. Managing the content of broadcast radio was thus vastly more important in Imperial Britain than in a democratic and non-imperial America where, in any case, white racism was much more socially acceptable among elites, especially in the South.

As indicated in chapter 4, as wireless technology matured in the early 1920s broadcast radio was the subject of national, then international conferences. At the level of American politics these forces were initially exacerbated by Herbert Hoover's poorly informed behavior as secretary of commerce. The switch following the 1920 election away from Wilson's robust interventionism and internationalism to the crude *laissez-faire* policies of Harding, Coolidge, Hoover and the Republicans had disastrous unintended consequences for America's geopolitical interests.

In Washington in 1922 Herbert Hoover, then secretary of commerce, convened the first of four national radio conferences, one a year for the next four years. At the First Radio Conference he sought to assign frequencies on the basis of both national interest and likely commercial viability. The navy pressured Hoover to allow it to retain control of the very longest wavelengths, still believed to be the only part of the spectrum capable of long-distance wireless telecommunications. On the basis, however, of the development of vacuum tube based radio telephony for the war in the air, its centrality to "Plan 1919," and the resulting vast release of surplus vacuum tubes onto the market once the war ended, it was quickly obvious that wireless in the medium wavelengths held a strong possibility for broadcast entertainment. Hoover thus assigned the medium wavelengths, the modern AM band, to the commercial sphere. In practice, given the rather simple early types of vacuum tubes then available, that defined for American broadcasters the lowest waveband usable by broadcast radio as the area around 200 meters and, because of the needs of the shipping industry for emergency wireless communications, the highest as around 560. The shorter waves and higher frequencies, below 200 meters, seemed inaccessible with the technologies extant in the early 1920s. This area, about to be made accessible by a vacuum tube technology that was continuing to improve rapidly, and thus increasingly the most valuable part of the radio spectrum, was consigned to the amateurs. Given that America was in the technological lead in radio by 1922 the rest of the world followed suit in terms of frequency allocation (Howeth 1963, 502; Hugill 1999a, 123).

Hoover's problems were not limited to those caused by poorly informed frequency allocation. AT&T's vision of broadcast radio was that they would provide studios and transmitters that anyone would pay to use, rather as they would a public telephone. AT&T called this "toll broadcasting" and it was a way station to commercialization of the airways. RCA's vision was of what they called "public service broadcasting," which was akin to what the BBC would develop in Britain. Since Hoover initially opposed "toll broadcasting" and leaned toward the BBC style outcome, the row between RCA and AT&T was compounded by Hoover's presence at the department of commerce. By 1922 RCA was thus "far more concerned with broadcasting and receiver sales than telecommunications and was no longer willing to expend its entire resources on becoming the American flag carrier in the international wireless field" (Sobel 1986, 54–61, quote from 61). Following the Third Radio Conference of 1924, at which he emphasized "his belief that broadcasting be left in private hands," Hoover began to back off his criticism of toll broadcasting, a concept toward which Sarnoff had warmed as he began to realize the potential that commercial broadcasting held for profit (Sobel 1986, 66). In 1926 the Supreme Court ruled against Hoover's controls and real chaos threatened as broadcasters began to switch frequencies, increase transmitter power, and abandon time-sharing agreements.

Given the minimal controls put in place at Hoover's First National Radio Conference over who could use which medium wavelengths when and for what purpose, the growth of broadcasting in America from 1922 on was rapid. America was a large country and there was enough distance between the urban markets where most early adopters of broadcast radio were located that the minimal controls over frequency allocations did not at first create much of a problem with interference. Adoption increased rapidly once the developing advertising industry began to realize that broadcast radio offered unparalleled access to potential consumers in the emerging consumer economy of the early 1920s. Despite the problems with Hoover's policies, internecine commercial fights, constant British harping on the "chaos" of American broadcasting, and the Supreme Court decision of 1926, a reasonable level of order was restored by the passage by Congress of the Radio Act in 1927 and the creation of the Federal Radio Commission as a regulatory board friendly to the broadcast industry. In 1934 Congress passed the more inclusive Communications Act, which merged the oversight of wired and wireless communications under a single agency, the Federal Communications Commission (FCC) (Messere 2004, 1202–1204; Sobel 1986, 73). After the 1927 and 1934 Acts commercial broadcast radio was thus able to mature in America in a coherent national political and technical form, although to British eyes cultural "chaos" continued to reign. In Europe, however, something much worse developed. Broadcast radio was no respecter of the lines on the map that delineated where the power of nation-states started and stopped and this led to serious geopolitical chaos. America was simply so large that it hardly had to worry about such problems.

Initial agreements for allotting European wavelengths were reached in Geneva in 1926 and accepted by the Council of the *Union Internationale de Radiophonie*, which became the International Broadcast Union (IBU) in 1927. In 1927 the Third International Radiotelegraph Conference in Washington sought to impose international order on all users of radio frequency, military and civil, broadcast and telecommunications alike, as well as to claw back some of the higher frequencies given away in 1922. However, it allowed signatories complete freedom to assign frequencies at the national level "upon the sole condition that no interference with any service of another country will result therefrom" (Codding 1972, 125). Given the ability of the higher frequency signals from medium wave stations to travel long distances at night, the small size of European states rendered such a "non-interference" clause moot, and several subsequent conferences sought to resolve the problem without success.

In October 1922 the British Broadcasting *Company* had been founded as a technically commercial entity but in 1927 it was re-configured as the British Broadcasting *Corporation* and as a *de facto* state monopoly that would restrict and increasingly control broadcast radio (Street 2006, 39, 44, 45, 77). At the end of WWI the Marconi Company controlled the same patent rights in Britain that the American government had to force into RCA in its attempt to ensure its continued ability to control global wireless telegraphy. Marconi's, which had also been the key company in the development of wireless telephony for the RFC and its replacement, the RAF (Burns 2004, 415), began commercial broadcasting from its headquarters at Chelmsford almost as soon as the war was over. On June 15, 1920, Marconi's broadcast a thirty-minute recital by the great Australian opera singer, Dame Nellie Melba, sponsored by an important and influential national newspaper, *The Daily Mail*. This created the same sort of interest in radio in Britain that the broadcast by RCA of the Dempsey-Carpentier fight would cause in America a year later. In postwar Britain, however, the GPO resumed its control over telecommunications, which the government now held to include broadcast radio because of its potential to interfere with wireless telecommunications. Marconi's broadcasts were shut down late in 1920 because of claims of such interference, but under public pressure the GPO allowed a team of Marconi engineers to resume experimental broadcasts for fifteen minutes a day in early 1922.

In the period between WWI and WWII the British economy remained the wealthiest in Europe. Since all radio owners in Britain were required to purchase a license to listen, which fees funded the BBC, we know with reasonable accuracy how many British households owned radios year by year from 1922 on, though some certainly evaded buying a license. In the earliest years of radio, before voice broadcasting began in the early 1920s, many adopters were men who enjoyed tinkering with a new technology, although Hilmes (1979, 132–133) notes that women were certainly not absent in America before voice broadcasting began since the use of Morse code ensured gender blindness. In the early years of voice and music broadcasting, radio listening was private, over headsets, since the technology for amplification of sound through loudspeakers was expensive and difficult, requiring much more electrical energy than a headset: "listening in a crowd . . . [was] a normal activity but the pattern of home listening was firmly established before 1927" (Pegg 1983, 222). Photographs from the period show whole families listening in on multiple headsets. By 1927 radios were in just over 20 percent of British households and by 1930 they were in 30 percent, by which time loudspeakers were rapidly replacing headsets. Thereafter the percentage of households with radios continued to rise steadily for the rest of the decade, the 50 percent mark being crossed in 1933 and 70 percent being reached in 1938 (Pegg 1983, 7).

In America by the late 1930s some 26 million households owed at least one radio. By 1938 radios were in over 80 percent of households and the typical household "spent an average of five hours daily listening" (Hilmes 1979, 183). Although the American market was different, clearly driven by the interests of advertisers, the data collected for commercial purposes showed that, as early as 1930, American audiences were predominately female, "ranging from 55–60 percent at night to more than 70 percent by day," when the female oriented serial, or "soap opera" tended to dominate the airwaves (Hilmes 1997, 131). Certainly the British market was different: the BBC's "monopoly" status meant the absence of "soap operas," since the "social composition of its employees and speakers meant that the BBC nearly always represented the voice of a ruling élite" (Pegg 1983, 223). Even in America there was considerably social tension over content, with the general assumption that daytime programming was, as Hilmes puts it, "Daytime Dirt." In 1934 the New York Continuity Acceptance Department was set up in an attempt to censor such material, usually unsuccessfully because of the demands of the advertisers for audiences (Hilmes 1979, 124–126). There was, in fact, always a market for more popular material than elites, especially in Britain, were willing to provide or put up with. As more sensitive vacuum tube sets using Super Heterodyne circuits came to predominate and transmitters improved, British and European listeners became quite capable of picking up signals from well outside the boundaries of any specific nation-state.

One of the prime examples of the problem posed for Britain and the BBC by European broadcasting is illustrated by the history of Radio Luxembourg both before and after WWII. As radio adoption boomed in the early 1930s the small state of Luxembourg found that its geographic position put the one long wave station it was allocated in 1927 within reach of pretty much anywhere in Western Europe. From 1933 on British entrepreneurs operating out of London persuaded Radio Luxembourg to focus on commercial broadcasting to the lucrative and rapidly growing British market since no British station was available to British companies wishing to use the new medium to advertise their wares as American companies were so successfully doing. After 1927 the BBC also became known for its imposition of rather staid middle class taste upon its listeners—it was not known as "Aunty BBC" for nothing! On Sundays, for example, the BBC broadcast only "serious" music and religious programs. In 1930 Captain Leonard Plugge founded the International Broadcasting Company in London as an advertising brokerage firm working with mostly Radio Normandie. Ireland played a lesser role, with Radio Athlone also focused on the English market. But in May 1933 Radio Luxembourg, "THE MOST POWERFUL IN EUROPE," started to focus on the English market under Plugge's tutelage and established a highly successful commercialized English language broadcast of popular music and light entertainment (Hilmes 2012, 91). Its normal market share in the years 1933 to 1939 was over 10 percent and could reach 50 percent on Sundays (Pegg 1983, 20). Both Normandie and Luxembourg were, of course, shut down by the Germans during WWII.

In 1948 a new plan to distribute radio spectrum in Europe was set up at the Copenhagen Conference, at which time the state of Luxembourg was allotted one high power long wave station and one high power station broadcasting on the medium waveband at 208 meters. As the British economy recovered after the war 208 grew into the most powerful privately owned radio transmitter in the world, using a focused antenna and 1,300 kilowatts of power to reach the crucial London area market, which it could only do after dark. The normal AM radio station of the period had a maximum daytime output of 500 kilowatts, reducing after dark to avoid interfering with other stations near it in the radio frequency spectrum. As the new popular teen oriented music known as Rock & Roll developed from the late 1950s on "Aunty BBC" refused to transmit it. Radio Luxembourg embraced it with its usual commercial enthusiasm. This was greatly helped by both the general increase in household income and the emergence of cheap, battery powered transistor radios. By around 1960 a teenager in the London area could easily afford a transistor radio and tune in to Radio Luxembourg to listen to the new American Rock & Roll stars and their British imitators.

After WWII and before the emergence of the "pirate" commercial radio stations Luxembourg had the lucrative British radio advertising market entirely to itself. The emergence of pirate radio, initially in the form of Radio Caroline on a ship moored just outside the three mile limit of British territorial waters, began to change things rapidly in the British market after 1964. Caroline could reach the London market by day, and focused heavily on Rock & Roll. Success spawned numerous imitators and finally pushed the British government into making the pirate stations illegal in 1967 and instructing "Aunty BBC" to relax its Sunday strictures and provide lighter forms of entertainment, including modern "pop" music, on BBC Radio One. The years from 1962 on were an era of great ferment in British popular music, with groups such as Liverpool's Gerry and the Pacemakers and London's Dave Clark Five generating a new, British sound. The Beatles and the Rolling Stones followed in their footsteps, but went further, adopting American Rhythm and Blues style music and building on the rock-and-roll revolution begun in America by such performers as Buddy Holly, Roy Orbison, Jerry Lee Lewis, and Elvis Presley as well as such important black artists as Chuck Berry and lesser known ones such as Howling Wolf to become immensely popular. Radio Luxembourg had catered to such tastes in late night broadcasts from the early 1960s on, but Caroline and the other pirates transmitted all day long. By 1967 a more relaxed and up-to-date BBC Radio One and new legislation had forced the pirates off the air. Many of the pirate disc jockeys, who had developed strong followings, moved to Radio One and in 1973 commercial broadcasting was finally allowed from within Britain itself. This would focus almost entirely on popular taste and represented the achievement of American cultural hegemony (Radio Luxembourg Wikipedia Entry).

The agreements reached at Geneva in 1926 proved increasingly difficult to change since the national broadcast interests squabbled incessantly among themselves about access to wavelengths. They also found themselves opposed by maritime and aviation interests unified across national lines. Most merchant ships, which had been forced to carry wireless by the International

Conference on the Safety of Life at Sea in 1914, were still using old-fashioned spark sets in the 1920s. These were "obsolete and unselective" and needed a great deal of space in the frequency allocation. Ship owners were thus particularly opposed to any reduction in their allotted wavelengths (Briggs 1965, 348). The international wavelength for marine distress calls had long been 600 meters. To avoid interference this effectively limited the upper wavelength for broadcast radio in the medium wavebands to 545 meters (Vyvyan 1933, 206). Even so the Washington Conference of 1927 agreed, "despite complaints of ship owners that extreme financial losses would be incurred by eliminating the old spark transmitters, . . . to establish a timetable for the elimination of all such transmitters" (Codding & Rutkowski 1982, 17).

In addition to the problem posed by the maritime interests, by the late 1920s governments, especially in an eastern Europe where the political problems caused by the break-up of the Ottoman, Austro-Hungarian, German, and Russian Empires had never been resolved, had come to realize that radio was a powerful propaganda tool. As the new nationalism took hold in the region it had to compete with Italian and, after 1934, German designs to rebuild their influence and control over the series of small, dysfunctional states that had emerged from the Versailles Treaty. Both Mussolini and Hitler were fully aware of the propaganda value attached to control of broadcast radio, and both were spellbinding orators able to use the "power in the realm of representation" the new medium gave them to great effect. This greatly intensified national squabbles over frequency allotments in the European environment (Briggs 1965, 342).

The Social Forces: the Impact of Written, Sung, and Spoken Communication

The Written Word

Although the new nationalism was percolating along under the surface by the late 1800s, until 1914 the political structure of the world-economy was still dominated by fourteen Empires, the global nature of which effectively globalized the demand for news. Developing alongside the continental and submarine cable networks from the 1840s on came a group of news agencies that created a substantial market for long-distance telecommunications services to feed the demand for written news material (Winseck & Pike 2007). This was most notable in the largest and most global Empire, that controlled by Britain, and the British press agency Reuters took an early lead in controlling the world flow of news, starting by providing European stock prices to a London Stock Exchange that became more and more central to British geopolitical interests as well as the global economy from 1851 on (Read

1992). As the literate population exploded in the later 1800s from the development of compulsory national education programs in the industrialized and industrializing world the demand for more dramatic news rose accordingly. One of the first massive disasters reported on a global scale and in near real time was the massive eruption in 1883 of the volcanic island of Krakatau in the Sunda Strait and the deaths of some 36,000 people in the explosion and resulting tsunami. The reporting was made possible by the British network of submarine cables, approaching truly global status by 1883, and the press agencies that used the cables (Winchester 2003).

Assessing the importance of communications to the British Empire in the Beit Lectures on Imperial History at Oxford University in 1950, the Canadian scholar, Harold Innis, noted "the effective government of large areas depends to a very important extent on the efficiency of communications" (Innis 2007, 26). But empire is also about something more than merely effective government, and the consequence of the development of the continental and submarine telegraph systems and then broadcasting was to extend

> the "space" of empire to facilitate the unification of territory into nation-states and to allow long-distance control of alien territories. . . . The telegraph, coupled with the printed word, altered political relations between the British government and its colonial territories, allowing more direct control. It also altered the autonomy of markets in the dominions of the British Empire, allowing London, rather than local, capital to meet demand. . . . [T]he empire became perceived as a community . . . [a vision that] was to recur throughout the early years of the twentieth century in arguments to decrease international telegraph tariffs and promote inter-empire traffic for purposes other than business. Such sentiments were what drove the first external broadcasting of the British Broadcasting Corporation in 1930s. (Hills 2002, 4–5)

The Sung Word

The rise of grand opera as a major art form culminating in the compositions of Giuseppe Verdi, Giacomo Puccini, Richard Wagner, and Richard Strauss changed the way the sung word interacted with national popular cultures from the mid-1800s on. "Operatic nationalism" became the order of the day in the emerging states of Italy and Germany. The chorus "Va Pensiero" from Verdi's *Nabucco* of 1841 became the unofficial national anthem for Italians as mid-century Italy struggled to unify and shed itself of the political control of the Austro-Hungarian Empire, which it would do in 1861. The final chorus of Wagner's *Die Meistersinger von Nürnberg*, premiered in 1868, is a ringing defense of "German art" before German political unification in 1871. In his later *Ring Cycle* Wagner went on to dramatize the legends and myths of Nordic and Germanic culture, works that were seen by Adolf Hitler as particularly valuable to the "new nationalism" of Nazi Germany. Related to this was the development of the heroic tenor who could sing to much larger theaters and hence larger audiences, and who quickly became the protagonist of so many operas in the late 1800s. By comparison tenor parts are relatively minor in Mozart's operas of the 1700s, such as *Julius Caesar*. Mozart's leads were usually written for castrati, now they have to be sung by countertenors. The heroic tenor emerged as an important focus for nationalist ideals and the "cult of the individual" that became strong in the Nazi geopolitical imagination. Unfortunately and presciently such heroism was often fatal.

Against such nationalist trends the great German composer Richard Strauss moved the operatic form toward modernism, thus toward visual rather than aural presentation on the stage and cinema. His first major work, *Salome*, debuted in 1905. It was performed once at New York's Metropolitan Opera in 1907 and then banned for its sexual content until 1934. His dark version of the Greek myth, *Elektra*, first performed in 1909, is sung almost entirely by females. The one tenor lead is Aegisth, the paramour of Elektra's mother, Klytaemnestra, who helped her murder Elektra's father, Agamemnon. The only other male lead is Orest, Elektra's brother, sung by a baritone. The opera ends with Orest's revenge killing of Aegisth and Klytaemnestra offstage and Elektra's collapse and death onstage. *Elektra* is riddled with intense sexual tension and violence that come from the heavy influence on Strauss of the Vienna Secession, led by such artists as Gustav Klimt and Egon Schiele (Kandel 2012), and the work of Sigmund Freud. Such work was never "popular" in the way that of Verdi or Wagner was and much of it would be condemned as "degenerate" in Germany in the Nazi period since it did not support nationalist goals. Strauss's work from this period did not always suit broadcast radio, although his 1896 tone poem, *Also sprach Zarathustra*, would eventually achieve massive recognition. It was used as the opening sound theme in Stanley Kubrick's 1968 film, *2001: A Space Odyssey* and then by Elvis Presley to introduce his late period concerts (1971–1977).

Nationalist composers such as Verdi also pioneered the popular treatment of the trials and tribulations of everyday life. The Parisian heroine of his *La Traviata* of 1853 sings herself toward her death from tuberculosis, but Austrian censors worried about rising Italian nationalism refused to allow it to be performed as a contemporary drama, insisting that it be set in the 1700s (Brett Cooke, personal communication, December 2014). Puccini's 1896 opera, *La Bohème*, set precisely the same theme in contemporary Paris, and the 1996 musical, *Rent*, followed *La Bohème's* storyline, but switched tuberculosis to AIDS. In Britain W. S. Gilbert and Arthur Sullivan satirized British politics, military life, culture, and *morés* in a series of comic operettas starting with *H.M.S. Pinafore*. Despite, possibly because of, their satirical genius Gilbert & Sullivan had a huge impact on shaping the emerging British national culture of the late 1800s. Such music was ideally suited to broadcast radio and opera and operatic solos from popular works were almost always featured in the earliest broadcasts as well as lighter concert material. One of the great pioneers of broadcast radio, Lee De Forest, actually attempted to broadcast New York's metropolitan opera featuring Puccini's *Tosca* and the great Italian tenor, Enrico Caruso, as early as 1910. Marconi's famous first broadcast in Britain in 1920 was of solos by the great Australian opera singer, Dame Nellie Melba. In December 1921 station WGX of San Francisco put on a pioneering performance of operatic selections so popular that "the entire Pacific slope was hushed" (Hilmes 1997, 37, 42).

The Spoken Word

The third crucial social force of the later 1800s was the increasing importance given to the spoken over the written word in first telecommunications and then popular culture. The development of telephone technology in the 1880s made this possible. Spoken messages simply carried very much more emotional information than the written messages delivered along telegraph cables, especially submarine ones, where commercial codes were commonly used to radically shorten message length and keep cost down at the expense, often, of any subtlety of meaning (Winkler 2008). The real key to this, both in telecommunications and, eventually, broadcast radio, was the development of the vacuum tube technology that enabled the transmission and reception of music and the spoken word.

THE ENABLING TECHNOLOGIES

In America in the early 1900s, AT&T had pushed hard in its newly created Bell Labs to develop long-distance voice transmission technologies for its telephone service in the early 1900s. Two were successful, the first of which was line-loading. Between 1900 and 1910 line-loading allowed the extension of long-distance telephony from the east coast of America to just west of the Mississippi river valley, but that was the limit of its range and it was not, in modern terms, "user-friendly." Because of the attenuation of the signal caused by switching losses, long-distance calls could not be switched into the regular telephone service that connected businesses and homes within all American cities by the late 1800s. These first-generation long-distance calls had to be made from specially designated long-distance booths connected to the city the caller wished to reach, and callers had to use a "telephone" voice pitched to the very narrow range of frequencies line-loading was capable of transmitting (Hugill 1999a).

Bell's second, and much more important long-distance technology, was the amplification of signals using Lee De Forest's *audion* vacuum tube. By 1915 this made possible the beginnings of a second-generation long-distance system. Coast-to-coast telephony on a wired network became possible, albeit at very high cost. More to the point, however, AT&T installed multiple vacuum tube amplifiers along their land lines so that calls did not attenuate, which effectively removed the switching problem. Calls could now originate and terminate anywhere in the system, at office desks and in homes wealthy enough to have telephones. In addition, vacuum tubes transmitted the human voice far better than line loading, which removed the need for callers to use a "telephone" voice (Hugill 1999a).

By 1915, driven in part by industrial espionage, similar developments in vacuum tube technology for telephony had occurred in Imperial Germany, which was expanding its geopolitical interests eastwards and was beginning to experience similar problems of geographic scale at the continental level to those of America. Within Britain proper, line-loading technology was adequate to the needs of the entire geographic space, but the telephonic needs of the Empire were well beyond a technology using vacuum tubes and land lines. Early British interest in vacuum tube technology thus did not focus on wired telephony but on the development of wireless telephony to meet the needs of military observers using airplanes to spot for field artillery, where voice transmission conveyed information more quickly and accurately than the primitive spark transmitters using Morse code that were deployed when war broke out in 1914 (Hugill 1999a). In their 1910 military maneuvers the British Army made the first air-to-ground wireless transmissions from an airplane using such transmitters "at ranges of up to a mile, although they were one-way only" (Hare 1990, 28).

Voice transmissions, especially if they could be two-way, were obviously likely to be much more useful in aerial observation for artillery than cumbersome Morse code. Marconi's chief engineer, H. J. Round, thus put a great deal of effort into developing vacuum tube technology for that purpose and by February 1916 the RFC was able to demonstrate wireless telephony from airplanes using vacuum tube sets at ranges of up to forty miles (Burns 2004, 416). However, these early sets used "soft" vacuum tubes that were hard to manufacture, use, and maintain in the rigors of military service. By 1917 "hard" vacuum tubes, so-called because they were more thoroughly evacuated than "soft" tubes, had reduced these problems. Sets using them required several more tubes to give the same performance level as those using "soft" tubes and still required considerable maintenance since the tubes were fragile. Manufacture remained a problem, but production on a huge scale ensured a plentiful supply of working tubes. This also required the training of a large number of radio engineers for military service. By late 1918 the mass production of "hard" tubes, especially in America, had reached a level where the British could promulgate what became known as "Plan 1919." This conceived 1919's spring war offensive as based on the advance of ground and air units coordinated by wireless and it required that "every two-place airplane at the front" be equipped for wireless telephony.

When WWI ended abruptly in late 1918 the millions of vacuum tubes manufactured for the war effort but not used were quickly released onto the surplus market at knock down prices. Vacuum tubes would otherwise have been by far the most expensive components in early radio receiving sets. This surplus of tubes made quickly possible the development of a broadcast entertainment industry using vacuum tube receivers operating in the medium wavelengths, today's AM band. In addition the military had trained a huge number of wireless operators and mechanics: by late 1918 nearly 7,000 were serving in the newly formed RAF. Many of these operators and mechanics would use the skills they had learned in the military to establish radio maintenance and repair businesses when they returned to civilian life and as the broadcast radio industry developed (Hugill 1999a). Charles O. Stanley, who built Pye into one of the main radio manufacturing companies in Britain in the early 1930s and was one of the main developers of television in the late 1930s, "volunteered for the newly created Royal Air Force and became a radio operator . . . in October 1918, just one month before the signing of the armistice that brought the Great War to its end" (Frankland 2002, 18).

A large number of radio enthusiast magazines sprang up in the early 1920s to cater to this rapidly growing audience. The Marconi Company had published *Wireless World*, initially as *Marconigraph*, from 1911, but it was intended as a trade magazine for the company. In the early 1920s, however, it published a significant number of circuit diagrams, many of which were of wartime radio sets designed for the war in the air. The magazines aimed at the new enthusiasts all began publication in 1922 or 1923. They usually sold for 3d (3 old pence), and all but the most professional had ceased publication by the late 1930s, which was the end of the era of home construction of radio sets. The third class of magazine was that aimed solely at the consumer of radio entertainment. The BBC began weekly publication of *Radio Times* in September 1923 for 2d (2 old pence). It was filled with chatty stories and, of course, a list of the week's programs. Regular newspapers boycotted such program lists, fearing they would lose readers to radio. Since *Radio Times* carried only BBC listings Plugge's International Broadcasting Company started to put out *Radio Pictorial* in the early 1930s to publicize the commercial broadcasts aimed at Southern England from European stations (Sterling Times Online).

In Britain especially, since distances to the transmitters were short, home constructors in the very early 1920s often used very simple crystal sets tuned with a cat's whisker that required neither vacuum tubes nor batteries. They were hard to tune, limited in geographic range, limited also in the range of sound frequencies they could receive, and allowed listening on headphones only. Charles Stanley recognized from his brief experience in the RAF that the future would lie with vacuum tube receivers. Stanley founded perhaps the most influential of the enthusiast magazines, *Radio for the Million*, late in the game, in December 1926, and sold it for 1/- (12 old pence), four times the price of the more common publications. Immediately WWI ended Stanley had left the RAF and moved into advertising. After founding *Radio for the Million* he took Mullard, one of the major British manufacturers of vacuum tubes, as one of his main clients. By the late 1920s the supply of war-surplus tubes was petering out and Stanley's use of *Radio for the Million* to advertise Mullard's vacuum tubes "won Mullard a quarter of the home radio construction market" (Frankland 2002, 22, 32).

As broadcast entertainment, especially of music, began in the early 1920s, demand rose for both better sound reproduction and radio sets that allowed the whole family to listen in. Such sets required both vacuum tubes and electrical energy from batteries since most homes in the early 1920s, even in America, lacked mains electricity. The enthusiast magazines advertised basic chassis kits for receivers that could be put together by enthusiastic radio amateurs to listen in their workshops to the experimental broadcasts of the early 1920s on headphones. By the late 1920s more sophisticated receiver designs offered amplified sound to power speakers. These kits were advertised as designed to use war-surplus tubes as both receivers and amplifiers. As such they were a jumble of wires, batteries, and antennae that consumed a substantial amount of domestic space and were often relegated by domestic authority to attics and outbuildings. But as broadcasting evolved by the late 1920s into the hands of large corporations such as the BBC and, in America, both RCA and NBC, manufacturers started to offer receivers complete with tubes and amplified speakers and fitted into furniture style cabinets acceptable to domestic authority in the living room. Between 1929 and 1931 the output of sets by the British radio industry almost doubled. These new sets used very much better technology and could be powered by the mains electricity being installed in almost all new houses in urban areas by the late 1920s. They quickly became a focal point for family life in better off households, and between 1927 and 1939 the number of households in Britain with radio licenses went up from 2 million to 9 million. Successful companies such as Pye prospered as the market evolved, but others fell by the wayside. In 1928 Britain had 107 radio manufacturers, by 1933 only 43 were left. Many more would fail before 1939 (Frankland 2002, 38, 44). In Britain, there were

> up to 1930 . . . probably more home made receivers than those manufactured by the trade. Receivers were constructed for relations, and boy friends built them for prospective parents-in-law. . . . From about 1929, the radio industry began to produce 'mains-only' models with indirectly heated valves. (Ramsbottom 1995, 117)

The tubes made for airplane use in WWI were battery powered, but by the early 1930s the supply of such tubes from war-surplus stocks was heavily depleted and most new houses were now being supplied with mains electricity. In America rural electrification came later, hand in hand with the New Deal in the late 1930s. Tube manufacturers thus began to introduce new tubes designed for mains operation. These mains sets no longer required the two sets of high and low tension batteries that earlier sets had needed, while longer broadcasting hours made frequent visits to battery recharging centers both impractical and expensive. The other key to the success of the mass-produced radio and its increasing centrality in the household was the development of superhet technology. This required complex circuitry and an extra vacuum tube, the expense of which did not seem justified to most manufacturers in the 1920s, RCA being the honorable exception. However, in the early 1930s the proliferation of radio stations and a more cluttered radio spectrum resulted in demands for more precise tuning. As radio broadcasts became a focus of household activity consumers also demanded better sound quality. Superhet circuits and automatic level control allowed the production of highly sensitive and selective radios without any of the instability in tuning and variations in volume level by station that plagued earlier receivers (Rosen 1995, 130). By the early 1930s, therefore, broadcast radio receivers had matured to the point where they were reliable, easily tuned and listened to, so designed that they were acceptable in living rooms as a focal point for a family's domestic entertainment, and affordable by most households. By the late 1930s mass production and marketing had ensured that almost no household in the developed world was without its radio receiver.

Britain's Odd Trajectory

On the face of it the British switch to government control of broadcast radio after 1927 might be seen as an unexpected outcome. British governments of the late 1800s and early 1900s, whether Liberal or Conservative, generally held to *laissez-faire* policies as much as or more than American governments. By the late 1800s, indeed, American governments had backed some distance away from pure *laissez-faire* policies. The Republicans in particular had become deeply invested in using Protectionism to promote the well-being of domestic entrepreneurs against foreign competitors, and such Protectionism only deepened from the turn of the century through into the 1920s. In Britain the Imperial Protectionist wing of the Tory Party began to push the Conservative Party in the same direction in the 1890s, but Britain remained heavily committed to *laissez-faire* policies until the demise of the Liberal Party and the events of the Depression pushed Britain briefly toward Protection in the early 1930s. That the British government intervened so aggressively to control broadcast radio by 1927 must thus be accounted for.

At the structural level WWI caused a huge backlash against classic liberalism in Europe. British leadership of the international system, established at the Congress of Vienna in 1815 but questioned increasingly since the rise of the new nationalism in the 1880s, now seemed threadbare. Two countries that had been on the Allied side in the war rejected liberalism utterly in its aftermath. Russia retreated into Bolshevism, Revolution, and Civil War: in Italy Fascism—National Socialism—reared its head. Although Britain emerged from the war with mandates to expand the Empire through its rule of former Ottoman and German territories in the Middle East and Africa, the Empire was showing serious signs of wear in two crucial regions. Following the Easter uprising of 1916 the long building demands for home rule in Ireland gave way to independence for the Catholic south in 1922. In 1915 Mahatma Gandhi returned from South Africa to begin to lead the Indian subcontinent toward independence in 1947. After WWI Britain itself was beginning, especially under the influence of an increasingly powerful Labour Party, to veer toward social democracy, and thus an increasing willingness for the government to regulate such public goods as broadcast radio. Both Labour and the Conservatives were much more willing to embrace Protection than the Liberal Party, which lost power rapidly in the aftermath of a war driven by nationalist ideals but in which a large swath of the working-class would die. Many Liberals had claimed, even as late as 1914, that Free Trade would prevent WWI. At the political level classical liberalism was thus in serious retreat in the early 1920s.

In Britain there was one overarching issue and four specific ones pushing for the abandonment of *laissez-faire* policies with regard to broadcast radio, even by those "antisocialists" who should have favored them. The overarching issue was created by

> the bureaucratic impulse of both national security organizations and post, telegraph, and telephone authorities who wished to keep radio in their hands. These antisocialists, as most of them were, shared with the left a belief in radio nationalization, and they probably understood the character of this proposal far better than did those socialist idealists who provided its rhetoric. (Pool 1983, 109)

British Concerns over National Security

Four national security concerns operating in the specifically British context can be identified as impacting the development of wireless technology in general and broadcast radio in particular in the period between the wars.

First came the internal political problems caused by the way the BBC had controlled access to broadcast radio during the Great Strike of 1926. During the strike the BBC had allowed the Conservative government to control radio messages and had refused to allow access to the new broadcast media by any other voices, whether they were the trades unions, the opposition Labour Party, or even the Archbishop of Canterbury, who was attempting reconciliation. This had an interesting result. Such heavy-handed censorship created a clamor against state control of the new media, which might reasonably have been resolved by shifting broadcast radio to the free market. However, that was not likely to happen for the reasons outlined below and the result was the insistence that the BBC become a publicly but not politically controlled monopoly and that it be made to play the same role as Britain's historically free press. The British Press, politically powerful at a national level unlike the American Press, also influenced the decision, since it lobbied strongly against commercialized radio, advertising being a major part of its revenue stream. In America's more free market system local newspapers simply developed their own radio stations.

Second, a major legacy of WWI for Britain was the emergence of the RAF out of the army's RFC and the RNAS as an independent third arm of the military. The RAF pushed hard to develop medium wave wireless telephony to serve the needs of "Plan 1919": the co-ordination of advances by airplanes, tanks, and armored cars, a slow-speed version of the form of tactical air war that the Third Reich would later perfect as "blitzkrieg" (Hugill 1999a). The RFC had resisted this because it saw its role as spotting for the artillery, and the RNAS had been far more concerned with strategic bombing than tactical air war. Despite the pressure after WWI from the army and navy to fold the RAF back into the older services from which it had been detached in 1918, the realization that air policing had the potential to significantly lower the costs of imperial control of such fractious regions as Iraq (Omissi 1990) ensured the support of Winston Churchill as chancellor of the exchequer. The survival of an independent RAF ensured its continuance as a powerful political actor demanding access to that portion of the radio spectrum, the medium wavelengths, that Hoover had assigned in America to the broadcast interests. This helped delay the onset of radio broadcasting in Britain.

Third, between 1914 and 1918 imperial transportation by sea had been seriously interrupted by the German U-boat campaign, and at the end of WWI the British government began to look seriously at the development of an imperial air transport system to move mail and significant civilian and military personnel around the Empire more quickly and safely (Hugill 1993). After a brief period of unsuccessful experiments with commercial air transport this emerged in 1924 as a state monopoly airline within the Empire, appropriately named Imperial Airways. Radio, in particular in the medium wavelengths used by the RAF, was critical to communications in the airways. For example, the first commercial broadcasts from the Marconi Company at Chelmsford were shut down in late 1920 because they interfered with "Croydon aerodrome's newly installed air traffic control system" (Street 2006, 41). Once again, this delayed the onset of radio broadcasting in Britain.

Fourth, the British Empire not only survived the wreckage of WWI, it was substantially enlarged geographically at the end of the war by the provisions of the Treaty of Versailles. This enlargement came because of the cumulative impact of the collapse of the Ottoman Empire before the war, the Austro-Hungarian Empire as the war started, and the German Empire at the war's end. Under the mandate system established by the League of Nations Britain was able to make substantial additions to its Empire, especially in the Middle East and Africa, from the ruins of two of those Empires, the Ottoman and the German. If imperial telecommunications had been an issue in 1914 they were even more an issue in the early 1920s. The only area where imperial power seemed in decline in the early 1920s was Ireland, connected by submarine telephone cables.

These four intersecting geopolitical and military concerns, along with the problems of spectrum allocation in the limited geographic space of Europe, pushed Britain further and further away from liberal policies toward the emerging technology of radio in the early 1920s. They had, however, no impact on America. In America socialism was weak and America lacked all four of the national security concerns that drove Britain's normally liberal politicians and decision makers away from liberal solutions to the challenge of how to structure broadcast radio. The retention of *laissez-faire* policies for broadcast radio might well have made sense to American nationalist politicians in the 1920s, but to those with a clearer sense of the building geopolitical issues of the interwar period American internal politics merely ensured the continuation into the wireless era of the British telecommunications hegemony begun with the submarine cables.

It was thus through its obsession with controlling global telecommunications to maintain the security of the Empire that Britain came most firmly into control of the new wireless technology, and here geopolitical concerns were clearly uppermost in the minds of British political elites, not broadcast radio. The shift by America's RCA to a primary focus on broadcasting and profit taking was not immediately obvious in Britain in the mid-1920s. Increasing fears of RCA's likely ability to dominate international wireless communications pushed the British government to force a merger between the British submarine cable interests, the Eastern and Associated Telegraph Companies, and Marconi Wireless in 1928. This created Imperial and International Communications, Ltd., which matured as Cable and Wireless, Ltd., in 1934. Although nominally a private company, Cable and Wireless often acted as if it were under state control. There were two reasons for this. First, the main predecessor of Cable and Wireless, the various submarine cable companies established by John Pender and amalgamated under the umbrella of the Eastern and Associated Telegraph Companies, had given the Royal Navy's intelligence arm unprecedented access to telegraphic signals in WWI's Room 40, and the British state was well aware of the likely value of continued access to such information in the event of another major war (Hugill 1999a). Second, the BBC had emerged as a state-licensed entity in 1927. Once the British state was, via the BBC, able to monopolize commercial free broadcasting for entertainment purposes Cable and Wireless could not enjoy in Britain the ability to generate profit through broadcast radio enjoyed by companies such as RCA and NBC in America. RCA simply evolved as a commercial company in a *laissez-faire* economic environment and left behind its geopolitical origins in a country trending toward isolationism. Cable and Wireless was first, last, and always a geopolitical construct in the interwar years.

BROADCASTING AND PROPAGANDA

The more subtle issue of the propaganda uses of broadcast radio remains. Three positions emerged during the 1920s and 1930s, especially in Europe. In Britain, for the reasons outlined, broadcast radio emerged as a public utility, noncommercial and politically neutral. American style broadcast radio as a voice of commerce, with programming created in England and paid for by advertisers, developed in some parts of Europe with broadcast access to the southern English market and to British relay systems. But over much of Europe broadcast radio developed as state propaganda. This third approach began alongside national socialist politics, and was ultimately epitomized by Nazi Germany.

At the level of geopolitical analysis the American impact was muted during the 1920s and 1930s. America itself was a huge but self-contained market, and its domestic transmitters had no ability to penetrate into European space. Some of the guiding principles of American broadcast radio did, however, most obviously in the development of a highly successful commercial radio station aimed at the British market in the form of Radio Luxembourg from 1933 on. Although Radio Luxembourg did not fully display the "alluring mass culture" that was developing in America in the period after 1900 it echoed it, most notably by transmitting American or American style popular music, especially of the "big bands" popular in the 1920s and 1930s. American records of popular music were freely available in Europe and sold well. There was considerable diffusion of popular culture across the Atlantic, especially between America and Britain where language barriers were nonexistent. American movies were also extremely effective carriers of America's "alluring mass culture." Language was not an issue in the silent era, when subtitles were easily added to suit the market. The "talkies" were initially a problem, and some quite hilarious versions of films exist with Anglophone stars mangling Spanish, French, or German as they remade their features for a foreign language market. Fortunately Western Electric developed what it called "re-recording" technology that allowed native speakers selected for appropriate vocal qualities to dub the appropriate language over the original English sound track. American made "talkies" and recorded music created effective demand for similar entertainment by British actors and bands. The BBC generally avoided such musical material as "lowbrow." Radio Luxembourg made good money catering to those "lowbrow" tastes.

Despite the origins of radio as state propaganda in fascist Italy, it was in Nazi Germany that it took its fullest hold in Europe as radio developed. "The most produced radio set models in Germany between 1933 and 1945" were sets called *Volksempfangers*. These radios were intended to receive only German stations and were available by 1938 for about 25 percent of the average monthly wage. The tuning knob was marked "Remember. The reception of foreign radio stations is a crime against the national security of our people. It is punished with several years of penitentiary by order of our leader." *Gestapo* searches that found a radio tuned to a foreign station had dire results (Old Radio World Online). From the late 1930s on that "foreign station" was almost always the foreign language service of the BBC. Despite being a state-licensed monopoly the BBC was relatively free from government control. Following the intense criticism of its lack of neutrality in reporting the Great Strike of 1926, the BBC steered an aggressively neutral course, emphasizing accuracy in its news reports, even when such accuracy might be construed as detrimental to British interests. This reputation for neutrality and accuracy stood the BBC in great stead during WWII, when its neutrality became a byword.

The other area in which radio became an arena of geopolitical contestation was what the BBC came to call Empire broadcasting, for which the Americas became a major market judging by the correspondence received by the BBC. Immediately the Nazis came to power in 1934 they began to invest heavily in shortwave broadcast stations designed to reach South

America in particular, but also some of the English-speaking settler colonies and dominions. By 1935 the *Reichs Rundfunk Gesellschaft* was receiving as much money from the Nazi state for its overseas shortwave foreign language services as it did for its domestic services. The BBC did not begin foreign language broadcasting, which it regarded as separate from Empire broadcasting, until January 1938. The first foreign language broadcasts by the BBC were in Arabic to counter Italian propaganda aimed at the British mandate states in the Middle East. The second foreign language broadcasts, which followed shortly after, were in Spanish to counter German propaganda stations aimed at Latin America. The third language added was German, starting in March 1938. The BBC monitored the foreign language broadcasts of other powers partly to ascertain quality and partly so that it could address propaganda. Palestine was a major market for such broadcasts, the British estimating in 1939 that there were "less than 100,000 wireless receiving sets in 47 Colonies—the Dominions were not included in this reckoning . . . [and that] Palestine accounted for more than a quarter of the receiving sets." A major emphasis by the BBC, in all its services, was that "when news was broadcast it would be 'straight' news and not propaganda." The very first news broadcast by the Arab language service concerned the sensitive issue of "the execution that morning of a Palestinian Arab on the orders of a British military court" (Briggs 1965, 393–410, quotes from 408–409, 400, & 404).

In the end the central issue comes down to the response of the three different models of broadcasting to the political and geopolitical events of the 1930s. The British followed the model of the relatively independent public corporation based on the notion that radio should function like a free press, but that as a "natural monopoly" the BBC had to be subject to public oversight. American broadcasting continued for most of the period with lowest common denominator entertainment that ensured high advertising revenue and high profits from an almost totally national market. But from 1934 on that began to shift with the election of the media savvy Franklin Delano Roosevelt to the Presidency and the victory of the Nazi Party in Germany. Many Americans became concerned about domestic anti-Semitism, peddled over the airwaves by such "homegrown bigots as Father Coughlin and the Reverend Gerald L. K. Smith" (Hilmes 1997, 238). In 1939 the National Association of Broadcasters implemented its "equal time" code to prevent the likes of Coughlin from buying up huge amounts of airtime. The Roosevelt administration had proven adept in the late 1930s at using radio to promote Democratic Party policies, but when America entered WWII it avoided creating any overt and centralized propaganda apparatus, working instead through "a multiplicity of government-related offices, bureaus and liaisons with private organizations to mold a more unified public response to the war without the appearance of direct authorization" (Hilmes 1997, 237).

Centralized propaganda found its home most emphatically in Nazi Germany where, as Reich Minister of Public Enlightenment and Propaganda, Dr. Joseph Goebbels

> fashioned radio into an effective instrument of propaganda. Each of the thirty-eight party regions (*Gaue*) had a *Gaufunkwart* or regional radio officer and under them were *Kreisfunkwarte* in each of the thousand districts into which Nazi Germany was divided. It was their business to see that when the *Gemeinschaftsempfang*, community listening, was ordered, every factory, public square, and school was fitted with receivers. The People's Wireless Set [*Volksempfanger*] was designed to receive German stations satisfactorily and to receive nothing else. (Briggs 1965, 475)

Goebbels was emphatically clear on the central importance of radio to the rise of Nazism, and he sought to focus all radio activities, including popular entertainment, on the Nazi worldview. In a speech opening the tenth German Radio Exhibition in 1933 he said

> the radio will be for the twentieth century what the press was for the nineteenth century. . . . We do not intend to use the radio only for our partisan purposes. We want room for entertainment, popular arts, games, jokes, and music. But everything should have a relationship to our day. Everything should include the theme of our great reconstructive work, or at least not stand in its way. Above all it is necessary to clearly centralize all radio activities, to place spiritual tasks ahead of technical ones, to introduce the leadership principle, to provide a clear worldview, and to present this worldview in flexible ways. (Goebbels 1933)

The BBC was slow to react to the shift of emphasis that was increasingly evident in Nazi broadcasting by the late 1930s. Even in the immediate run-up to WWII the BBC continued to broadcast escapist fare and the lighter forms of entertainment, all the while being attacked by Britons for what seemed to some pro-German propaganda, as though "talk about the 'Prussianism' of the BBC was an excuse for not understanding what was happening in Prussia" (Briggs 1965, 475). But as the European crisis began to unfold more fully in 1938 it revealed for Britain what would be throughout WWII "the double role of broadcasting: first for the home audience—to give orders and to maintain morale; and second for the world—to spread reliable news and views" (Briggs 1965, 646). The experience developed with the Spanish and Arabic language services was applied in 1938 to developing a German service, despite the problem "that British broadcasts lacked adequate power to be picked up" by a *Volksempfanger* designed to minimize reception of non-domestic broadcasts (Briggs 1965, 649). There was, in the last few months before WWII broke out both considerable preparation for war by the BBC and considerable discussion as to what the proper role of the BBC should be. In the event R. T. Clark, the home service news editor, summed up cogently what would become the very successful policy of the BBC throughout WWII.

> It is very difficult for me to put up any suggestions on the matter of improving the morale of the country. I am afraid that at the present the majority of people would admit that the main items of news are, in themselves, depressing. . . . It seems to me the only way to strengthen the morale of the people whose morale is worth strengthening is to tell them the truth, and nothing but the truth, even if the truth is horrible. (quoted in Briggs 1965, 656–657)

The events of WWII certainly bore out the advantages of the British approach to broadcast radio at that time and in Europe, although it would also be fair to note that the BBC broadcast a reasonable amount of commercial style entertainment on the domestic front, which it saw as morale building at a time of crisis. The advantage of this deliberately neutral approach to news is that, to this day, "the BBC remains an important element of British public diplomacy because the independence of BBC News enhances British soft power" (Mirchandani & Abubakar 2014, 15).

The ultimate success of the American model of broadcast entertainment, today including television as well as radio, had to wait until the liberalization of the global economy that took place in the 1960s and 70s, and much of that success emerged from the huge representational power of America's increasingly "alluring mass culture." This was epitomized at the level of radio broadcasting by the fact that, from the late 1950s on, American popular music, whether performed by American groups or British ones that had adopted American style Rock & Roll, completely swept the world stage, creating a massive demand for radio stations that would broadcast such material. After 1945 there were certainly attempts to establish American stations in Europe, notably the propagandist and rather heavy-handed "Voice of America." But what went farthest in the establishment of American hegemony and a uniquely American form of imperialism that did not depend on "the materiality of spatial organization" was the carrying of America's "alluring mass culture" on European radio stations, especially British ones, that confirmed America's ability to project its remarkable "power in the realm of representation."

Chapter 6

The Oil War

The Struggle to Control the Middle Eastern Oil Spigot after World War I

THE OIL WAR AND THE MULTIPLE STRUGGLES BEHIND IT

The third arena of struggle foreseen by President Wilson at Versailles was over oil. In this he believed America held the advantage over Britain, which suggests that, as was the case with transportation, where he thought the reverse, he was not well informed. Much has been written about this arena of struggle between America and Britain, and it comes down to two quite contradictory arguments: American "oil exhaustion" versus the "search for markets" for American oil. Much of the modern writing, exemplified by scholars such as Bromley and Stivers emphasizes the American "search for markets" for American oil, the need for bunker fuel for American ships in global trade, and the like. Such scholars reject the idea that fears of American "oil exhaustion" drove American companies to seek foreign oil.

Most oil scholars tend to ignore what seems to them a secondary struggle between America and Britain over the Middle East. In the expanded British Empire post-Versailles, the Middle East loomed large for a series of reasons: "the (anti-Hashemite) caliphate movement among Indian Moslems, the Turkish menace to Mosul, the French suspicion of the motives of Britain's Middle Eastern policy, . . . the disturbance of the Middle East by the Wahhabi movement led by Ibn Saud . . . [and] Britain's strong imperial interest in safeguarding the acquired strategic and economic assets against any future interference, be it by Arabs or her imperial rivals at large." A key issue in this was the "purpose of making US recognition of Iraq conditional upon prior acknowledgment of the British mandate in Palestine . . . to consolidate the political security of the strategically important trans-desert motor and air route from the Mediterranean via Amman to Baghdad and the Persian Gulf" (Mejcher 2007, 71).

America's relative international naivety as it emerged onto the world stage driven by Wilsonian idealism and the hangover effects of a belief from before WWI that "dollar diplomacy" could trump geopolitics tended to cause America and Britain to act at odds over oil in the early 1920s. The Turks made peace over the Mosul boundary issue in 1925 and most of the diplomatic, economic, and geostrategic issues between America, Britain, and France were resolved by the time of the Achnacarry Agreement over oil in the Middle East signed in 1928. As Secretary of State for the Colonies in 1922 Churchill realized that King Faisal of Iraq could never be recognized as representing the caliphate without inflaming Moslem opinion in India. India was far more important to the Empire than Iraq, but the problem of Moslem politics in muted form and the rise of Wahhabism were destined to remain as issues throughout the interwar years.

To observers in the early 1920s, however, the struggle between America and Britain, Standard Oil and the anglicized Royal Dutch Shell, to control global oil resources, especially those of the Middle East, seemed clear enough, and they saw that struggle in two lights: "oil exhaustion" and geopolitics. Modern historians have analyzed the diplomatic struggle between Britain and America that swirled around these themes. Unfortunately the diplomatic documents rarely directly identify these themes as such and much of the American diplomacy seems to have been aimed at keeping foreign sources of oil, whether apparently controlled by Britain or other powers, accessible to American companies (Venn 1982, 66). The diplomatic documents thus support, at least in part, both the "oil exhaustion" and "search for markets" theses.

The geopolitical part of the oil struggle the writers of the 1920s saw was caused by the rapid evolution of maritime power in the early 1900s and thus related to at least the maritime component of the transportation struggle (see chapter 2). To Wilson, in 1919 the issues of transportation and petroleum would have seemed clearly linked. Although Admiral Mahan had made it clear in his analysis that hegemony was derived from the possession of the dominant blue-water navy, it was also clear in Mahanian reasoning that possession of the dominant blue-water merchant marine was equally important: much of Mahan's geostrategy lay in the use of the blue-water navy for commerce-raiding and blockade duty to interdict the wealth generation made possible by the merchant marine (Mahan 1890).

Although it was coming into focus in the first decade of the 1900s, by the end of WWI the impact of oil-fired navies on military struggles, geopolitics, and geostrategy had become blindingly clear. As a closed world-economy structured increasingly around nation-states began to replace the relatively open global trading system created by relatively interlocking imperial structures, and as autarky began to increase in the early 1920s, the consequences, in particular for Britain, seemed likely to be negative. In *The Oil War* (1926) Anton Mohr, a political geographer at the University of Oslo, was clear about the military need to control naval fuel supplies.

> For Great Britain especially, whose whole fleet, and therefore security, is dependent upon abundant and safe supplies of petroleum, it is a matter of vital importance to secure control of the necessary oil fields. Thus the struggle for oil has developed into a quarrel among the Great Powers, comparable to the struggle for colonies in the preceding century. There can be no truce to hostilities; the battle must go on as long as there are any oil fields left to fight for. (Mohr 1926, 139)

But there was also an economic struggle with subtle geopolitical undertones under way after 1918 as the world-economy restructured around oil fuel for nonmilitary transportation: shipping; automobiles, trucks, and tractors; and aviation. By the end of WWI the likely consequences of the coming nonmilitary struggle over the emerging oil economy seemed clear to non-Americans, as well as intimately connected to the issue of transportation that Wilson was also concerned about. Chapters 2 and 3 cover some of the material on these three transportation arenas. This chapter addresses the issues of the four main types of fuels needed in the main arena of transportation: heavy oils for oil-fired steamships; refined heavy oils for diesel engine motorships; leaded gasoline fuels to increase the performance of automobiles; and high octane gasoline fuels for flight. Shipping was, however, central. In shipping there was an experimental period roughly from 1900 to 1918 and an effective implementation period for heavy oil fuels thereafter: "as a means of raising steam oil was about twice as economical [as coal], and was four times so if used in internal combustion engines" (Mejcher 2007, 8).

Effective implementation of oil fuels for gasoline-engine vehicles had to wait for several improvements in distilling technology in the late 1920s and again in the late 1930s. These were not needed in the production of the heavy oils that were used in shipping although diesel fuel was more demanding. As the demand for automobile fuels rose as mass adoption got under way in the first decade of the 1900s distilling technology changed quickly to allow the production of more of the more volatile components of petroleum, although to begin with there was no quantifiable way of measuring octane. By 1927 research was showing fuels being marketed with octane ratings ranging from forty to sixty, the latter being the new leaded fuels that were just being introduced (Hamilton & Falkiner 2003, 62). Until that breakthrough all but the most expensive automobiles with the largest-engines had limited acceleration and low cruising speeds. By the late 1920s leaded fuels began to allow cheap vehicles to carry the extra weight of closed steel bodies, which made them useful in a much greater range of weather conditions. These rapidly replaced the lightweight touring bodies with soft tops and poor weather protection of the teens and early 1920s. By the late 1930s octane ratings for automobile fuels had improved to the upper sixties.

Although such progressive increases in octane rating helped improve aviation the aircraft industry always sought more power from less engine weight. Initially aircraft engines ran on the same fuels as automobile engines, although in WWI it was quickly recognized that fuels from some parts of the world performed better than others without any understanding of why. Shell's fuels from Indonesia were later shown to have high natural octane ratings. The leaded fuel technology introduced by the mid-1920s made it possible to produce better fuels in the production process whatever the petroleum being distilled, but as late as 1938 the USAAC recognized only two octane ratings for its aviation fuels, sixty-eight and ninety-two (Strauss 2003, 89). Higher octanes allowed higher compression ratios, which gave higher power output. Chemically complex and very expensive high octane racing fuels began to allow very high outputs in such arenas as the Schneider Trophy seaplane races in the 1930s, and Rolls Royce in particular capitalized on the experience they gained in this arena. The Rolls Royce Merlin, which would power such iconic fighter planes as the Supermarine *Spitfire*, was designed to give optimum performance on the 100 octane fuels used in the Schneider Trophy races, fuels that were not available in commercial quantities when the Spitfire entered service in 1938. Running on eighty-seven octane aviation fuel the first generation Merlin gave just over 1,000 horsepower. When 100 octane fuels became available in late 1939 the same engine gave, with minor adjustments, just over 1,300 hp. In the Battle of Britain those extra horsepower mattered, making the Spitfire the equal of the German fighters, which used a much larger engine. By the end of WWII refined and redesigned Merlins running on fuels as high as 145 octane were capable of just over 2,000 hp. It was the breakthrough to mass produced 100 octane leaded aviation fuels in the very late 1930s that finally allowed lightweight, high performance engines, and by that time their military use mattered far more than any possible civilian applications.

Oil-Fired Shipping and Geopolitical Maneuvering

After 1918 a huge technological shift got under way in shipping, although it had begun in the experimental period as oil firing of steamships became increasingly common. Heavy oil rapidly became the fuel of choice for merchant as well as naval shipping because it was substantially more caloric than coal, far more easily bunkered, and removed the need for large numbers of stokers. Used direct in diesel engines oil could power blue-water merchant ships that had much longer range, were just as fast, and were much more fuel-efficient than the coal-fired ships Britain had built in such vast numbers in the late 1800s. Where much more power was needed than that era's relatively low-powered, slow speed diesels could deliver, such as in fast liners and warships, coal-fired boilers were easily converted to burn oil (Delaisi 1922, 17–20). Compared to coal, oil smoked much less, making oil-fueled warships much less visible to their opponents. By the end of WWI the oil-fired warship reigned supreme. By the beginning of WWII about half the world's merchant ships were oil-fueled: "of these ships, half were steam and half diesel powered" (Winkler 2003, 145).

The fear for some in the 1920s was that, if one country monopolized global oil output, as America indubitably did at the time, it would be able to use that fact to achieve economic hegemony in the post-WWI economy rather as Britain had used its vast output of coal to continue its hegemony in the last half of the 1800s, as steamships using highly efficient coal-fired triple-compound engines replaced sailing ships as the dominant merchant vessels (Hugill 1993).

> Suppose that one great country supplies by itself eight-tenths of the new fuel; the ships of other nations will soon be unable to sail without recourse to its oil depots, and if that country creates a powerful merchant fleet, it will be at once the controller of ocean commerce. Now the people which obtains the world's carrying trade levies a tithe upon all those for whom it provides transport, and thus adds to its capital. New industries arise round its ports, its banks become clearing houses for international payments. The controlling centre of credit is displaced: we saw it pass from Amsterdam to London in the eighteenth century with the growth of British shipping. Is it not going to move to New York? Thus, under the pressure of economic facts, arises one of the great *financial* problems of to-morrow. (Delaisi 1922, 2, emphasis in original)

Before WWI America had no blue-water merchant fleet of any consequence. American merchant ships came in two types: those that catered to American coastal trade and those that catered to the Great Lakes traffic. Britain owned most of the world's blue-water merchant ships and many of its blue-water liners, despite German inroads into the latter market in the late 1800s and the intense competition that developed over the "Blue Riband," the unofficial title given the liner that made the fastest transatlantic passage. In 1897 the crack German liner *Kaiser Wilhelm der Grosse* wrested the Blue Riband from Britain. Other great German liners followed, *Deutschland* and *Kronprinz Wilhelm*, the sister ship to *Kaiser Wilhelm der Grosse*.

Britain worked hard to get the Blue Riband back with the new Cunard liners, the turbine powered *Lusitania* and *Mauretania*, which both took the record in 1907. *Mauretania's* 1909 record was unbroken for twenty years and the British would hold the Blue Riband with brief breaks until it was taken permanently for passenger ships by the American liner, *United States*, in 1952. Such liners proved technologies later applied to the new generation of all-big gun battleships: *Lusitania* and *Mauretania* were laid down the year before the equally revolutionary turbine powered British battleship, *Dreadnought*.

But WWI shifted the balance. The German merchant fleet was devastated by British commerce raiders and by its internment in neutral ports, especially those of America, which took German ships as war prizes when America entered the war in 1917. Most of the High Seas Fleet was scuttled at Scapa Flow in 1919. Second, many British merchant ships were sunk by U-boats. Third, America began mass-producing replacements for those British merchant ships, so that by the end of WWI the British and American merchant marines were about even and the German gone (Delaisi 1922, 20–21). By 1920 the American potential was even greater than that of Britain: the *New York Times* could boast that "if we are using 751 ways and can average three ships a year per way, we should turn out in one year 13,518,000 tons" . . . more than Britain, "heretofore the greatest builder of ships, has completed in any five years of her history" (*NYT* June 11, 1918).

> For six years Britain waged an exhausting war, with the main object of ruining German shipping for ever. At the cost of enormous sacrifices, which will weigh upon her for half a century, she gained her end. The Kaiser's magnificent battleships now lie at the bottom of Scapa Flow, and the fine steamers of the Hamburg-America Line and the North German Lloyd have been shared as booty among the victors. The only maritime rival she had to fear is humbled to the dust. Yet now, from the very war which destroyed that competitor, a new one has arisen, twice as formidable as the old, for, in addition to a superiority in tonnage, it enjoys the practical monopoly of a fuel of which Britain has none. The burning of American oil in the boiler-rooms of the great liners may be the downfall of the British Empire! (Delaisi 1922, 21)

Many German and Austrian liners had been interned in America when the war broke out. Sixty-six were reported interned in American ports in early 1915 with nine more in Hawai'i. The *New York Times* described thirteen of these as large liners (*New York Times* February 3, 1915). *Kronprinz Wilhelm* was taken as a war prize when America finally entered the war in 1917 and re-named *Von Steuben*. Many German crack liners, such as *Kaiser Wilhelm der Grosse*, were used as armed auxiliaries and some were lost in combat. As it turned out, however, possession of even these great German liners did post-WWI America little good for four clear reasons. First, there was no attempt to create the extra ships that would have been needed to create an effective, regular, transatlantic service when the war was over; second, shipbuilding technology had improved rapidly during the war and prewar ships were "dated" both technologically and stylistically as Victorian and Edwardian aesthetics were abandoned; third, the emigrant trade all such ships had depended on collapsed after America passed the Johnson Quota Act restricting immigration in 1921; and fourth, American flagships could not attract the luxury trade since they could not serve alcohol after the implementation of the Volstead Act in 1920.

Accounts of *The Oil War*, Contemporary and Modern

The issue of fuel for naval vessels, merchant ships, the burgeoning automobile fleet, and airplanes represents the demand side of the emerging oil economy. By the early 1920s fears were growing rapidly that the supply side could not keep pace and that, in particular in America, most of the known reserves had already been consumed. These fears drove *The Oil War* and are thus of the greatest concern here.

Modern understanding of the size and depletion rate of oil reservoirs began with the innovation of well-head logging by the French company, Schlumberger, in 1926, and the development using the laboriously accumulated data from well-head logging of the concept of "peak oil" in 1956 by the Houston based geoscientist, M. King Hubbert (Deffeyes 2006). Neither Schlumberger's data nor Hubbert's model were available to inform those worried about the possible collapse of American output in the period 1919–1930, hence the peculiar potency in that period of the "oil exhaustion" thesis to American observers before the discovery of the massive East Texas field in late 1930. Although the Alaska field is larger East Texas is still the largest producing field in the history of oil in America. During the period in which the oil exhaustion thesis held sway, the Norwegian political geographer Anton Mohr described the likely coming global conflict over oil resources as *The Oil War* (1926) and noted the potential of the Middle East to rise to dominance over the global oil supply. Mohr's work is one of a substantial series of polemics and academic accounts from the 1920s and 1930s that follow this theme. The polemics are exemplified by Ludwell Denny's *We Fight for Oil* (1928), which constructed the fight as between Britain and America, but the polemics include a substantial number of French and German works, which took an explicitly anti-British stance (Delaisi 1922; Fischer 1926; Hanighen 1934; L'Espagnol de la Tramerye 1924; Marcosson 1924; Zischka 1933; 1939). One polemic attempted to deny that it was a struggle at all (Davenport & Cooke 1924). The first of the polemics was

Delaisi's book, published initially in Paris in 1920 as *La Pétrole*. There are two "odd men out" in this list: Anton Zischka and Frank Hanighen published their books much later than the rest, after the Great Depression had bitten. Zischka was a Viennese born journalist who later took up with the Nazis and whose writing focused heavily on the role of energy, commodities, and technology in world power, usually from a strongly German perspective. His first book, *La Guerre Secrete pour le Pétrole* (1933) was republished in German in 1934 as *Der Kampf um die Weltmacht Öl* and a second book on oil followed in 1939. Zischka's work is intertwined with that of Hanighen, and hard to disentangle. Hanighen was a classic muck-raking American journalist, famous for co-authoring *Merchants of Death* (Engelbrecht & Hanighen 1934), which blamed the outbreak of WWI on arms manufacturers. In *The Secret War* (1934) Hanighen was responding to the rise of fascism in general, Hitler in particular, and fears of a new war in which oil would be an even more crucial resource than it had been in WWI. Hanighen notes that his "Chapter 1 is based mainly on Antoine Zischka, *La Guerre Secrete pour le Pétrole* (Paris 1933)" (Hanighen 1934, 289), and in 1935 Hanighen's 1934 book appeared in London under the expanded title of *The Oil War, The Secret War* with Zischka listed as coauthor.

There has been a significant amount of modern scholarship on *The Oil War*. Kent's account, *Oil and Empire: British Policy and Mesopotamian Oil 1900–1920* (1976), is the first to be driven by British official documents. Kent's thorough analysis of these documents just after their release is her study's very great strength, but its exclusive focus on them is also its weakness. As her subtitle indicates she says little of events after 1920. In general she underplays the likely scope of American involvement in the period and, as an economic historian, she pays no attention to the scientific developments that led to petroleum geology and engineering and the technological developments that made oil increasingly the fuel of choice for the Second Industrial Revolution.

Other scholars have interpreted past events through modern eyes. Bromley has attempted to debunk the extent to which contemporary fears of a domestic American oil shortage pushed American companies into the Middle East, arguing that American companies merely sought in the Middle East "access to control over production in order to complement their already dominant marketing position outside the Western Hemisphere, and they also wanted to compete in the supply of the bunker trade for the merchant marine and the Navy," the classic "search for markets" position (Bromley 1991, 111). Yet a looming potential shortage was real enough to contemporary authors, as was the struggle between Britain and America for access to Middle Eastern oil that looms large in British official documents after the end of WWI. On May 13, 1919 "M. L. Requa, a former director of the US Fuel Administration . . . warned the undersecretary of state that . . . no matter how domestic reserves and growth are extrapolated . . . 'our petroleum reserves . . . will not last over seventeen years'" (Black 2006, 270). Such a clear "oil exhaustion" statement is all the more interesting in light of the use by Stivers of Requa to support his "search for markets" argument.

> Even when exhaustion fears ran high, the expansion of U.S. petroleum exports was a live concern among government officials. In fact, wartime oil administrator Mark Requa was so eager to increase American oil exports that, shortly after the armistice, he instructed the American petroleum mission in London to determine whether France would require large quantities of fuel oil during the reconstruction period and to offer advice and material assistance to help the French set up oil-burning equipment. (Stivers 1982, 197)

Stivers reading of Requa here is odd. Requa could just as easily have been concerned that high French demand would impact increasingly exhausted American supplies, especially as the output of America's pioneering Texas field at Spindletop was in serious decline by late 1918. In the light of Requa's comments in May 1919 such an interpretation seems far more likely than a "search for markets" interpretation. It is all very well to say that "with the entry of US interests into the Turkish Petroleum Company the way was open for a general agreement throughout the international oil industry: this was the Red Line Agreement of 1928" (Bromley 1991, 111). The critical struggle was in the events that led up to the Red-Line Agreement, fueled by such fears as those of Requa, and the concerted efforts by Britain to exclude American interests as much as possible until that point. In any case the Red-Line Agreement did not produce comity: companies such as SOCAL would make serious end-runs around it and help radically reduce the power of Shell and Standard Oil in the process.

A problem for all early petroleum engineers—to use the modern term—most of whom came out of hard-rock geology and Schools of Mines, was that they had no idea of reservoir capacity. Their "estimates of domestic reserves were based on exceedingly conservative assumptions as to prices and the state of technology: indeed, it was assumed that technology would hardly improve at all" (Stivers 1982, 196). Both drilling technology and, thanks to the Texas Railroad Commission's rulings, reservoir management would improve substantially by 1925, when Spindletop was revitalized, but that could not have been easily predicted when its output began to decline in 1918. Early petroleum engineers could only measure output and the rate at which it changed. The economists among them also tended to use price as a critical variable, and oil prices, driven by falling supply, "rose sharply from

June 1919 to July 1920 and remained high until January 1921" (Stivers 1982, 195). Such rises were because American output, especially from the massive Spindletop field, was turning down. In 1921, and in response to fears of a serious downturn, the government went so far as to create strategic oil reserves for the USN at Teapot Dome in Wyoming and at the Elk Hill and Buena Vista oilfields in California.

Since the invasion of Iraq in the Second Gulf War (2003–2012) more polemics have appeared dealing with the Middle East, though these seem much less connected to reality than the polemics of the 1920s and 1930s. For example, William Engdahl's *A Century of War* (2004) updates work first published after the First Gulf War (1990–1991), draws on Hanighen and Zischka, though none of the other writers of the 1920s, adds nothing and confuses much. Engdahl is both a peak oil denier with very strange views on how oil is formed and a global warming denier.

Two academic books from the 1920s stand out: Edward Mead Earle's dissertation at Columbia University on the Berlin to Baghdad Railway (1923); and Anton Mohr's *The Oil War* (1926). Earle would spend much of his later career at Princeton's Institute for Advanced Study and is best known for editing *Makers of Modern Strategy: Military Thought from Machiavelli to Hitler* (1943). Mohr was a political geographer at the University of Oslo. Mohr's work was reviewed twice in the British international affairs literature of the period: once, rather negatively, when it was first published in Norwegian as *Kampen om Oljen* (Gathorne-Hardy 1926) and later, rather positively and in conjunction with Fischer's book after it appeared in English translation (Slade 1928).

As these books indicate, by the early 1920s *The Oil War* had become a *cause célèbre* and the cause deepened over time. The first polemic to appear, in French in 1921, was *Le Pétrole*, by French socialist Francis Delaisi, the English translation of which was published as *Oil: Its Influence on Politics* in 1922. In Delaisi's opinion British businessmen who accepted the likelihood of American oil exhaustion were clearly engaged in an undeclared economic war with America.

> In Britain certain men foresaw the danger from America even before the United States had appreciated it. They made no appeal to Parliament or the public. They silently drew up statistics of the world's oil supplies, noticed that their rival's reserves would soon be exhausted, set about obtaining a monopoly of all available sources, and carried out their plan almost in entirety by the adroit use of financial combinations and diplomatic pressure. Their fellow countrymen as well as their opponents only heard of the manoeuvre when it had succeeded. Thus a problem vital to the existence of a nation was solved solely by business methods, unknown to the peoples interested or their official representatives. (Delaisi 1922, 3)

Pierre L'Espagnol de la Tramerye followed Delaisi's lead in attacking Britain for its anti-French actions in the Middle East in a book published in France in 1923 and available in English translation as *The World Struggle for Oil* by 1924. Neither he nor Delaisi saw the San Remo Agreement of 1920 in a positive light even though it resulted in France gaining 25 percent of Iraq's output. Kent, however, argued from the British official documents that France used its claim to the Mesopotamian oil region of Mosul as "a lever to further her territorial ambitions with regard to Syria" and gained geopolitically rather than lost economically at San Remo (Kent 1976, 14). As Kent notes, at the time of the San Remo Agreement Mosul's oil was "commercially as hypothetical as ever" (Kent 1976, 157). For France, control of Syria was a tangible geopolitical gain.

The San Remo Agreement followed on from the Sykes-Picot (-Sazonov) Agreement of 1916 between Britain, France, and Russia on how the three powers would divide up the former Ottoman territories in the Middle East at the end of WWI. In 1916 the Foreign Ministers of the three main Entente Powers allied against Germany, Sir Mark Sykes for Britain, François Georges-Picot for France, and Sergei Sazonov for Russia, secretly agreed to a division of the territories of the collapsed Ottoman Empire adjoining Russia once the war was over. Italy was also to be given a share. This initial agreement was, in fact, for a radical redistribution of the lands of the old Ottoman Empire, leaving Turkey as a rump state. Italy was to get most of southern Turkey. Russia was to get the Bosporus and most of Armenia. Britain and France were to split the rest, either as British controlled states or spheres of influence, the British state extending well south along the Persian Gulf through Kuwait and what is now Saudi Arabia as far as Qatar. The continued strength of Turkey and the surrender of Russia in 1917 and her exit from the war reduced the agreement to Sykes-Picot and left out any concessions to Italy and Russia. As implemented Sykes-Picot was an Anglo-French carving up of former Ottoman possessions.

The San Remo Agreement of 1920 emerged from arguments at Versailles over the Sykes-Picot division now that the Russians were out and how to divide up the probable oil output of the new state of Iraq. At the time France had a minimal domestic oil industry and less need for oil than the other industrial powers. San Remo assured France would share Iraq's oil with Britain. From a British perspective this made it much more difficult for France to create an industry that would search for oil elsewhere and begin to commercially compete with Britain. Further, from a geopolitical perspective it cast France as in alliance with Britain and opposed to America.

There were other contemporary perspectives. Isaac Marcosson's *The Black Golconda: The Romance of Petroleum* (1924) was dedicated "to the wildcatter, whose faith and courage made the American oil empire possible."

Marcosson left readers in no doubt of the severity of the American struggle with Britain although he omitted mention of the internal struggles within America between Standard Oil, with its conventionally trained mining geologists, and the wildcatters of places like Texas. Louis Fischer's *Oil Imperialism* (1926) took an explicitly pro-Bolshevik stance and accused all the western powers of having imperial designs on Russia's rich reserves. In the case of Britain such an accusation is clearly supportable. Britain had both occupied the Baku oil region at the end of WWI, albeit briefly, and encouraged the attempts of General Anton Denikin and his White Russian forces to control the Caucasus region and keep open the pipeline to the Baku oilfields and the Black Sea oil port of Batum.

E. H. Davenport & Sidney Cooke's *The Oil Trusts and Anglo-American Relations* (1924) was one of the few books of the period that tried to pour oil on these troubled geopolitical waters: the authors described Delaisi as a publicist whose "malice [was] winged with imagination" and blamed much of the conflict on the comments of British banker E. Mackay Edgar, who saw British control of the global oil supply as the solution to the huge costs incurred by Britain to fight WWI (Davenport & Cooke 1924, viii–x). Davenport & Cooke advocated a classically liberal perspective, claiming that American and British oil interests competed strongly, that the market had ensured Britain of adequate supplies at decent cost even at the height of WWI. For them, this meant that there had been no need for the British Government to have bought in 1914 majority ownership of the Anglo-Persian Oil Company (APOC), which Winston Churchill had pressed for to ensure a geostrategically secure source of oil fuel for the British navy. They clearly had no understanding of the implications of the Treaty of Washington of 1871.

Ultimately, Davenport and Cooke's argument collapses on the reefs of politics and treaties. They note that, as allied forces faced a serious fuel shortage in mid-1917, Lord Northcliffe, the British Commissioner in America, "called up the Standard Oil head man" to plead for oil. Standard Oil delivered, with no price gouging, and when thanked simply replied "its our war as well as yours" (letter of Northcliffe to Walter H. Page, August 13, 1917). Allies reasonably supply each other. Neutral powers are under no such obligation and, as Churchill would have been well aware in 1914, under the terms of the Treaty of Washington of 1871 that ended the dispute between Britain and America over the "Alabama Claims" from the American Civil War and that created the first international law on the subject, a neutral power could not legally supply war materiel to a combatant. In such circumstances the Treaty of Washington would have effectively suspended normal British-American commercial relations as it would have for Dewey in Hong Kong harbor in 1898. By mid-1917 no one could have seriously doubted that oil was war materiel and any intelligent politician had to be ready for a war without allies who could supply oil. Russia had proven unreliable and the Treaty of Washington controlled supplies from America.

Recent Histories—What *Oil War*?

In conventional recent political histories of the Versailles Treaty and its impact on the Middle East it is argued that only a few visionaries saw the potential of the Middle East to be a major oil region before WWI. Winston Churchill is always advanced as one of the visionaries, and the naval fuel issue is duly mentioned. This conventional understanding has a clear view of traditional geopolitical maneuvering in the region, mostly by Britain, but also by France, although it has America playing an almost nonexistent role because few of its vital interests were at stake. President Wilson is usually depicted as a poorly informed but honest broker whose ideals of self-determination were brought to naught in the Middle East by Machiavellian Europeans, venal local warlords, and his failure to persuade American politicians to sign the Versailles Treaty and support the League of Nations (see, for example, MacMillan's fine 2001 history of the Versailles Treaty, *Paris 1919*). The struggle to control the oil of the Middle East that occurred at Versailles is little mentioned.

In one of the best histories of the impact of the Versailles Treaty on the Middle East, David Fromkin's *A Peace to End All Peace,* oil appears as an afterthought, with American companies briefly upset at being frozen out of the hypothetical spoils of war by the 1920 San Remo Agreement between Britain and France that split Iraq's output in Britain's favor. Fromkin concludes his three-page section on oil by merely noting that by 1922 "the dispute with the United States was resolved" (1989, 534–536, quote from 536). It was not, at least until the Achnacarry Agreement was signed in 1928. A recent cultural history of the Middle East, Michael Oren's *Power, Faith, and Fantasy. America in the Middle East, 1776 to the Present* (2007), looks at developing American obsessions with "The Holy Land" in the 1800s and the increasing role of Christian missionaries and academics in the region. As Oren notes, the academic interest eventually spilled over into the geology and oil exploration of Saudi Arabia.

Recent British military histories of WWI in Mesopotamia, which was run largely out of India with Indian Army troops, not out of Britain, are, however, clear that the military struggle there was about oil as much as about protecting the route to India: "securing and protecting the supply of oil from southern Persia was a perfectly straightforward and realistic aim" (Townshend 2010, 519). Given the well-understood need to secure the oil supply for the navy, such an emphasis is logical. Initially the British, believing "the oil pipeline was threatened . . . ordered the reluctant Indian Government to send reinforcements to the theater" (Barker 1967, 59). But as control shifted to India the commander sent from India, Sir John Nixon, came to regard the safety of the Mesopotamian oil wells as "subsidiary to control of the . . . Basra vilayet" and that "Persian Arabistan should be cleared of Turks and pacified, so that pumping could be restarted at the oil wells" (Barker 1967, 60, 61). Summer 1915 through summer 1916 was the crucial period. German agents in Persia attempted to cut off the flow of oil to Britain, promising the local population all the oil revenue rather than the share they had from the British. By mid-1916 unrest in the region was endemic, but the tribes who controlled the oil region "knew the German fleet had been swept off the seas [at Jutland] . . . the Germans had no ships . . . and unless it could be sold, the oil was valueless" (Barker 1967, 136).

Kent, working from the political documents, argued that the initial advance into Mesopotamia was not about oil at all, but about maintaining Britain's prestige in the region (Kent 1976, 119). This, of course, was hardly a viable war aim and certainly not a grand strategy. Controlling the flow of oil, as the military historians have argued, made sense as a war aim, but the problem of the turbulent Mesopotamian front was, of course, that absent a grand strategy, serious "mission-creep" set in rapidly, with the usual disastrous results. Although one can see the Sykes-Picot (-Sazonov) agreement to carve up the Ottoman Middle East when the war was over as something approximating a grand strategy, even that was initially flawed. As Kent notes, the British took forever to decide where boundaries should be drawn and seemed initially little concerned with oil, conceding Mosul *vilayet* to the French. Sykes however, seems to have understood the importance of the oil region, arguing that, "even though the fields of Flanders might decide the battle, what Germany was fighting for was the Middle East" (Kent 1976, 124). In Sykes-Picot (-Sazonov) as signed the British claimed "well known oil seepage areas" and had the French agree that both sides should honor in their respective regions each other's "existing concessions, rights and privileges" (Kent 1976, 121, 124). Much of the later British claim to making Mosul part of the Iraqi state was based on this. After the Turks had signed the Armistice of Mudros on October 30, 1918, the British advanced on Mosul. This was ostensibly "a deployment of the occupation garrisons as agreed in the armistice" but as was clear from "the discussions that took place beforehand, and the succeeding negotiations with France" it was really all about Britain gaining control of what was now understood to be a potentially major oil region (Kent 1976, 126–127).

In July 1918 Admiral Slade delivered a report to the British Admiralty entitled *The Petroleum Situation in the British Empire* in which he argued that "the power that controls the oil lands of Persia and Mesopotamia will control the source of supply of the majority of the liquid fuel of the future" (quoted in Kent 1976, 125). British foreign secretary Arthur Balfour suppressed this report and substituted his own with the comment that Slade's argument was "imperialistic." The Cabinet was divided. Prime Minister David Lloyd George was clearly in favor of an "imperialistic" oil policy and in the run-up to Versailles persuaded French premier Clemenceau to cede to Britain all of Mosul *vilayet*, not just the oil-bearing parts. By 1919 the Admiralty was, in any case, convinced that Persian and Mesopotamian oil was the solution to its fuel supply problems. An unstable Iraqi state was thus "invented" to cover the oil regions of Mosul, Baghdad, and Basra *vilayets* and the British imposed on the area a Sunni ruler of the majority Shia population, King Faisal. Given that they had anointed him the British thought they could control him, but Faisal turned out to have a mind of his own, and Iraq achieved independence, albeit with the retention of a strong British military presence, in 1932 (Townshend 2010, 519–522).

It has been in business history that *The Oil War* has been most recognized in the modern literature. The most important work is Daniel Yergin's *The Prize: the Epic Quest for Oil, Money and Power* (1991). Though useful, it betrays its origins and, because it focuses on the personalities of the businessmen who ran the oil companies as they developed, it lacks any real sense of the geopolitics of the struggle to control the Middle East and the complex history of that struggle, despite paying lip-service to it, or of the underlying geological science and petroleum engineering technology needed to find and extract oil. Yergin notes only some of the accounts published in the 1920s and 1930s, referring to neither of the academic books by Earle (1923) and Mohr (1926), although he does refer to Earle's important 1924 article on the Turkish Petroleum Company. Yergin refers to only four of the eight polemical accounts in his bibliography (Delaisi 1922, Denny 1928, Fischer 1926, and L'Espagnol de la Tramerye 1924).

The most significant recent work from the perspective of geopolitics is David Harvey's *The New Imperialism* (2003). Harvey expresses the modern importance of the Middle East in classical geopolitical terms, invoking the ghost of Halford Mackinder when he states "whoever controls the Middle East controls the global oil spigot and whoever controls the global oil spigot can control the global economy, at least for the near future" (Harvey 2003, 19). The evidence in the British National Archives confirms that, at the level of policy, Britain, Germany, America, and to an extent, France, understood the central importance of the Middle Eastern "oil spigot" well before WWI broke out. Just before WWI broke out APOC and *Deutsche Bank* put together a deal to create the Turkish Petroleum Company (TPC) to replace the Berlin-Baghdad Railway concession. Once WWI was over the British took the German TPC shares as a war prize which resulted in a row with France and America, first for control of Germany's TPC shares, then more generally for access to the region's oil. By the 1920s most observers could see clearly the reality of Mohr's *Oil War* and the centrality of the Middle Eastern "oil spigot" to a world in which the American supply seemed to be faltering and the Russian supply to have receded behind closed doors.

Oil Companies and *The Oil War*

Three companies were involved in *The Oil War* of the 1920s and 1930s, one American, one British, and one that had become British. Although the American Government certainly intervened to support Standard Oil in its struggle, it did not control it. Of the two British companies one, APOC, the progenitor of today's British Petroleum (BP), was controlled by the British Government from 1914 on. The other began as a Dutch company, Royal Dutch, but morphed slowly into a British one after it merged with the British company, Shell. In the normal pattern of the form of British capitalism that had emerged in the late 1800s to deal with the increasingly complex relationship between the commercial interests of transportation, communications, and energy companies and the geostrategic and geopolitical interests of the state, Royal Dutch Shell (Shell) accommodated the British Government, so that even if it was never "government owned" it generally did what the British Government wanted it to, as did Pender's cable companies.

To the contemporary observers of events in the mid-1920s it seemed obvious that the building struggle over oil between America and Britain was a very serious one. Those contemporary observers tended to characterize the struggle in human terms as between two powerful personalities, John D. Rockefeller of Standard Oil and Henri Deterding of Shell: "Mr. Deterding is Napoleonic in boldness, and Cromwellian in depth" and is now "the strongest personality in the oil world." Further, Deterding was supported by such men as the Rothschilds and Calouste Gulbenkian, the "Talleyrand of Oil" (L'Espagnol de la Tramerye 1924, 84). Although there was no question that the various Standard Oil companies still contributed from their American production the lion's share of the world's supply of petroleum in the early 1920s, those who held the "oil exhaustion thesis" believed that America's resources were nearly tapped out and that Shell had accumulated drilling rights to the lion's share of potential resources elsewhere in the world. It was also increasingly clear by the early 1920s that there were now four real or potential major oil regions:

1. the Gulf of Mexico, where American interests predominated, although interest in Mexico was dialed back after the Mexican Revolution of 1910, and by 1939 Venezuela had become the region's main exporter (Bromley 1991, 95);
2. the Trans-Caspian region around Baku in what had been Imperial Russia, after the end of the Russian Civil War (1917–1923) securely within the grasp of Russia's new Communist State;

3. Indonesia, where Royal Dutch had started out and which Shell continued to control; and
4. The Middle Eastern oil region, Persia and Mesopotamia.

Of the four major known oil regions of the early 1920s the Middle Eastern region thus logically became the major arena of struggle between American and British oil interests since it was the only one into which American interests could expand. Despite attempts by Lenin in his "New Economic Policy" to reach out to the West once the Revolution seemed secure, neither of the two great western companies got much traction in the new USSR, not least because Stalin, who had started out in 1901 as the chief socialist organizer in Batum, the main Black Sea port for the Baku oilfields, was opposed (Yergin 1991, 130, 237–240). Shell continued to dominate Indonesia. Despite some inroads into the Gulf of Mexico, mainly Mexico, by Shell, which also penetrated California, American companies continued to dominate the output of the region. It was, however, becoming clear by the mid-1920s that "Standard Oil" was no longer synonymous with "American oil." A new breed of "wildcatters" had arisen in Texas, where the geological conditions in which oil had been formed were very different than those in Western Ohio where Standard Oil had its start. Compounding the problem was the fact that Standard Oil, which had started as largely a distribution and sales company, began to move into production in the 1920s under the leadership of its President Walter Teagle, but took very conservative geological advice. Meanwhile, the developing city of Houston, without help from Standard Oil, was busy making itself the center of what would become the American pipeline network (Melosi & Pratt 2014). Within the larger British–American struggle a further struggle thus arose among American interests between Standard Oil, now headquartered in New Jersey, and the aggressive new companies emerging in places like California, headquartered in Los Angeles (Elkind 2014), and Texas, headquartered in Houston. What was initially a domestic American struggle driven by regional rivalries and different technological understandings driven by petroleum geology and engineering would then spill over into the Middle East in the 1930s.

The origins of Standard Oil lay in the development by John D. Rockefeller of the oil region of western Ohio and the incorporation of Standard Oil of Ohio as a company in 1870. Standard initially focused on lubricants and the medium weight, not very volatile distillates, mainly kerosene for lamps. The seemingly useless heavy oils and the much more dangerous volatile distillates were usually discarded, although the volatile distillates found some use as cleaning fluids. After 1900 or so internal combustion engines, mostly operating on the Otto cycle, started to use the volatile distillates, and ships started to burn the heavy oils to make steam. Distillation technology improved and better cracking allowed higher yields of fuel for Otto cycle engines from the same barrel of crude. Standard Oil recognized early that the production of oil was in many ways less reliably profitable than its transportation to market and retail sales. After its early use of whiskey barrels from the Appalachian mountains of western Ohio to transport oil via the railroad system, Standard Oil started to invest in pipelines. Although Standard Oil was broken up into a series of smaller companies in 1911 by the anti-trust movement of the Progressive Era the combined companies retained almost total control of American sales and distribution. By 1920, however, Walter Teagle, who had taken over as President of Standard Oil of New Jersey in 1917 and was part of the new breed in American business of professional managers, commented that the "policy of the Standard Oil Company is to be interested in every producing area no matter in what country it is situated" (Yergin 1991, 199). In that regard Standard Oil was now beginning to follow Shell's lead.

The origins of Royal Dutch Shell are much more complicated. Royal Dutch was founded in 1890 with concessions from the Dutch Government in Sumatra to exploit the Indonesian oil reserves. Since it needed to move oil to market in Europe but had few ships Royal Dutch found itself involved with a British trading company in the Far East owned by Marcus Samuel. Samuel's company was initially set up to trade in exotic shells, from which the company took the name Shell, but it found shipping Royal Dutch oil from Borneo far more profitable, and Samuel invested in a rapidly growing fleet of tankers.

In 1898 Standard Oil attempted a hostile takeover of Royal Dutch, without success, and its use of predatory pricing in markets where the two companies competed left a bad taste in Dutch mouths. In 1902 Royal Dutch therefore negotiated a merger with Shell to form what became known as Royal Dutch Shell or, more generally, just Shell. A third of the capital for the 1902 merger came from Royal Dutch, a third from Shell Transport, and a third from the Rothschild banking family. In 1907 they transferred their assets to two companies: one, the *Bataafsche Petroleum Maatschappij* (Batavian Oil Company), dealt with production; the other, Anglo-Saxon Petroleum, dealt with transport and sale (L'Espagnol de la Tramerye 1924, 59–63), but the company was still Shell.

At Shell Deterding pursued a series of very different policies than Rockefeller, four in particular.

1. "the most skilful part of the policy of the *Royal Dutch* was to *establish itself anywhere there was any oil*, while the *Standard* confined itself almost exclusively to America. This was a great mistake. . . . It became an error from the day on which important oil deposits were discovered in other parts of the world."

2. when undercut by Standard in the market for lamp-oil in China and forced into substantial losses, Deterding made up his losses by shipping oil to America, where Standard Oil maintained high prices.
3. Shell's origins had its production well away from its markets and it needed ships to get its oil to market. It was thus was obsessed with shipping costs and sought production sources to minimize those costs. Once he had started to market oil in America Deterding therefore sought American wells, first in Oklahoma and then in California. In 1918 he also acquired British owned Mexican Eagle, which had hitherto sold its output to Standard Oil.
4. Deterding followed "the commercial example of Great Britain. *He has stations at all the strategic points in the world*," such as at both ends of the Panama and Suez Canals. By the early 1920s Standard Oil produced "no more than 17 percent. of the oil of the United States—it is true that it has always disdained the extraction of petroleum—but it refines only 49 percent., which is a much more serious matter" (L'Espagnol de la Tramerye 1924, 65–83, quotes from 79, 72, 81–82, emphasis in original).

One problem shared by both Standard Oil and Shell was that an older breed of geologists, both American and British, usually trained in traditional Schools of Mines, advised the older companies. These geologists knew that oil was to be found in sandstone and limestone deposits, in anticlines or trapped by faults. They did not initially understand how wildcatters along the Texas and Louisiana coasts of the Gulf of Mexico found oil in salt domes. Slowing their understanding, the first salt dome strike, in 1901 at Spindletop east of Houston, seemed depleted by 1920 and was not revitalized by deeper drilling and more careful spacing of wells until the mid-1920s. There was thus little or no scientific understanding of salt dome oil deposits along the Gulf until much more research was done on the region's geology by the region's universities. The University of Texas founded its Bureau of Economic Geology in 1909, but it took until the 1930s before well-trained geologists used to the geology of the Gulf Coast started to appear. The Gulf Coast's two major Morrill Act Colleges, the Texas Agricultural and Mechanical College (now Texas A&M University) and Louisiana State University were even further behind the curve, establishing their programs in oil geology and petroleum engineering only in the late 1930s. Spindletop, its production in decline by 1918, looked like a fluke by 1920, the Texas wildcatters seemed out on a limb, and the various companies of Standard Oil seemed likely to remain in firm control of American resources for the foreseeable future. The exception was SOCAL, which, founded in the middle of the break-up of its parent Standard Oil company by anti-trust legislation, had gone its own way and forged significant early links with the Texas Fuel Company, later Texaco, which got its start at Spindletop in 1901and understood salt dome deposits at the empirical level.

BRITAIN'S GEOPOLITICAL AND GEOSTRATEGIC CONCERNS IN THE MIDDLE EAST, A COMPLEX HISTORY

Britain's geopolitical and geostrategic concerns in the Middle East did not, of course, begin with oil in the early 1900s: they started in the late 1700s. When Britain came into control of the Indian subcontinent at the end of the Seven Years War in 1763, it had to confront what all rulers of India have had to confront, the threat of attack from Central Asia through the Khyber Pass. This led Britain into what it came to call the "Great Game," the long confrontation with Russia over Afghanistan that the Russians called "The Tournament of Shadows" (Meyer & Brisac 1999). The problem was that Britain's lines of communication to India, which rapidly became by far the most valuable part of the British Empire, were long drawn out, and had to pass either around Africa or through the Mediterranean and the Middle East. The problem of secure passage around Africa was dealt with during the Napoleonic Wars by the annexation of South Africa in 1806 to serve as a way station for ships bound from the southern Atlantic into the Indian Ocean.

British control of the shorter, easier route, by sea through the Mediterranean, then overland to the Persian Gulf, then by sea across the Indian Ocean, was a much thornier and more complex problem. Britain had acquired Gibraltar in 1713 in the settlement of the War of Spanish Succession, thus controlling the crucial maritime entryway into the Mediterranean from the Atlantic. Control of the land passage through the Middle East and the exits into the Gulf was much more difficult but became a critical component of British geostrategy in the early 1800s. In the event, much of that geostrategy was initially left to a private company, the British East India Company (BEIC), albeit one in which the elected government played an increasingly significant role. The BEIC grew substantially after the Napoleonic Wars were over, and more and more women found their way to India—"as late as 1770 there had been only 39 British women in Bombay." The women particularly complained that the sea route to India via South Africa was much too long, taking between five and six months, and much too uncomfortable: "living conditions were cramped, the ships verminous." The emergence in the 1830s of steamships that could handle shorter routes, such as through the Mediterranean from Britain to Egypt and across the Indian Ocean, cut the voyage to two months. In 1838 the BEIC acquired Aden at the Gulf end of the

Red Sea as a coaling station for its growing steamship service in the Indian Ocean (Gardner 1971, 179–185, quotes from 185, 180).

A British government policy of shoring up the decaying Ottoman Empire logically followed, for two ends: it facilitated British movement through the Ottoman territories between the eastern Mediterranean and the Persian Gulf; and it ensured that, with a friendly power in control of the Dardanelles, no Russian fleet could exit the Black Sea into the Mediterranean whereas a British fleet could, if necessary, enter the Black Sea and interdict Russian expansionism via the Crimea. In the Crimean War (1853–1856) British and French naval forces did exactly this, using the Russian defeat of Ottoman naval forces as a *casus belli* and entering the Black Sea. In the Treaty of Paris of 1856 that settled the Crimean War the Russians agreed not to establish any naval or military presence on the Black Sea coast (Gourdon 1857), thus ensuring that no Russian fleet would be able to challenge British control of the Mediterranean.

By the 1850s British concerns about the long sea passage to India around Africa meant that British interests were becoming heavily invested in railroads that crossed Egypt from the Mediterranean to the Red Sea (Ayrton 1857). The British initially opposed France's building of the Suez Canal, completed in 1869, but once it proved successful they quickly reversed course and, when the Egyptian shares in the canal became available in 1875, then Tory Prime Minister Benjamin Disraeli purchased them, albeit with considerable criticism from the leader of the Liberal opposition, William Ewart Gladstone (Lee 2005). If British geopolitical thought was concerned with the passage to India before the acquisition of the canal shares, it became totally obsessed with it thereafter. Here, British domestic politics intruded dangerously to muddle its previously coherent Middle Eastern geostrategy. Abandoning Disraeli's realist policies, Gladstone's Liberal Government of 1880–1885 reacted in classic idealist fashion against Ottoman corruption and "withdrew its protection and influence from Constantinople" (Fromkin 1989, 30).

With British protection withdrawn the Turks, who had no love for the Russians, turned increasingly for support to Bismarck's Germany. By the end of the 1800s German geopolitical designs were clashing with increasing regularity with those of Britain, and the Middle East was no exception. As early as 1889 George Curzon asked "are we content to see a naval station within a few days' sail of Kurrachi and to contemplate a hostile squadron battering Bombay?" (Curzon 1889, 378). Curzon had just returned from Russia and was concerned about Russian railroads thrusting toward the Persian Gulf. When Curzon became Viceroy of India (1899–1905) threats to Britain's control of the Persian Gulf and the route to India concerned him even more. In 1903 Germany obtained a concession from the Ottomans to build the Berlin to Baghdad Railway. This clearly offered the possibility of a German railhead supplying and maintaining a naval base on the Persian Gulf. A German base on the Gulf would have threatened Britain's imperial lifeline to India as clearly as the Russian base Curzon opposed in 1889. In the light of such behavior Halford Mackinder's famous recasting of global geostrategy in favor of powerful continental states using railroads and against maritime power in his 1904 paper to the Royal Geographical Society (RGS) could have done nothing to reassure the Committee for Imperial Defence, which had been established to worry about such things that same year (Mackinder 1904; Woods 1917, 33–34; Fromkin 1989, 31). Imperial Germany obviously posed an increasingly serious threat to the main artery of the British Empire.

The messy geopolitics of the Middle East were not helped by uncertain British domestic politics or by the ongoing and increasingly serious collapse of Ottoman power. The costs of the Crimean War pushed the Ottoman Empire into increasingly uncontrolled debt. Poor fiscal policy resulted in bankruptcy in 1875 and the need to be bailed out, which let in more foreign bankers and interests (Clay 2000). Internal revolts developed and the European powers saw the chance to chip away at the Empire territorially. In 1875 Serbia and several other Balkan *vilayets* declared their independence. In 1878 Britain acquired Cyprus, and in 1882 de facto annexed Egypt. In fact, Egypt had never been firmly held by the Ottomans, was briefly conquered by France in 1798, and only partially restored to Ottoman control when the French were ousted from the region by the British following the Battle of the Nile later that year. From then on, British interests in Egypt increased steadily for three reasons. First, during the "cotton famine" of the American Civil War British cotton textile mills had found in Egypt an important source of raw cotton. It quickly became the second main supplier after the American South and became even more important as the boll weevil began to ravage American production after 1892, especially of the sort of long staple, high quality cottons that grew well in Egypt. By the first decade of the 1900s Britain was beginning to look to the Ottoman area of the Sudan for even more cotton, believing it might equal the America South in its potential, as it would later look even less successfully to Mesopotamia (Hugill forthcoming 2020). Second, the importance of the Middle Eastern route to India continued to deepen. Third, the increasing collapse of the Ottoman Empire was dangerous to British interests as other European states seized opportunities. France occupied Tunisia in 1881 and in 1912 Italy occupied neighboring Libya. After the Bosnian Crisis of 1908 most of the rest of the Balkan *vilayets* revolted. Within the Empire the Young Turks seized power in 1908 and overthrew the Ottoman sultanate. The Ottoman Empire was effectively ended by that point, although the fiction that it existed continued until 1922. To protect its cotton sources, its route to India, its oil and its geosecurity, Britain had little choice but to become increasingly involved in the politics of the region.

The origins of the complex struggle between Rockefeller and Deterding for control over Middle Eastern oil, Mohr's *Oil War* (1926), lie mainly in the mess left as the Ottoman Empire collapsed into rubble and the struggle to pick up the pieces intensified. Before WWI the Ottomans and their Turkish successors, effectively bankrupt and in the grip of massive corruption, had leased, or promised to lease, various concessions in the collapsing Empire for railroads and oil to American, British, French, and German interests without rhyme or reason. Many concessions were at odds with each other. This internal, corruption driven struggle was compounded by the need at the end of WWI for separate treaties with the different combatant powers, the Treaty of Versailles dealing only with Germany. The rise of Turkish nationalism under the Young Turks from 1908 on did not improve matters. The Young Turks questioned the validity of all Ottoman concessions and led the resistance to the carving up of former Ottoman possessions that the victorious allies had sought to impose on them from the Sykes-Picot (-Sazonov) Agreement of 1916 through the draft Treaty of Sevres of 1920.

In any case, the triumph of the Young Turks in the Turkish War of Independence led to the annulment of the Treaty of Sevres and its replacement in 1923 by the Treaty of Lausanne, a treaty in which the Turks were strongly supported by America. At Versailles in 1919 the British claimed the German Ottoman oil concession, restructured as TPC in 1914, as a war prize, at which point an Anglo-French squabble broke out with the French demanding their share of the spoils. No sooner than that was settled at San Remo than the Americans chimed in, demanding a share based on their claimed prewar Ottoman oil concession. Following the Treaty of Lausanne the new Turkish Government ratified the prewar concession that American interests claimed had been awarded to America, but not those claimed by Britain or the concession that actually had been awarded to Germany and taken over by Britain. The refusal of the American Senate to ratify the Treaty of Lausanne reduced the impact of Turkey's behavior somewhat, but it gave Standard Oil a strong negotiating position against the British in Iraq. The British diplomatic correspondence on this is voluminous, irate, and frustrated.

In 1907, as part of the restructuring of its geostrategy to cope with the rise of Imperial Germany, Britain had settled its old arguments with Russia and the two powers agreed to partition formerly powerful and independent Persia into British, Russian, and neutral zones of influence (Fromkin 1989, 31). When WWI broke out in 1914 Britain used Indian troops to occupy the three Ottoman *vilayets* that made up Mesopotamia: Mosul, Baghdad, and Basra, although the initial focus was on the two southern *vilayets* and not until late 1918 was Mosul invested. After 1915 Britain embarked on a policy, personified by the work of Lawrence of Arabia, designed to encourage the Arab populations of the region to seek independence and revolt against Ottoman rule. By the early 1920s, along with a host of lesser states, five major successor states had emerged in the Middle East, four from the rubble of the Ottoman Empire: Egypt, Iraq (formerly Mesopotamia), Saudi Arabia (formerly Arabia and the Hejaz), and Turkey; the fifth state was Iran (formerly independent Persia). The final factor complicating British geostrategy in the region was the desperate need in the early part of WWI to keep the main Entente powers, France and Imperial Russia, satisfied. This led Britain to negotiate the secret Sykes-Picot (-Sazonov) agreement of 1916 on how the former Ottoman *vilayets* would be assigned to the victors when the war was over.

Oil and (the British) Empire

Europeans were well aware by the late 1800s that there was oil in the region from the Red Sea to the Caspian. Members of the RGS regularly reported such evidence in the Society's proceedings and journals, and in this time frame the RGS was effectively a semiofficial arm of the British Foreign Office. In 1886 Colonel Stewart, after his posting to the Persian/Afghan border, reported the regular use of oil from the Baku region of Russia to fuel Caspian Sea steamers and Central Asian railroad locomotives. That same year Colonel Ardagh reported on "The Red Sea Petroleum Deposits" (Stewart 1886; Ardagh 1886).

Two lines of geographical reasoning about potential Middle Eastern oil sources developed, one focused on Persia, the other on the Ottoman Empire. In 1894 Captain Maunsell read a paper to RGS on "Kurdistan," describing substantial oil deposits in the Ottoman *vilayet* of Mosul. Maunsell's 1897 paper, "The Mesopotamian Petroleum Field," mapped twelve "petroleum springs" in Mosul and Baghdad *vilayets*, suggested that "the present political troubles" might encourage the Ottomans to grant concessions, and noted that "a line of railway from Baghdad through Kifri, Tuz Kurmatli, and Kirkuk to Mosul could bring the produce of these places to the river [Tigris]. If such a line were extended from Mosul to the Mediterranean, communication towards both seas would be complete" (Maunsell 1897, 532). In 1901 a German technical commission described Mesopotamia as a "lake of petroleum" which it would be advisable to develop to break the stranglehold on world supplies of the American company, Standard Oil (Earle 1923, 15). In 1903 the Germans negotiated a concession from the Ottomans to build the Berlin to Baghdad Railway with a ninety-nine year lease on mineral rights in a twenty-kilometer swath on either side of the line in Ottoman territory.

In 1907 Britain and Russia agreed to partition Persia into a British zone of influence, a Russian zone, largely excluded from the D'Arcy concession, and a Neutral zone. By early 1908 D'Arcy and his backers had sunk over £500,000 ($66 million in 2017), into his concession, but not until mid-1908 did his company finally strike oil. The problems in Ottoman territory were political, and two forces prevented any concession in Ottoman territory being worked before WWI. One was the corrupt nature of Ottoman and Turkish politics. The other was the involvement of the American company, Standard Oil of New Jersey, via the shadowy figure of Admiral Colby Chester, who seems to have effectively helped derail all German and British attempts to negotiate workable concessions.

In a "Very Secret" Treasury memorandum of 1916 entitled "The Future Control of Oil-Supplies" the British Board of Trade (BOT) expressed considerable concern about oil, noting that while America, Russia, and Germany (in Hungary and Rumania) had ample oil, "the British Empire has to rely almost entirely on supplies drawn from foreign countries." The BOT noted that "an 'all-British' company or combination would be inadequate . . . [since] the nucleus of such a company would . . . have to be Mexico, and the political position and conditions of that country and the quality of the oil produced there render it unsuitable." The BOT then laid out for the British Cabinet the case for an "Imperial Oil Company" built around Shell, but without including the APOC. The report concluded by noting "the proposed arrangement . . . was only provisionally obtained by long and difficult negotiations with the Shell Company" and that the arrangement should not be made subject to further negotiation by the Cabinet for political reasons. In 1916 Shell's considerable holdings in the Russian oilfields (46 percent of its 1915 output of 3.3 million tons) were still accessible, as well as its traditional holdings in Asia, mostly Indonesia (24 percent), and its recently acquired holdings in the Americas, mostly in California (27 percent) (BNA/T 161/738, August 12, 1916). Russia's exit from the war in 1917 devastated this rosy picture.

Despite the intense chaos in Russia caused by military collapse and the series of revolutions and Civil War that followed the British did not immediately give up on Russian oil. In late 1917 a hand-picked force of troops from the Mesopotamian theater was assembled near Baghdad, its goal "to get through to the Caucasus and help the Armenians, Georgians, and Azerbaijanis, evict their Turkish masters and form a buffer state between revolt-torn Russia and Turkey." Under the terms of the Treaty of Brest-Litovsk the Russians had been forced to agree to "supply Baku oil to Germany while the Peace of Bucharest gave the Central Powers control over Rumania's oil wells" (Bromley 1991, 93). In chaotic conditions a small British force under Major-General Lionel Dunsterville pushed through to Baku by early August 1918, its "declared object . . . to prevent the Turks establishing a base on the Caspian, to protect the Armenians, and—probably the most important reason of all—to lay hold of Baku's oil supply" (Barker 1967, 389–393). Townshend suggests, however, that "Britain's grand strategy now had two new objectives: to prevent the Germans from accessing the oil and mineral resources in the Caucasus and to turn Persia into an effective British ally" (Townshend 2010, 419). Dunsterville's orders, which he never seems to have even made the slightest attempt to carry out, were to do as much damage to the Baku oilfields and oil infrastructure as possible. Although Dunsterville was a fluent Russian speaker and believed he could rally local resistance to the Turks the only disciplined troops in the area were the Red Army, which refused any help to Dunsterville's small force and abandoned the city. Unable to rally the chaotic remnants of the resistance to the Turks, Dunsterville had no choice but to withdraw: "farce . . . [became] . . . tragedy when Baku fell on 15 September and the irregular forces pillaged and burned it. The entire population of the Armenian suburb was massacred" (Townshend 2010, 426–431, quotation from 430).

This was by no means the end of it. The Turks in turn were pushed back and the Russian Civil War raged in the region. Denikin's "White Russians" held onto the oil regions during the early part of the war, and he proposed to ensure oil shipments from Baku to the West from the Black Sea port of Batum for a fee of £1 per ton "to strengthen his financial position" (T 161/738, November 19, 1919). British forces pushed hard across the Black Sea from 1918 to 1920 to keep Denikin supplied, but after the revitalization of the Red Army by Leon Trotsky and a series of defeats in November and December of 1919, Denikin gave up the uneven struggle and went into exile in April 1920. The British evacuated the remaining White forces to Constantinople that November (Mawdsley 1987). British naval activity in the Black Sea had ensured that the head of the pipeline at Batum remained accessible, but with the White Russian army defeated and the oilfields at Baku now securely under Soviet control, the British had little choice but to withdraw.

In the aftermath of WWI the Treasury thus had little option but to return to the model of a British controlled Shell company, but one now clearly focused on the Middle Eastern oil spigot and which thus had to be merged with the APOC. Shell asked both that "the new Mesopotamian State, whatever it may be, shall recognise their Turkish Government concession and that the Shell holding shall be increased from 25 to 34%" and that "clauses protecting Dutch shareholders" from British taxes be agreed to (T 161/738, January 5, 1919). The British Foreign Office in reply expressed concerns about "the validity of existing concessions in the Middle East" that turned out to be all too prescient (T 161/738, n.d, but February or March, 1919). Whatever the Foreign Office's concerns, from this point on British geostrategy in the Middle East was as much about access to oil as access to India.

The Rise of Race and Nationalism in the Middle East, and the Increasing Role of Oil

As the Ottoman Empire disintegrated three powerful new forces emerged in the Middle East: race, nationalism, and oil, where they sat uneasily with the preexisting force of religion. Racialized geopolitics were, as Agnew so clearly shows, one of the dangerous new trends of the late 1800s (Agnew 2003). In Empires such as the Ottoman, ethnicity had been ignored or suppressed, but as such Empires broke up previously minor ethnic differences became dangerously elevated by the theories of racial distinctiveness that bloomed in the late Nineteenth Century. As Ferguson makes appallingly clear, such forces would then proceed to rip apart the fabric of global culture throughout a short but devastatingly genocidal twentieth century (Ferguson 2006). One of the first, if not the worst, of many such rips was the genocide perpetrated against up to 1.5 million Armenians by the Turks as the Ottoman Empire disintegrated and the ideal of a Turkish, Muslim nation elevated ethnic and religious difference into a drive for racial purity that resulted in the mass murder of Armenian Christians as well as a host of others—Coptic Christians, Greeks, Cypriots and the like (Bloxham 2005; Fisk 2005, 316–355; Oren 2007, 293). Other newly important racial divides also ripped at the former Ottoman possessions. Arabs predominated in the Arabian peninsula, but modern Iran was never Arabized and Iraq only partially so, with ethnically distinct Kurds remaining important to the north. Shiite Muslims fought Sunni Muslims, especially in Iraq. Linguistically and ethnically non-Arab Berbers dominated western North Africa.

In Egypt and Turkey religion and, to a lesser extent, race, was relatively suppressed by the new nationalism, but neither state had oil. Although Ibn Saud was able to forcibly unify the relatively small but ethnically homogenous Arab population of much of the Arabian Peninsula, including the Hejaz, under an austere form of Islam, Wahhabism, oil was not known there until the early 1930s. Iran's version of Islam was geographically distinct, since it was in Iran that the Shia form of Islam had taken strongest hold. In addition to being homogenously Shiite, Iran was relatively homogenous ethnically, Persian not Arab, and known from the early 1900s to have oil. This left Iraq to emerge as a state that was known to have oil by the early 1900s, but in which three ethnically and religiously disparate former Ottoman *vilayets* were forcibly united at Versailles to suit British geostrategic interests in oil, cotton, and control of the route to India. The British, especially Winston Churchill, believed that under British tutelage—and against most of the evidence—nationalism could be used to create an Iraqi state (Catherwood 2004). To the south and east, in Basra *vilayet*, the population was ethnically Arab but the Shia form of Islam had diffused across the border from Iran. To the north and west, in Mosul *vilayet*, the population was ethnically Kurdish but practiced the dominant form of Islam, Sunni. Between Basra and Mosul *vilayets* was Baghdad *vilayet*, ethnically Arab and religiously Sunni.

The American role in the Developing *Oil War*

Although Anglo-American relations were improving steadily in the early 1900s by the 1920s "Anglo-American [oil] competition centered on . . . the Middle East and Britain resigned herself to effective exclusion from Latin America" (Bromley 1991, 95). This was a consequence of the fact that British attempts to exploit Mexican oil had been prevented by America.

> During 1912 and 1913 Britain [via Mexican Eagle] had succeeded in obtaining the grant of important oil concessions from the various Governments in Costa Rica, Colombia, Venezuela, and even Ecuador. This would have given it, in point of fact, a monopoly of the supply of oil to all the shipping which passed through the Panama Canal. Were the United States going to let slip the absolute control of this world-highway which had cost them so dearly in treasure and in toil? There followed . . . a demand by the Washington Government for the cancellation of these concessions as an infringement of the Monroe Doctrine. (Delaisi 1922, 24)

Kent notes the tenuous nature of the mineral rights promised to the Berlin to Baghdad Railway Company. They were hedged around with many conditions that were never met: the German company's own internal correspondence admits these failures. Such failures opened possibilities for others, notably D'Arcy's APOC, which sought and received many promises of concessions from the Ottomans, promises correctly regarded as dubious by the British Foreign Office at the time (Kent 1976, 29). Between 1908 and 1911 the Americans also saw the possibilities and one Anglo-American and one American company applied for concessions. The Anglo-American firm was J. G. White & Company and was fronted by Bruce Glasgow, who the British Foreign Office regarded as a rather shady character. The American company was fronted by another shady character, Admiral Colby Chester. Chester's group had "strong commercial backing and the support of the State Department" and effectively removed the Glasgow group by offering better terms (Kent 1976, 26–28, quote from 27). Kent makes the point that German and British interests did not sit idly by and that "their (especially the German) opposition contributed largely to its failure. . . . As one way of disquieting the Turks about the project, they carefully fostered the rumour that the Standard Oil, the great oil bogey, was behind it" (Kent 1976, 28). DeNovo argues, however, that the Germans did actually believe that Chester was attempting to cut into the German railroad concessions on behalf of American railroad interests, though really acting for Standard Oil so that it could prevent the development of competition. The Head of the Deutsche Bank in Constantinople, Arthur von Gwinner, paid a visit to the American ambassador in Berlin, David Hill, in June 1913 to make precisely that point (DeNovo 1963, 70–71). Given the behavior of Standard Oil after WWI DeNovo's account seems likely. Chester's grasp on the problems of exploiting the region's oil also seems to have been more tenuous than that of the Germans and British. Whereas Chester only sought railroad concessions, and only in part of Mosul *vilayet*, D'Arcy's APOC sought "the sole right to build and work pipelines in the vilayet . . . there was no use in [Chester's] obtaining a concession to produce oil that could not be transported to its markets" (Kent 1976, 28).

Chester remains a shadowy, self-promoting, rather unscrupulous figure in the historical record—for example, in his 1922 article in *Current History* he claimed that his "concessions" predated the German and British concessions when the only formal concession ever granted by the Ottomans to anyone was to the Germans in 1903. Chester never received any formal concession before 1923, and that was not from a defunct Ottoman Empire but from a Turkish state that no longer controlled Mesopotamia. Even so, *Time* magazine reported on November 19, 1923 that Chester had rights to "the famous Mosul and other oilfields," reporting at the same time substantial dissension among the directors of the company backing him. Earle, who knew the region well and traveled extensively there in the late teens, comments that the Chester concession of 1923 from Turkey simply replicated one Ottoman and one Turkish concession: the Baghdad Railway concession given to German interests in 1903 and the Black Sea Railway concession given to French interests in 1913 (Earle 1923, 333–335, 339–345). Chester's "concessions" existed mostly in his mind, and "a common feature of this bizarre enterprise throughout its strange history was the poor business planning, the flimsy financing" (DeNovo 1963, 228).

In retrospect one must look at the Chester "concessions" after WWI in the light of two sets of forces. On the one hand, and supportive of the "oil exhaustion" thesis, DeNovo notes that "the postwar need for petroleum in the United States provided a favorable environment for the revival of Admiral Colby M. Chester's ambitious plans for economic development in Asia Minor" (DeNovo 1963, 210). Interestingly, whereas the state department seems to have wanted little to do with him in the post-WWI environment, the Navy became enthusiastic, clearly worrying about a controlled source of bunker fuel outside America's boundaries. Chester seems to have become self-delusional by this point, however. The petroleum expert advising the state department, Arthur C. Millspaugh, noted that "the Admiral is now an old man, the Turkish project has become an obsession with him, and he seems to be oblivious to recent changes in the Near East and to the fact that Near Eastern diplomacy centers now in London and Paris rather than in Constantinople" (quoted in DeNovo 1963, 215). The second set of forces revolves around the international condemnation of Turkey's role in the Armenian genocide. Chester very publicly supported Turkey's claim that the genocide was a fiction (Chester 1922). Chester's comments are still cited on pro-Turkish websites to this day as "evidence" that the Turks did not commit the Armenian genocide (Chester on Armenia Online). But Chester is almost absent from the contemporary accounts. Earle makes only the briefest mention of Chester's presence in the prewar Ottoman Empire (Earle 1923, 15) and then only by reference to another article in *Current History* (Woodhouse 1922). Woodhouse's account of 1922 seems the most coherent one of Chester's activities in the Ottoman Empire and Turkey before WWI, but was clearly written after the fact.

Earle's thorough account of the formation of the TPC focuses on one area of Chester's involvement, his likely role in preventing the 1914 agreement being signed. As negotiations were reaching a conclusion in the summer of 1914 the Turkish Grand Vizier inserted a clause that the company would have to "indemnify any persons who could establish a claim to participation in the concession," a clause so nebulous that the British and Germans dug in their heels and refused to sign, but which Earle footnotes may have referred to "certain unsubstantiated claims of the American Admiral Chester or to the personal property rights of the Sultans" (Earle 1924, 270–271). Negotiations with the Turks over concessions were always fraught with such twists and turns, and this insertion prevented the signing of the TPC concession before WWI broke out and Turkey lined up with Germany.

The British and German Roles in the Developing *Oil War*

In 1912 Churchill, as first lord of the admiralty, committed to a radical new class of oil-fueled battleships, the *Queen Elizabeths*, heavily armored and designed to outgun and outperform any other battleships in the world. That same year he appointed the Royal Commission on Oil Fuel and Engines, chaired by Admiral Jackie Fisher, which reported favorably on oil as a naval fuel but expressed concern about the security of supplies. In 1913 Churchill announced another major naval expansion and a contract with Anglo-Mexican Petroleum Products to fuel them: "this . . . crystallised the American belief that British policy in Latin America was largely directed by oil" (Venn 1986, 21). President Wilson became increasingly hostile to British oil concessions in Mexico and American intervention in the Mexican Revolution intensified. There was little hope of the Anglo-Mexican contract bearing fruit and Britain turned to other sources.

In 1913 The Royal Commission on Oil Fuel and Engines established the Slade Commission under the leadership of Rear-Admiral Edmund Slade. The Slade Commission sent experts such as John Cadman to examine the APOC's operations in Persia. Cadman's background was in coal mining in northern England, but he spent the years 1904 to 1908 in Trinidad, where he organized the mines and petroleum department. As professor of mining at Birmingham University from 1910 on he organized the first department of petroleum technology in the British Empire (Smith 1941). After the positive report of the Slade Commission on Persia's oil in 1914 Churchill, no longer willing to rely on American supplies in the event of war, persuaded the British government that same year to take a controlling share in the APOC (Massey 1991, 782–785; Yergin 1991, 139–141). As Cadman noted in the Slade Commission report, in Persia oil was found in "well-defined anti-clines of tertiary beds of a clearly established nature, namely the so-called Asmari Limestone" (BNA/CAB/24/202, 241). Cadman became director of the Petroleum Executive of the British Government during WWI, moving thereafter to the APOC, where he became chairman in 1927. Cadman was also very clear in the Slade Commission report that the oil-bearing Asmari Limestone continued into Iraq.

When they obtained the concession to build it in 1903 the Germans took care to route the Berlin to Baghdad Railway through the major oil regions around Mosul and Baghdad, although in the event they were never able to either develop the oilfields or finish the railroad, which depended on Turkish mileage payments and was therefore chronically underfunded. In the discussion at RGS following Woods' 1917 paper on the development of that railway the barrister Sir Edwin Pears noted that a German oil expedition to the Mosul region that passed through Constantinople around 1903 vanished, probably done away with by Mahmoud Yahia, the local warlord. Pears was a pillar of the British community in Constantinople from 1873 on, an expert on Turkey, and the head of the European Bar there. Pears cast doubt on the likely value of Mesopotamian oil: although "a great German expert who had gone out just before the above-mentioned expedition went so far as to declare that Baku and the great oil district round that famous city on the Caspian were not in it with the fields of the Upper Tigris" a Standard Oil expert had convinced Pears that "no company could work [the Mosul] field profitably against the great American field" (Woods 1917, 56). Given Standard's likely use of Admiral Chester in Turkey before 1911 to attempt to obtain an Ottoman concession and their concerns with gaining access to Iraqi oil in the early 1920s it is likely that the Standard Oil expert was simply attempting to reduce interest in Mosul by others. As a barrister and expert on Roman law and jurisprudence (Sir Edwin Pears Wikipedia Entry), it is unlikely that Pears would have had the technical education to understand that he was being fooled by Standard Oil's expert.

The second, more important, German foray came in 1912 when Deutsche Bank partnered with the Anglo-Saxon Petroleum Company and the APOC (25 percent, 25 percent and 50 percent) to create the TPC, but it was the British who would have been the main beneficiaries (BNA/FO 93/110/74). None of the three groups involved had a particularly coherent claim to their real or presumed concessions, so the logic was to co-operate (Kent 1976, 29). According to Kent this agreement was particularly beneficial to the Germans, since it assured the German Navy "the same share of Mesopotamian oil as its British counterpart" and the Germans had signed off on the TPC by April 25, 1914 (Kent 1976, 97). The TPC agreement seemed to have worked its way through British Government channels by March 19, 1914, just before the massive Government investment in the APOC that was completed May 11, 1914, with the Government holding 2,001,000 APOC shares out of four million total (BNA/CAB 37/119). By mid to late 1914 the British thus reasonably expected to have control of all Iran's and three quarters of Iraq's potential oil, which experts such as Cadman had identified as a huge potential field. According to British Admiralty documents from November 1918, Appendix A, the formation of the TPC was necessary because:

> It will be remembered that Mr. D'Arcy (Anglo-Persian Group) received promises from successive Turkish Grand Viziers of oil rights in the Vilayets of Mosul and Baghdad, but was unable to secure written confirmation, and that German interests claimed mineral rights along the route of the Baghdad Railway under the Convention of 1903. As there seemed no prospect of either party obtaining sole rights over the whole area owing to the vacillation of the Turks between claims supported on the one hand by the British and the other by the German Government, it was decided to effect a compromise, and . . . The Turkish Petroleum Company was reconstituted to prospect for oil in the Vilayets named. The Vilayet of Basra was subsequently included. (BNA/ADM 116/3452)

Following WWI Standard Oil in particular sought to negate the TPC, noting with accuracy that the TPC was "granted *not a concession* but *the promise of a concession*" (Earle 1924, 271, emphasis in original). Even so, as Earle noted, the promise was substantial, negotiations were almost complete and "had no war supervened, and had Mesopotamia remained till now under Turkish rule, the exploitation of these oil deposits would long since have begun. . . . If the Turkish Petroleum Company does not possess a definitive concession, it may at least be said to enjoy priority over other competitors for the Mesopotamian oilfields" (Earle 1924, 271).

Kent's analysis of the reason the deal was not consummated before WWI broke out in August 1914 runs counter to the argument that it was Turkish deviousness and Chester's maneuvering that held it up. She focuses on what can only be called British confusion. Three British government agencies were involved, the Foreign Office, the Admiralty, and the BOT which was supposed to act as mediator. None could quite agree on what their aims were, and "the Admiralty . . . was, as often happened, almost pathologically unable to send an opinion reasonably quickly" (Kent 1976, 97). The Foreign Office disagreed with the Admiralty on the proper role of APOC in the TPC and the BOT failed to find common ground throughout the fall of 1914, and had not done so when Turkey entered the war on the side of the Central Powers on November 5 (Kent 1976, 98–99). Two weeks later the Admiralty and the BOT finally sent their views to the Foreign Office, which replied simply, "Wait" (Kent 1976, 103).

Before WWI Britain obtained 62.3 percent of its oil from America, 11.6 percent from Rumania, and 7.7 percent from Russia. The last two supplies were lost during the war and for most of it "over 80 per cent of our supplies came from the United States" (BNA/ADM 116/3452). Before the war Britain saw the APOC as its alternative to Standard Oil. Although Shell was developing fast and beginning to challenge Standard Oil, the British initially worried that, as a Dutch company, it might lean toward Germany. Once the war began Shell quickly allayed those fears, becoming increasingly a British company. Henri Deterding, Shell's managing director, became a British citizen in 1915 and was knighted in 1920 for his services to the state (Earle 1924, 273). Churchill continued to have concerns over the price of oil, worrying as late as November 14, 1923, that Shell's merger with the APOC would result in the Royal Navy having to pay the market cost of fuel, not the much lower contract price arranged with APOC (BNA/ADM 1/8658/55).

The Post-WWI Anglo-French-American *Oil War*

At the end of WWI Shell's foreign shareholders were promised freedom from British taxes: a report to the Treasury noted "Mr. Deterding is prepared to come under British control provided he does not pay British income tax" (BNA/T 161/738 January 5, 1919). In early 1919 a report of the British Government's Petroleum Executive to the Admiralty and Cabinet noted the need to merge Shell with APOC using as bait the Deutsche Bank share of the TPC taken by Britain as a war prize. The report further noted that "the French had made up their mind to acquire an interest in oil-bearing territory, and that proposals had been made with a view to their co-operating with the American interests." As the Petroleum Executive put it, "the intervention of the United States would not be desirable" (BNA/ADM 116/3452). The British Government was well aware that the deal with France would raise American hackles. The British ambassador in Washington was warned by the prime minister on July 23, 1920 that the San Remo Agreement was about to be made public and was "more calculated to produce bad feeling between England and America at the present time than Ireland" (BNA/FO 141/144).

A Treasury *aide memoire* of mid-1921 made crystal clear the need to include Shell in British oil geostrategy, the incapacity of the APOC to achieve British goals, and the need to include the French in the division of TPC interests:

> In view of the increasing dependence of the world on oil, and the small proportion—only about 3 per cent—of the world's production which before the war was in British hands, it has been urged that it is of vital national interest to secure control over one of the great Oil Trusts. The Standard Oil Company being out of the question the Shell is the only great company with which an arrangement for British control can be made. . . . The possibility of any Government Department being able to work the oil fields of Mesopotamia on Government account or to market the oil when won, is of course exceedingly remote. Similarly then the prospect that the French can be kept out of participating is very small, and we may lose control of the Shell Company. Negotiations have been continuing with the Shell Company on the basis of the Anglo-French Agreement ultimately going forward and the Shell being given a share in the Turkish Petroleum Company rights. (BNA/T 161/738 June 27, 1921)

One of the problems for observers in the 1920s was that it seemed that, since France had been promised Mosul *vilayet* as part of Greater Syria in the secret Sykes-Picot (-Sazonov) agreement of 1916, French premier Georges Clemenceau's giving up Mosul to British prime minister David Lloyd-George at a private meeting in December 1918, along with Jerusalem, meant that France was giving up access to the region's oil (MacMillan 2001, 382). The solution to this apparent loss was the San Remo agreement of 1920 and the assignation to France of the formerly German rights to Mesopotamian oil via the *Deutsche Bank* shares. Books such as Denny's *America Conquers Britain* (1930) saw this as a "sweetheart deal" in which the San Remo Agreement bought off the French. In addition Denny quoted the Report of the American Federal Trade Commission in 1923 to the effect that "denial of reciprocity of treatment to citizens of this country appears to exist with regard to the petroleum industry of Australia, British Borneo, certain African colonies, British Honduras, British Guiana, and Trinidad; France and French possessions; Italy, and the Netherlands and its dependencies" (Denny 1930, 238). The San Remo Agreement allowed French and British companies equal access to each other's territories, but French oil companies amounted to almost nothing in 1920. In conjunction with British moves to exclude non-British—and by virtue of San Remo, French—companies from the British Empire it was clear that this was intended to freeze out American companies from a very large portion of the world and thus no surprise that it sparked a massive British/American struggle over Middle Eastern oil.

However, more recent work, especially in the French archives, has shown that Sykes-Picot gave France much less in, and of, Mosul than such contemporaries as Denny thought and this casts a slightly different light on British aims at San Remo. The French sources show that the French were only ever to have the northern part of Mosul *vilayet* and that the British were to retain both control of the crucial oil region around Kirkuk and drilling rights throughout the *vilayet*. Fitzgerald (1994) thus argues that the original cession of northern Mosul was, for the British, about two things: gaining French support for the British policy of creating anti-Turkish Arab states in the region, which did not sit well with French ambitions in Syria; and creating a French buffer state between British controlled territory and that of Imperial Russia. The French went along because Britain offered them a much freer hand in the creation of Greater Syria than they had expected. For the French to give up in 1918 that section of Mosul they were granted under Sykes-Picot did not mean they were giving up oil rights, since these had never been part of the initial bargain (Fitzgerald 1994). In addition, by 1918 Imperial Russia was history, Russia was riven by civil war as the Bolsheviks struggled for control, and the need for a buffer state was thus seemingly lessened. At San Remo Britain not only assigned to France the German shares in the TPC, but also confirmed France's claim to a League of Nations mandate for much of what they sought as Greater Syria. The real *pro quo* for Britain's generosity at San Remo was thus French support for the effective exclusion of American oil companies from both the Middle East and the French Empire.

Despite San Remo, French relations with Britain remained strained in the Middle East throughout the 1920s and 1930s for a variety of reasons that had little to do with oil. Territorial claims in the region conflicted, and the British support for a French controlled Greater Syria still caused problems because its boundaries remained unclear. The Sykes-Picot agreement had assigned much of Palestine to France as well as the northern part of Mosul *vilayet*. Lloyd-George had based his case to Clemenceau that Britain should have all Mosul and Palestine, Jerusalem in particular, on the grounds that it was British troops who had occupied and secured the region against German influence during WWI. Lloyd-George was also mindful of the need to create a Jewish state in Palestine based on the Balfour Declaration of 1917. As a consequence of the Lloyd-George/Clemenceau agreement Mosul *vilayet* was attached to the Mesopotamian *vilayets* of Baghdad and Basra that had been assigned to Britain under Sykes-Picot, and Palestine and Transjordan were separated from what the French had intended as Greater Syria and emerged as separate political entities.

A thornier problem was that before Britain agreed to a French Greater Syria at San Remo, Syria was briefly ruled by King Faisal (Faysal ibn Husayn) of the Hashemite dynasty. Faisal was a nationalist who supported the British-led Arab revolt against the Ottomans during WWI, led the Arab delegation at Versailles, and was elected King of Syria in 1920 with British backing. When France came into power in Syria he was then overthrown and forced into exile in Britain. At the Cairo Conference of March 1921 Churchill then pushed Faisal as a logical candidate for King of Iraq, a position he agreed to take with British support, although he subsequently demonstrated that the reputation for unreliability the French had ascribed to him was correct, frequently posed problems for Britain, and was always implacably opposed to French rule in Syria (Catherwood 2004). This last would later cause major problems in the construction of the oil pipeline from Iraq to the Mediterranean, which the San Remo Agreement required pass through French mandated territory.

This complex struggle between Britain and America over the Middle Eastern oil spigot actually had two other causes than the one seen at the time, that of Britain's need to secure its naval fuel supply. The first was the state of the geological science and petroleum engineering technology of the period. The second was Britain's parlous fiscal condition following WWI. British foreign reserves had been run down almost to zero by the costs of the war, on top of which it had incurred a substantial debt to America in war loans.

The State of Oil Geology and Petroleum Engineering in the 1920s

The geological science and petroleum engineering technology of the 1920s could do little to relieve the perceived coming decline in American oil output. Geologists and fledgling petroleum engineers of the period simply had no proper understanding of reservoir behavior. The allies had fought WWI on the massive output of several American regions: California, Oklahoma, Louisiana, and Texas in particular. But Texas' 1901 Spindletop field looked like a boom and bust field, the first ever drilled in salt dome conditions. Such conditions were completely unfamiliar to the pioneers of the oil industry. Geologists believed at the time that oil was found in porous sandstones or limestone, trapped in anticlines or by faults. Yergin's often excellent business history of oil, for example, misses this distinction entirely (Yergin 1991). States such as Oklahoma and, to some extent, California conformed to this early geological understanding of where oil would be found: Oklahoma lacks salt domes (Hudson 2002, 253). This lack made it easy to develop oil there in the early period, since traditional geological techniques worked well. From the perspective of most formally educated geologists as late as the early 1920s Texas's Spindletop was a bizarre fluke. Such a conclusion was supported by the fact that, by 1918, Spindletop's production was dropping fast. Most of those who initially moved into what is now petroleum engineering were geologists, usually trained in hard-rock mining at Bureaus of Mines, who did not understand how salt domes were oil-bearing. As a result, companies staffed by traditionally educated geologists were late to enter Texas and the Gulf Coast, effectively leaving the region to its local wildcatters, who were pure empiricists.

In 1909 the University of Texas established its Bureau of Economic Geology and began to study local conditions, but it took time to train the needed cadre of engineers from scratch there and even longer at the two major Morrill Act Colleges serving the Gulf region: the Texas Agricultural and Mechanical College (TAMC), now Texas A&M University (TAMU); and the Louisiana State University (LSU). At TAMC most of the oil men went through Geology and were fairly conventionally trained. Petroleum Engineering did not develop as a department at TAMC until the late 1930s. The initial department of geology at LSU collapsed in the Depression and had to be rebuilt from scratch, a program that began in 1931 when geology and petroleum engineering were combined with geography and anthropology. Much early training on where to find oil actually concentrated on topography, the study of earth surface features taught by Geography departments. The LSU program, though small, seems to have been the first to pay careful attention to salt domes (Louisiana State University Geology online) that were, initially, usually identified topographically.

The initial consequence of this lack of formal academic knowledge of salt domes was unrestrained drilling and massive initial depletion of the upper reserves of the Spindletop field. Deeper drilling after 1925 renewed output, but was still purely empirical and not based on any formal understanding of salt dome geology. Unrestrained drilling was caused primarily by the "rule of capture" derived from British "common law regarding migratory wild beasts and game" and which allowed a landowner to remove all the oil in a given reservoir even if it extended well under the property of his neighbors (Yergin 1991, 221). The only defense was to drill harder and faster on one's own land. The problems with Spindletop led the State of Texas to use the Texas Railroad Commission in 1919 to control well spacing in an attempt to conserve output. One oil man of the period, Henry Doherty, argued that to conserve oil "the rule of capture" should be abandoned and fields "unitized," with royalties pro-rated on the basis of acreage owned and the field managed to maximize output over the long haul. Doherty's ideas began to take hold in the late 1920s (Yergin 1991, 221–222). After the massive East Texas Field came online in 1930, the Texas Railroad Commission imposed controls on depletion to maintain prices as massive oversupply combined with the reduction in demand induced by the Depression caused prices to collapse (Texas Handbook Online). Both spacing and depletion controls were, however, driven by economic concerns, not scientific understanding. Until the East Texas field came on line it looked throughout the 1920s as though the American supply was about to dry up while demand for petroleum from automobile owners was increasing rapidly.

By the early 1920s one in three American households had an automobile. Most of these were basic Model T Fords, their suspensions designed to cope with the appalling "roads" of the period, usually little better than muddy tracks. Speeds were low and automobiles lacked closed bodies and all-weather equipment, which kept weight down. Given the low octane fuels of this early period of automobile adoption, engine output was poor. Low speeds and low weight meant fuel economy was relatively good. The leaded, higher octane fuels developed in the mid-1920s changed the way automobiles were built by the late 1920s. Higher octane fuels allowed even relatively basic automobiles to carry enclosed steel bodies and all-weather equipment, but fuel consumption rose alongside weight. The rapid construction of a network of all-weather concrete roads starting in the early 1920s also greatly encouraged the adoption and diffusion of these heavier enclosed automobiles, which attained much higher speeds on the new concrete roads, again increasing fuel consumption. Electric starters and better transmissions made such automobiles much easier to drive and ownership exploded. Chevrolet began to market its much improved Model K in 1925, rapidly overhauling Ford's ageing Model T in the sales charts, and pushing Ford to respond with its massively improved Model A in late 1927. New entrants appeared at the affordable end of the market, such as the Chrysler brothers' Plymouth. Demand for automobiles climbed inexorably through the 1920s until it collapsed when the Great Depression hit (Hugill 1982; 1988).

Britain's Fiscal Problems and its Debt to America

Perhaps the most serious underlying cause of the 1920s *Oil War* between Britain and America, unmentioned by authors such as Mohr and Denny, was that by 1919 Britain had to pay for the appalling costs of WWI, a war that had caused it to both spend down almost all the huge foreign reserves with which it had entered the war, after which it had to borrow massively from America (Ferguson 1999). Denny recognized at the time that American fiscal policy with regard to the war debts of its European allies was geopolitically bizarre: "cancellation of the British debt amounted (at a 5 percent interest rate) to 30.1 percent of the total"; that to fascist Italy, a country that had made only token payments on its debt, and those under duress, amounted to 80.2 percent.

Yet Britain was "the first foreign nation to fund its American debt and the only one to do so voluntarily" (Denny 1930, 22).

Coming into control of the Middle Eastern oil spigot thus made a great deal of economic sense to Britain. Canadian born British merchant banker E. Mackay Edgar put this more than forthrightly in an interview published in the *New York Times* in 1919: "Sees Oil Control Passing to Britain. In Ten Years American Imports, He Predicts, Will Be Restoring Sterling Equilibrium." Edgar accused America of "recklessly and in sixty years run[ning] through a legacy that, properly conserved, should have lasted a century and a half" and went on to detail America's lack of success in acquiring leases in the rest of the world. Edgar concluded that British interests "hold secure control of the future of the world's oil supply" and estimated that within ten years American imports would be running at "£200,000,000 per annum, most, if not all, of which will find its way into British pockets" (*New York Times*, September 21, 1919). "Prominent British spokesmen continued to make similar unrestrained remarks . . . while . . . the policies of the British Government added weight to the arguments of those who thought they saw a conscious British plot directed against American security" (DeNovo 1956, 860). Others tried to pour oil on such troubled waters: Sir John Cadman "Denies That Britain Seeks Oil Control . . . 'It is not true that the British Government is engaged in the oil business and controls the Anglo-Persian and Royal Dutch Companies, or that Great Britain is trying to secure a monopoly of the oil resources of the world.' He also added that it is not true that the British Government is shutting out American citizens and American capital" (*New York Times*, January 8, 1922).

Cadman's denial was partially correct, if disingenuous. Although the British Government held no shares in Shell it was the majority shareholder in the APOC and thus in the Middle Eastern oil spigot and, as Cadman well knew, since he had chaired the British Government's Petroleum Executive (Smith 1941, 918), had been more than willing to shut out American citizens and capital from that region. Not until the Cabinet meeting of March 13, 1922, did Winston Churchill, by then secretary of state for the colonies, admit that "neither the United States nor France would be sorry to see the Turks back in Mosul and in a position to give their nationals oil concessions which are at present claimed by His Majesty's Government for the Turkish Petroleum Company." Churchill therefore took the position that the only solution was to cut the Americans in for twenty-five percent, assuming they were willing (CAB 24/134). Cadman, a senior adviser to the APOC by 1922, helped make half of the APOC's shares in the TPC available to American interests to buy them off when the Americans began to demand a share of the Iraqi oil (Marcosson 1924, 22–24). With some justice the British found this hypocritical:

> One observes that the high-sounding note of the principle of economic equality had now sunk into the lower note of the principle of "sharing the swag." How had the mighty fallen! The United States had originally set a fine example of charity [at Versailles] by virtuously declining to take a share of German reparations or a square mile of the German colonies, but after four years was found making an exception to its self-denying ordinance in the case of oilfields in Mesopotamia! And that, too, in hardly logical circumstances. What, indeed, are the usual consequences of association in war"? Surely that one should be associated in making a treaty of peace. This, however, the United States has avoided. So the astonishing spectacle is presented of the Washington Government refusing to make war with Turkey, or to make peace with Turkey, or to take the Armenian mandate, yet claiming equal privileges in exploiting the oilfields in former Turkish territory. (Davenport & Cooke 1924, 120–121)

Despite such British ripostes, the noise generated by American writers continued. Denny, the chief editorial writer of the Scripps-Howard Newspaper Alliance, perhaps the best-informed and certainly the most vocal, argued that, as the price of oil inexorably rose, crisis was likely "unless the two empires seek through mutual sacrifice to reconcile their many conflicting interests. This would involve sharing raw materials and markets, and dividing sea supremacy. . . . If some such miracle of diplomacy is achieved oil may cease to be an international explosive" (Denny 1928, 274).

When the British began to move in on Mexican oil during the Harding administration (1921–1923) the potential for explosion apparently increased:

> An American note was sent to London. The note was so strong, the diplomats decided it was "not fit to print." Members of the London Government who considered its function of maintaining friendly relations with the United States more important than its functions as an oil company, insisted on a general oil compromise . . . the British Government fell to bribing Standard Oil; the bribes were to be paid in the oil of Persia and Mesopotamia. (Denny 1930, 239)

Denny, like most of the other American writers who commented on *The Oil War*, paid little attention to Edgar's financial comments and focused almost entirely on oil. In *America Conquers Britain: A Record of Economic War* Denny notes that his chapter on "oil diplomacy" is, in part, taken "verbatim from the author's *We Fight for Oil*, 1928, the purpose being to condense and bring down to date that study of the earlier battles of oil diplomacy" (Denny 1930, 418). The section taken verbatim argues that "the struggle is not alone between American and British capital. It is between American capital and the London Government. Of the two dominant British companies, the London Government has close unofficial relations with one and has direct controlling ownership of the other" (Denny 1928, 18; 1930, 226). Given the tenor of Denny's 1930 book, and especially the chapter "Dollar versus Pound" what he misses is the major thrust of Edgar's comments, which is that oil—and oil alone given the state of British industry—would have the capacity "in ten years . . . [to] be restoring sterling equilibrium" (*New York Times*, September 21, 1919). Denny was certainly well aware of the parlous state of both Britain and British industry by the mid-1920s, if rather hyperbolic in his comments.

> Britain has too little food, America has too much. Britain has industrial strife, America has industrial peace. Britain has diminishing coal supplies, America has coal in abundance and the more important energies of the future, oil and hydro-electric power. Britain has virtually no domestic raw materials, America has many. Britain has an obsolete plant and technique, America leads all in industrial equipment and organisation. Britain is losing world markets, America is gaining them. Britain is declining in sea power, America is rising. Britain is on the defensive, America is on the offensive. (Denny 1930, 404)

The Turkish award of "concessions" in a Mosul it no longer governed to American Admiral Chester in 1923 following Chester's denial of the Armenian genocide and American support of Turkey at the Treaty of Lausanne muddied the *Oil War* considerably. In 1924 the American state department official, Alan Dulles, told Standard Oil that it did not recognize as valid the 1914 negotiations on which the TPC based its claim to Iraqi oil (Yergin 1991, 201). The 1923 Chester "concessions," however dubious, plus massive American diplomatic pressure involving the nasty exchange of notes noted by Denny (1930, 239), finally persuaded the British to accept what became known as the "open-door policy" in the Middle East. This allowed American oil companies to purchase a twenty-five percent share in the TPC. These shares were provided by the APOC, whose share of the TPC established when the TPC was founded, was fifty percent. By this point the TPC had been cut into four equal parts: one owned by the APOC, one by Royal Dutch Shell, one by French interests, and one by American interests focused on Standard Oil of New Jersey. Calouste Gulbenkian, whose good offices in the region were considered necessary, and who in 1928 drew the infamous "Red-Line" around the former Ottoman region, including Iraq and all of Saudi Arabia, skimmed five percent off the top of the TPC. The Ottomans had controlled only the Red Sea and Persian Gulf coasts of the Arabian Peninsula, leaving the interior to the desert tribes. Even after this new division, however, Britain still controlled, through the APOC and Shell, all Iran's oil and almost half Iraq's.

The British issue with Turkey, which hinged on a disagreement over the nature of the Iraqi-Turkish border and the award of Mosul to Britain at Versailles was finally settled in 1926 when

> In return for complying with the League's boundary determination, the Turks agreed to accept 10 percent of the royalties for a twenty-five year period. . . . From this point on, Turkey would once again support Britain in the Middle East as a buffer state, throwing her weight against a resurgent Russia, serving as a stabilizing force in a turbulent region. . . . The very commodity that had made the Anglo-Turkish conflict so intense became the means of bringing that conflict to an end. (Stivers 1982, 171–172)

Other Complications of *The Oil War*

There were, as it transpired, three other complications of *The Oil War* in the 1930s, one contributing to the impending crisis predicted for America by writers of the 1920s, two ameliorating it. The one that might have deepened the crisis was the emergence of complications simply getting Iraq's reserves into the market. The ameliorations, both of which came from increased output from hitherto undiscovered fields, were the renewal of domestic American production from the vast East Texas field, and the exploitation of sources of Middle Eastern oil not firmly under British control.

The "battle of foreigners"

> [Many oil resources] can be tapped only by long pipe-lines across mountain, desert, or jungle to the sea. That is the situation in Persia, Mosul, or Colombia, and less important fields. Often a second weak country or territory is the only practicable outlet for otherwise inaccessible deposits; as the outlet for the south Persian field through the Baktari tribe region, the Russian Caucasus gateway for the north Persian field, the projected pipe-lines across Syria or Palestine to tap Mosul in Iraq, and the Venezuelan passage out of the east Colombian pool. Thus the *battle of foreigners* for one field may extend from the producing territory to the transit country. (Denny 1930, 226, italics added)

Getting Iraq's oil to market proved problematic. In Persia a short pipeline to the Persian Gulf had sufficed, but Britain was reluctant to commit the Iraq supply to the same costs and potential constraints. Britain had no mandate for Persia, so oil production could be reasonably easily shut down by political disruption there. But oil for Britain taken to the Persian Gulf coast, whether from Persia or Iraq, was subject to the tolls on the Suez Canal as well as possible closure of that canal, despite the British presence in Egypt. The British general staff were well aware of this and argued in late 1919 in the run-up to the San Remo agreement that Britain needed rail and pipeline facilities between the Persian Gulf and the Mediterranean to lessen "our dependence on the Suez Canal, which is a vulnerable point in our line of communications with the East" (Kent 1976, 151). But pipelines were as subject to disruption as canals, and the only geostrategically secure pipeline would be one passing through securely held territory.

For geostrategic reasons Britain thus wanted the pipeline from Mosul to pass entirely through British mandated territory in Iraq, Transjordan, and Palestine to the Mediterranean port of Haifa, but the San Remo agreement with France clearly called for a pipeline from Iraq to the Mediterranean through French controlled territory. France therefore wanted the pipeline to pass from Iraq into the French mandate of Syria and to the Lebanese port of Tripoli. The *battle of foreigners* began. King Faisal of Iraq, having been thrown out of Syria by the French, heavily resisted the passage of this pipeline through French territory. Since Faisal had been installed as King by the British the British also resisted a French route (CAB 24/195; CAB 24/202). Further, a route through Syria was too subject to French control in the event that Britain had to fight France, a hitherto not uncommon occurrence. The eventual solution, which both involved a long diplomatic row between Britain and France and was expensive, was to bifurcate the pipeline at Haditha, with one branch to Haifa and the other to Tripoli (Cadman 1934). It was American intervention, in the person of the president of Standard Oil of New Jersey, Walter Teagle, that eventually persuaded the British to give in to French demands for bifurcation (Fitzgerald 1993).

Amelioration of *The Oil War* Through Increased American Supply

In 1930 American domestic oil output took a dramatic turn for the better as the huge East Texas oilfield centered on Kilgore came on line. Although other American fields are larger more oil, over five billion barrels, has flowed so far from this field than any other in American history, and it was easily connected into the vast pipeline network that had been built out of Houston to Spindletop and that ran north into Oklahoma through east Texas. In the late 1930s it also became clear that there were very substantial reserves in West Texas, in the huge Permian Basin oilfield centered on the cities of Midland and Odessa. These reserves would take longer to tap because of the length of pipeline needed. After WWII Texas oil companies, aided by President Harry Truman's extension of America's territorial limits to the edge of the continental shelf, well beyond the historical three-mile limit, would push into the Gulf of Mexico. Although his oil career started in the Permian Basin, in 1954 George H. W. Bush commissioned the first jack-up rig for his company, Zapata Oil, to operate in the waters of the Gulf of Mexico. It was designed by R. G. LeTourneau using technology initially used in his earth-moving machinery. This Gulf oil also flows into the Houston centered pipeline network.

Amelioration of *The Oil War* Through Increased Middle Eastern Supply

The second amelioration was the discovery and development of Middle Eastern oil that was not under British control, not geographically part of *The Oil War*, and not in a region with serious domestic political problems. America was able to achieve success in the ethnically and religiously homogenous state of Saudi Arabia in the 1930s. Churchill had reluctantly accepted the rise of the Saudi monarchy to power on the Arabian peninsula, a region notable before the discovery of oil there for being almost entirely composed of seemingly valueless desert. King Ibn Saud had repaid the British for their support by evicting the British supported Hashemite monarchy from control of that region of the former Ottoman Empire known as the Hejaz. The Hejaz was a thin strip of coastal land south from Suez along the eastern coast of the Red Sea to the Persian Gulf that was supposed to fall to Britain under Sykes-Picot (-Sazonov). Britain then maneuvered the Hashemite Kings onto the thrones of first Syria, then Iraq, then Transjordan—now Jordan—in the 1920s. Ibn Saud wanted control of the Hejaz, then the only valuable part of the whole Arabian peninsula, for its substantial revenue from pilgrimage to the two holiest sites of Islam, Medina and Mecca. Ibn Saud captured Mecca in 1925, was declared king of the Hejaz in 1926, and was recognized as such by the British in 1927.

After forcibly uniting the Arabian peninsula under his monarchy between 1902 and 1932, three forces pushed Ibn Saud away from the British, even though the British had not resisted his seizing power. For the most part he embraced the Americans, although he flirted with Germany. First, he had come to appreciate and need the services of American medical missionaries. Second, he was engaged in a long-running row with the British telecommunications company, Cable and Wireless, that stemmed from his annexation of the Hejaz. Third, he has generally been considered by historians to have distrusted the British, although the Arabian peninsula was clearly within the British sphere of influence as defined at Versailles. To begin with, therefore, oil had little to do with events in Saudi Arabia.

The first American success in Saudi Arabia was that Ibn Saud had learned to appreciate American medical missionaries, a process that began in 1911. In 1914 he sent relatives stricken with malaria for treatment at the American clinic in Kuwait, and in 1923 he summoned an American physician from that same clinic to treat him for facial cellulitis (Oren 2007, 409).

The second force pushing Ibn Saud into the arms of the Americans was the argument over external telecommunications. Just before Ibn Saud's invasion of the Hejaz the governments of the Hejaz and Sudan had given Britain's Eastern Telegraph Company twenty-five year monopoly rights to telegraphic communications in the region on the basis of their installation of a major link in Britain's global system of telegraph cables across the region. In a letter of December 15, 1926, Ibn Saud recognized that arrangement in a letter to the governor general of the Sudan (BNA/FO 141/716/6 290/1/35). By the mid-1920s it was, however, becoming evident that the short wave, beam antenna wireless technology pioneered in the early 1920s by the British Marconi Company was a much cheaper solution to the problems of long distance telegraphy than the capital-intensive cable systems used by Eastern. In 1928 the British Government forced, for largely geostrategic reasons, a merger between Marconi Wireless and the various Eastern cable companies (Hugill 1999a, 130–131). The new company, Eastern's legal successor, Cable and Wireless, held that the 1926 agreement between Eastern and Ibn Saud over the Hejaz included wireless telegraphy. Ibn Saud objected and, although Cable and Wireless agreed to negotiations at a meeting in London with Saudi officials in September 1934, he attempted shortly thereafter to evade the contract by establishing "wireless communication between Riyadh and Basra, without reference to what had passed in London" (BNA/FO 141/716/6 290/7/35). At that point Cable and Wireless refused to negotiate any more. Ultimately, in 1944, the American oil company Aramco stepped in, chartered a boat that it equipped with wireless communications equipment and then parked it in international waters in the Persian Gulf to provide Ibn Saud with better telecommunications service (Hart 1998, 24).

Third, it has been generally believed by historians that Ibn Saud did not want British geologists and engineers surveying his kingdom. Allowing in even non-British members of the British dominated Iraq Petroleum Company (IPC) that was the successor company to the TPC in 1929 meant effectively allowing them all in. Fitzgerald (1993) contests this, noting that Ibn Saud did considerable business with various British interests. As the Depression bit, however, Ibn Saud's income from pilgrimage dropped both abruptly and precipitately and he began to look for other sources. In 1930 116,000 pilgrims had come to the Hejaz: by 1933 that figure had collapsed to 30,000 and it had recovered only to 48,000 by 1939. In that time period most of the Saudi state's revenues came from customs revenues, taxes, and tithes and in 1930 the estimated income from pilgrims was about $20 million. But the state was spending $13 to $15 million yearly on imports, some 70 percent of which were "essential food products and textiles." By the mid-1930s Saudi Arabia, whose King "was used only to a budgetless financial system," was sliding rapidly toward bankruptcy (Mejcher 2007, 243).

In 1929 King Ibn Saud had allowed an American, Charles Crane, a long-standing supporter of Arab nationalism and funder of Middle Eastern studies, to fund a geological study of the country. Crane's surveyor, Karl Twitchell, was commissioned to look for aquifers to service new pilgrimage routes across the desert (Fitzgerald 1991, 447). Although the survey found and resuscitated an ancient gold mine, it found little water and the gold mine failed to generate enough revenue to cover the lost pilgrimage revenue (Oren 2007, 412–413). Twitchell was not searching for oil, but in his final report he mentioned topographic evidence of what seemed to be salt domes along the Red Sea coast in Bahrain (Fitzgerald 1991, 447). This report did not set off any immediate rush for oil in Saudi Arabia.

Traditionally trained geologists did not believe salt domes held oil. In addition, by 1930 Standard Oil of New Jersey was part of IPC, which claimed a monopoly on oil exploration in the Middle East under the "Red-Line" Agreement of 1928. This agreement included all of the former Ottoman possessions except Kuwait, and Kuwait was, in any case, a British preserve. However, even within the IPC there was dissension over the "Red-Line," with some companies wanting to expand out of Iraq and Iran, Gulf Oil in particular, and the more conservative companies refusing to.

Most of the members of IPC were "crude long." Gulf was not, and needed supplies of oil to bring to market. Gulf saw the possibility of oil in Bahrain and made a clandestine deal with SOCAL, which was also "crude short," wishing to hold its California supply for its California customers. Before WWI, "in an effort to ward off German penetration in the Gulf region, Britain had made agreements with the local sheikhs, including those of Kuwait and Bahrain, that oil development should be entrusted only to British concerns" (Yergin 1991, 283). SOCAL therefore created a Canadian based subsidiary, the Bahrain Petroleum Company (BAPCO), to make it appear to be British so that it could operate within the Red-Line area. In 1928 BAPCO acquired the right to drill in Bahrain from Gulf Oil. Gulf, like SOCAL's partner, Texaco, had started in the Spindletop field in Texas and understood salt domes. Both Gulf and SOCAL needed overseas production to feed their overseas sales, SOCAL being particularly "crude short" outside the Western States of America (Fitzgerald 1991, 444, 464).

In 1932 SOCAL petroleum engineers working in Bahrain under their BAPCO cover found oil in salt dome conditions in the Persian Gulf just across from the Saudi Arabian coast. Looking across to Saudi Arabia the SOCAL geologists could see what they believed were salt domes. A great deal of early oil exploration focused on interpretation of surface topography. Oil in California and the Gulf of Mexico was found a far greater variety of conditions than in Ohio and Oklahoma, Iran and Iraq. Salt dome deposits were common around the Gulf of Mexico and in California, but these were conditions which British geologists and those who worked for Standard Oil were unused to. A former Chevron employee and TAMU Professor of Petroleum Engineering, Dick Startzman, remembers standing on the same Bahrain salt dome in the 1960s and seeing the same apparent salt domes in

Saudi Arabia, although he notes that the initial find there turned out to not be a salt dome after all. With the Bahrain well proved, Ibn Saud agreed to lease the Persian Gulf coastal region for drilling. After much maneuvering, SOCAL outbid IPC, changing its name for Middle Eastern operations to the California Arab Standard Oil Company (CASOC), and again in 1944 to the Arabian American Oil Company (ARAMCO). In the event CASOC nearly bankrupted itself before finding oil: Ibn Saud demanded and received from SOCAL and CASOC payment for his leases only in gold, and exporting gold from America in the 1930s was illegal.

IPC lost Saudi Arabia for several reasons that Fitzgerald comprehensively lays out (1993). Although it was certainly the case, as Yergin notes, that IPC was interested in keeping competitors out of the region and that managerial decisions had an impact on IPC's eventual withdrawal (Yergin 1991, 291), Fitzgerald draws a much more complete and complex picture. As a large consortium of disparate companies with different goals IPC members frequently had problems coming to a clear decision and they always took a great deal of time to do so. SOCAL was highly focused, tightly managed and could make crucial decisions quickly. SOCAL was "crude short," most IPC members were "crude long." Ibn Saud demanded payment in gold, which SOCAL went to considerable lengths to obtain, whereas IPC offered payment only in sterling. But the main problem for IPC was exemplified by the conservative behavior of Shell, which refused to participate in negotiations in Saudi Arabia since it simply "did not regard the prospects as favourable from either a geological or a political point of view" (Fitzgerald 1991, 458). SOCAL was certainly more willing to take risks than IPC, but it also understood salt dome geology a great deal better and, operating as BAPCO, it had found oil under such conditions in the region. After much investment, it finally hit commercial quantities of oil in the coastal region of Saudi Arabia in 1938. This oil was easily shipped using short pipelines to the coast. Nowhere in Saudi Arabia was it possible for a *battle of foreigners* like that which had erupted in Iraq to occur over pipelines. Finally, Ibn Saud put the rest of his kingdom up for lease, and again CASOC was the highest bidder, this time not only against IPC but also the Fascist governments of Germany and Japan, desperate for oil for the war they knew was imminent (Oren 2007, 413–415). Oren probably overstates the role Germany could have played. The Nazi Government had played a much more "hands-off" role in oil development than the governments of America, Britain, and France. Indeed, after Achnacarry, a strong governmental role was expected in those country's oil industries. Ibn Saud was well aware that the Nazi government was unlikely to assist in German investment in Saudi oil, and that it had geopolitical concerns about Middle Eastern oil: "in the event of a war in Europe and the Mediterranean with Britain as one of the adversaries, German oil supplies from the Middle

East would not be defensible" (Mejcher 2007, 245). Even oil assets in Central America would be more defensible, although most valuable as a bartering device. Nazi oil policy thus focused on an Eastern Europe it was confident of taking diplomatically or militarily and on synthetics it was confident of developing technologically (Mejcher 2007, 266–267).

By 1939 Saudi oil output was five million barrels a year, rising fast, and exports had started. Once WWII broke out Ibn Saud's anti-British leanings caused him to seem to tilt to the Nazis, although in 1941 he appealed to Washington for an emergency lend-lease loan of $10 million, which was turned down. In part this was because lend-lease was to aid the fight against tyranny, and Saudi Arabia was "backward, corrupt, and non-democratic." In part it was because the American state department continued to regard Saudi Arabia as within the British sphere of influence and "bankrolling the king as Britain's exclusive obligation" (Oren 2007, 418). As the war intensified Saudi Arabia began to suffer. Oil exports were shut off and the pilgrimage trade to Mecca collapsed completely. Serious food shortages developed, and CASOC, worried that it might lose its concessions to the British, who were providing financial aid, began to use its political connections to pressure Washington. In 1943, "at the urging of Harold Ickes, secretary of the interior and wartime oil administrator, Roosevelt declared Saudi Arabia vital to the defense of the United States and therefore eligible for financial aid" (Lippman 2005, 4).

Ibn Saud was an absolute monarch of a state with a tiny population. Personal relations ruled all his decisions. In September 1943 two of his sons visited America and returned home to report very favorably on their treatment. Ibn Saud expressed a strong wish to meet President Franklin Roosevelt, assuming their meeting would be "monarch to monarch." Following the Yalta Conference in February1945 FDR arranged to meet King Farouk of Egypt, Emperor Haile-Selassie of Ethiopia, and King Ibn Saud before returning to America. Contemporary reports suggest that Ibn Saud and FDR got on well on a personal basis, although no formal record of their discussions exists. Finding at Yalta that FDR was to meet these three, Churchill pushed for similar meetings. Ibn Saud was not impressed and found that "Mr. Churchill speaks deviously, evades understanding [and] changes the subject to avoid commitment, forcing me repeatedly to bring him back to the point. The president seeks understanding in conversations; his effort is to make the two minds meet, to dispel darkness and shed light upon this issue" (quoted in Lippman 2005, 11). In most ways the long-standing relationship between America and Saudi Arabia began at the crucial meeting of Ibn Saud and FDR, and Aramco became the principal beneficiary.

American control over Saudi oil had to deal with one more problem. It could not be said to be complete before "the abrogation of the pre-war Red-Line Agreement which had been preventing Exxon and Mobil's expansion in the region. In 1946 SOCAL and Texaco (Aramco), with their large Saudi concessions, were crude-long, capital- and outlets-short and unconstrained by the Red-Line Agreement, and accordingly they successfully invited Exxon and Mobil to join the Aramco consortium" (Bromley 1991, 112).

CONCLUSION

As part of the long economic struggle between two trading states, with Britain sensing a slow loss of hegemony and America a steady rise, *The Oil War* was crucial. Control over the global oil spigot would clearly be central to the oil-powered Neotechnic world-economy that almost all contemporary observers saw coming not long after the turn of the century and assuredly by the 1920s, and that they saw replacing the British, coal-powered Paleotechnic world-economy. These observers clearly understood the implications of the shift away from coal. But embedded as they were, in particular the Germans, French, and Americans, in the new, territorially structured and tariff-protected nation-states which began to emerge with the French Revolution and which intensified in the late 1800s after America embarked on a program of increasingly heavy protection of its industries, they could not see what was likely to happen if the world-economy reverted to liberalism. Britain sat uneasily between the old, globally structured, capitalist world-economy of Manchester Liberalism that it had done so much to create in the 1800s, where it sought to control nodes and flows, and the new Protectionist world-economy of more territorialized states. For many, revamping the British Empire as a territorialized state seemed to be a better response than maintaining its traditional trading state position based on Manchester Liberalism. After WWI that response began to gain the upper hand in the British polity, although the Imperial Protectionist wing of the Tory Party had moved public opinion a long way in that direction even before WWI.

The Oil War is best understood in these terms. It occurred in a period when the world-economy was riven by uncertainty about whether liberalism or protectionism would prevail. Britain had risen to global power in part because it had plenty of coal, the main fuel source for the Paleotechnic phase of the Industrial Revolution, securely within its territorial boundaries. By the end of WWI it was obvious that America had risen to be a global power of at least equal proportions, and that it had risen on the basis of a highly protected domestic industry serving a huge domestic market and with exceptional domestic resources of a new fuel, oil, as well as coal. If the new world-economy was to be driven by oil, and if it was to be politically constructed as a system of largely territorial states using tariff walls for protection of domestic industry and markets, Britain needed a geostrategically secure source of oil, for it had none at home. Germany and France had faced exactly the same problem in the period before WWI, but France was less industrialized than Germany and the French had therefore paid less attention to the monumental technological shifts occurring in the late 1800s in oil and electrical energy, and in organic chemistry. The formation of the TPC in 1914 by Britain and Germany made perfect sense in this evolving geopolitical scene, as did the replacement of Germany with France after WWI was over. The constant attempt to exclude America also made perfect sense both before and after the war. As late as 1944 Lord Beaverbrook was arguing that Middle Eastern "oil is the greatest single postwar asset remaining to us. We should refuse to divide our last asset with the Americans" (quoted in Mejcher 2007, 57).

What observers saw clearly by the 1920s was true enough. Britain and America were embarked on a full-blown struggle to control what we now see as the global oil spigot, the massive oil reserves of the Middle East. What they did not quite see was why, though the first to comment, Delaisi, got much of it right when he echoed a British prime minister that Britain had no particular claim on its hegemony and that just as the world's financial center had moved from Amsterdam to London it might well now be en route for New York (Delaisi 1922, 2). What no one fully grasped, not even Denny, who clearly knew more of the financial struggle than most of his contemporaries (Denny 1930), was that the British/American struggle was both far more serious than he imagined and also far less likely to lead to military confrontation. Denny (and Mohr, and Delaisi and all the other commentators in the 1920s and 1930s on the row in the Middle East over oil) was a product of his time: he saw a world in which the economic struggle he described between Britain and Germany had seemingly led inevitably to military struggle.

But as Mackinder noted in his caustic analysis of WWI, Imperial Germany was by no means a single polity: it was two polities in conflict, one driven by the elites of the German trading and banking cities and one by those German elites, the *Junker* class, who believed wealth came from landowning (Mackinder 1919, 154). For the Hamburg bankers expansion into the Atlantic basin to engage in economic struggle with the other great trading states was the norm. This expansion had part of its expression in the development of the great German steamship lines, *Hamburg-Amerika* and *Norddeutscher-Lloyd*, as well as in a foreign policy that stressed naval power and a geosecurity brought about by the construction of a High Seas Fleet with which Britain would not risk war. On the other hand, the state most central to the emergence of Imperial Germany as a great power in 1871 was Prussia, a state controlled by a highly territorial, *Junker* elite that was particularly obsessed not only with the ownership of landed estates as the route to wealth, but also with a very different sort of geosecurity. Military struggle to control territory was not only a thinkable option but one hard-wired into the culture of Prussia as a marcher state. Prussia sought Mackinder's heartland in part because that territory was valuable for the people and resources it contained, but also because it needed geosecurity. The memory of the massive Prussian loss at the Battle of Jena (1806) and of repeated French incursions from the west during the Napoleonic Wars was strong, and by the late 1800s Russia possessed a huge army and posed an obvious threat from the east.

Once the Napoleonic Wars were over and French power securely defeated, Prussia, Britain's main ally by 1815, was able to rebuild. Initially Prussia had strong ties to liberal Britain, reinforced by the Germanic ancestry of Britain's monarchy. This eroded slowly during the remainder of the nineteenth century, since Britain's German connections were, through the various Georges, to liberal Hannover and, via Victoria's husband, Prince Albert, to Saxe-Coburg-Gotha, a part of Germany that, embedded in Bavaria, did not like the increasing Prussian dominance of the German state that would finally emerge at Unification in 1871. In part because of Jena, Prussia committed early to the emergence of a new form of state, in which science was turned to the service of the state by its universities. In the 1830s the University of Berlin invented the modern university with its curricula set by its faculty, emphasizing research, and using seminars to educate its graduate students. Previous universities had had curricula set by Boards and were usually linked to religious groups that emphasized scholarship over research and rarely approved subjects beyond divinity, law, and medicine. Faculty control allowed new subjects to emerge quickly and become professionalized. This was particularly important for the development of science, often heavily resisted by religious interests.

As the nineteenth century wore on Prussia moved away from liberal Britain and toward American style protectionism. As new technologies emerged the Prussian state sought to link them to its research universities, and in the emergence of organic chemistry in the later 1800s its technocrats came to clearly value hydrocarbon resources because they could be manipulated scientifically to provide resources that could not be obtained either within Imperial Germany or the limited suite of colonies it had put together from the Berlin Conference on (Hugill & Bachman 2005). In that regard the Berlin to Baghdad Railway project and the follow-up TPC concession both made perfect sense to a Prussian dominated state since they both extended territorial control and gave access to an obviously valuable source of hydrocarbons, both for energy and for the resources into which organic chemistry could convert such hydrocarbons. A Russian railway to the Persian Gulf merely posed Britain a geopolitical challenge to its lifeline to India. A German railway to the Gulf posed both a much more serious geopolitical and an industrial threat.

Contemporary observers such as Denny could see how this sort of struggle might work out, since a version of it had obviously worked out in their lifetime in the military conflict of WWI. They logically projected that sort of conflict into an increasingly uncertain economic future, especially after the collapse of the world-economy marked by the Great Depression (Denny 1930; Hanighen 1934). What they could not see, both because they had no historically informed model of transitions in power and because the only example of a transition based only on economic struggle was buried in a struggle between Holland and Britain that had receded into the distant past of the 1600s, was that Britain and America were both trading states in which the power of elites committed to trade was strong and for which economic struggle was normal. British behavior at the time did not help, however, since the geostrategically driven attempt to control the TPC, however understandable because of the insecurities of a state which depended heavily for its security on its navy, was atypical in that it was like that of a territorial state, seeking to control territory for the resources in contained. One consequence was that neither Denny, Mohr, nor Delaisi could see how, for all its intensity, Mohr's *Oil War* (1926) was an economic struggle, not a war. When Denny claimed *We Fight for Oil* (1928) he could not see that, ultimately, the fight was about access to flows, not control of territory. In that regard America and its companies, even though clearly willing to seek and take political help, focused on control of flows and behaved like a true trading state where, for once, Britain did not.

Chapter 7

Multi-Polarity to Hegemony

The Switch from British to American Hegemony through Australian Eyes

Australia offers an interesting case study of the transition from British to American hegemony. In many ways Australia exemplifies all the problems that Britain had to deal with as it declined, especially vis-à-vis America. Australia was pretty much as far from Britain as it could be. This imposed very high energy transportation costs by sea, at least after steam and motor ships developed, and the time-cost of distance its antipodeal location imposed was always high. Air travel promised some reduction in time costs by the late 1930s, but the economic costs were still very high. Telecommunications offered a bright spot with near instantaneous communication, and were of crucial importance in connecting Australia to Britain, whether by submarine cable or, later and even more importantly because cheap, Marconi's beam wireless. Although Australia had some important minerals, such as gold, as well as good supplies of coal and iron ore, it lacked, in the period covered by this book at least, easy access to fuel oil and natural gas. On top of that, by the 1930s it was clear that it lived in an increasingly dangerous neighborhood. In 1902 Britain and Japan signed an important naval alliance, but the defense of Australia by Japanese naval vessels in WWI never sat well with a rather racist Australia that saw Japan as part of the "Yellow Peril" and imposed a strict "White Australia" policy on immigration. When that naval alliance ended in 1922 it became increasingly obvious that Japan had expansionist aims in Southeast Asia, and that those aims included Australia. Australian politics adjusted accordingly, becoming strongly anti-Japanese.

As British hegemony slipped away after WWI, Australia lost its privileged position as a Dominion of what had been the world's most powerful Empire. Britain looked—and began to act—more and more remote. Australia thus found itself subject increasingly to all of the three sorts of boundary problems that circumscribe analysis in international affairs: geopolitical, economic, and politico-military (Hugill 2009, 33–37). In geopolitical terms Australia could no longer count on being a protected and privileged member of the British Empire when a rapidly expanding Japan saw it as a potential solution to Japanese resource problems. In economic terms Australia made its living as a massive exporter of agricultural surplus, but over a global supply line that was clearly subject to disruption once the Germans had demonstrated the capabilities of the U-boat during WWI. In politico-military terms, until 1902 Australia had depended entirely on the British navy. Its replacement from 1902 to 1922 by elements of the Japanese navy was not culturally acceptable. From 1922 on Britain was in retrenchment to try and cope with the massive costs of WWI, its navy was ageing rapidly in technological terms, and it chose to "defend" Australia with a fortress at distant Singapore, which Australia had to garrison. In the 1920s and 1930s defense became a major issue for Australian politics, and Australian politicians of all stripes coped poorly with it.

Structurally, Australia was, at least until WWII, part of the periphery of what Immanuel Wallerstein (1974) described as the World-System, a provider of agricultural surpluses and raw materials. It did not, of course, meet Wallerstein's too constrained definition of a periphery since it did not use a controlled cost labor system such as slavery or sharecropping and actually had an aggressive and effective set of labor unions. From the perspective of Wallerstein's procrustean analysis Australia would fit more into his semi-periphery than anywhere else, but even that would be somewhat forced. Finally, because of what it saw as its privileged position within the British Empire, Australia saw itself as culturally much more part of the core. Effectively, Australia brings Wallerstein's model into considerable question since it was structurally peripheral, culturally core, and effectively probably in his semi-periphery.

Because of its cultural sense of itself as a core nation, in the run-up to WWII Australia associated itself heavily with the declining core, Britain. The Australian Labor Party (ALP) recognized this rather clearly and made it a central part of an economic policy that would move Australia away from excessive dependence on Britain:

> social justice could be achieved only if governments took power to regulate the workings of capitalism within Australia and to protect as far as possible Australia's position in the international world of trade and finance. The experience of the 1930s suggested that . . . Australia must reduce its dependence on overseas borrowing and overseas technology and develop the capacity to generate its own capital and its own industrial skills. (Bolton 1990, 84)

There was also a down side to the cultural understanding that made Australia want to claim something approximating core status: Australians suffered "the belief that Australian culture was inferior to that of the metropolis" (Alomes 1988, 219). Even as a "white" and self-governing Dominion of the British Empire Australia had still refused to ratify the Statute of Westminster of 1931 and accept effective independence from Britain.

The explosive push by the Japanese into the Pacific in December 1941 thus came as a massive shock to Australian *amour propré*. Until that point and despite Australian suspicions of Japan, Japan had been generally regarded as weak: a potential nuisance, not a dangerous enemy, an idea shattered by the Japanese assaults on Pearl Harbor, Singapore, and Burma. Singapore was, in British defense doctrine, the naval fortress supposed to protect Australia. For Australia the body blow came with the sinking by bombers of the Japanese Naval Air Force of two British capital ships attempting to relieve Singapore, the loss of that fortress to the Japanese, and the surrender of the Australian Division garrisoning it. Making matters worse was the fact that one of the two ships, the *Prince of Wales*, was one of Britain's very latest and finest capital ships, one that had recently helped sink one of Germany's finest battleships, the *Bismarck*.

Even the most basic reading of *Documents on Australian Foreign Policy, 1937–1949* published by the Australian Department of Foreign Affairs and Trade (DFAT) indicates, however, that Japanese naval aggression in the Pacific was scarcely unexpected by many in Britain, Australia, or America: it was merely a matter of when and where the Japanese would choose to strike and how effective they would be. Like too many of their peers in Britain with regard to Germany, notably the Conservative Party's Cliveden Set, Australian governments before John Curtin became Prime Minister had a strong commitment to appeasement, both of Germany and Japan.

Churchill, in his secret session briefing of the British House of Commons on April 23, 1942, noted that, before the fall of Singapore, the performance of the Imperial Japanese Army against China, a weak, ill-armed, and ill-supplied opponent, had been poor. Although Churchill did not mention it, against Russia on the Mongolian border—in the Nomonhan Incident at the battle of Khalkin Gol—Japanese military performance had been disastrous.

> I did not think that the injuries Japan would certainly inflict on us in our ill-guarded and even denuded eastern theatre would be too heavy a price to pay for having the immense resources of the United States bound indissolubly to our side and to our cause. . . . But I frankly admit that the violence, fury, skill and might of Japan has far exceeded anything that we had been led to expect. (Eade 1946, 47)

Churchill noted, however, that he did not believe Australia was threatened. Any attempt at invasion would commit Japan to

> a very formidable campaign at a great distance from home with American sea power, as it regains its strength, operative on their communications. No doubt the Japanese will do their utmost to threaten and alarm Australia . . . [but] neither Britain nor the United States must be drawn into immobilising in Australasia undue numbers of the limited forces we can transport across the sea within any given period. (Eade 1946, 59)

Despite the remarkable accuracy of Churchill's reading, when the Japanese strike did come it hit Australians hard, even those who knew it must come. The conventional reading of Australian history has it that the ALP Prime Minister in late 1941, John Curtin, realizing that British power had waned, then turned to America for support. Only three weeks after the Japanese struck Pearl Harbor, Curtin's New Year's message to the Australian people contained perhaps the most quoted sentence by any Australian Prime Minister. The sentence is generally claimed to have re-set Australian foreign policy and has led to the whole being usually referred to as the "without pangs" message: "Without any inhibitions of any kind, I make it quite clear that Australia looks to America, free of any pangs as to our traditional links or kinship with the United Kingdom" (Melbourne *Herald*, December 27, 1941).

Australian historians have generally interpreted this as a classic piece of *realpolitik*: Australia, her back to the wall, turning from the protection of a friend whose powers were failing to one whose powers were rising, and to Curtin as the simple embodiment of that *realpolitik*. Such historians fail to note that when Curtin's message was published the fortress of Singapore had not yet fallen and the massive scale of the Japanese push into the Pacific was by no means clear. On the other hand, American troops were already fighting the Japanese hard in the Philippines and it was clear that Americans would not back away from the war Japan had thrust upon them at Pearl Harbor.

Curtin was not alone among Australian politicians in seeing America as a logical ally. Supporters of Robert Menzies, Prime Minister of the United Australia Party (UAP) Government that preceded the Curtin Government, point to stated support by Menzies, despite his being a notable Anglophile, for a similar alliance. While there is truth in this reading, Menzies also subtly supported appeasement of Germany and Japan well after the war with Germany had broken out, which Curtin and the ALP did not. The Australian historian, Michael Day, makes Menzies' willingness to appease Germany crystal clear (Day 1993). The case for Menzies' appeasement of Japan is less obvious. However, Menzies strongly supported S. M. Bruce, the Australian High Commissioner in London, and on September 21, 1939, Bruce wrote a long memo to R. A. Butler, then Parliamentary Under-Secretary for Foreign Affairs, suggesting that Britain and America should provide Japan the finance to develop China, that Britain should provide needed raw materials, that restrictions on Japanese trade with the British Empire be removed, and that the Chinese government should be made to "recognise either permanently or for a period of years the existing status quo in Manchukuo" (DFAT II, 239). Menzies had earlier intervened to ensure shipments of iron ore to Japan when anti-Japanese Australian dockworkers struck to protest such shipments and at a time when America was refusing to make such shipments to Japan.

Curtin's anti-Japanese sentiment, on the other hand, ran deep. Curtin's numerous writings as a newspaper correspondent, his career for many years before he successfully entered politics, and his early speeches in the Federal Parliament, reveal a deep distrust of Japan. Given Australia's racist immigration history to that point and its constant fear of the "yellow peril," such a reading was by no means unique among Australians of all political persuasions. But where Menzies appeased, Curtin opposed. Curtin's distrust of Japan deepened throughout the 1930s. What the Japanese threatened most in December 1941 was not so much Australia's physical security as Australia's continuation as a culturally core if economically peripheral state. As a cultural claim on Australia's continued membership of the core of the world-system, Curtin's willingness to jump from attachment to a declining British core to a rising American one thus has much more resonance than as a simple piece of *realpolitik*.

Naturally, the story is deeper and more complex than this brief summary suggests. Given the centrality of Curtin to Australian history, the ground has been well worked over by historians with interests in Australian foreign policy. To re-set this discussion in its broader context is, however, worthwhile because it casts light on the complexity of events in periods of multi-polarity. The late 1930s was both the most complex period of multi-polarity in the entire history of the modern world-system and the one that marked America's final, delayed rise to hegemony. Re-setting requires recapitulation of world events of the period 1935–1947 as Australians saw them. The dates here are relatively arbitrary, but are intended to encompass the period in which Australian politicians began to realize they would have to deal with the rise first of Fascism, then of Communism.

The period from 1919 to 1947 began as one of serious multi-polarity but ended with clear American hegemony. The simplest way of structuring a recapitulation of world events in periods of multi-polarity is to analyze it in terms of three sorts of boundary problems: geopolitical; economic; and politico-military (Hugill 2009, 33–37). In the Australian case geopolitical and politico-military boundary problems were starting to loom very large indeed by 1935 because of the building global struggle against Fascism that would eventuate in WWII. Most of my concern here is with Fascism, but once that struggle was over the three problems were reshaped by a second global struggle, this time in what became the long Cold War against Communism.

Although it had not started out with them, Australia found itself with geopolitical boundary problems in both struggles. Three individuals were keys to these two struggles in the Australian context: John Curtin, ALP Prime Minister from late 1941, just before the Japanese attacks, until his death in 1945; Herbert Vere Evatt, minister for external affairs in Curtin's ALP Government and the Chifley Government that followed it; and Robert Menzies, failed leader of the UAP governments of 1940 and 1941 and successful founder of the Liberal Party that would both succeed the UAP and go on to take power from the ALP in 1949. Menzies then served as prime minister of a succession of Liberal Governments from 1949 until his retirement in 1965, strongly supporting America throughout. It is around these three key individuals that Australian foreign policy of the period revolved.

AUSTRALIA'S PROBLEMS

One of Australia's main problems was that she was at once geographically much too far away from and culturally much too close to Britain's declining core. To make matters worse, in the first half of the 1900s Australia's universities were under-developed and far too reliant on those of Britain for intellectual capital. It is true that British decline was a growing concern to the British by the first decade of the 1900s, but that concern was among only a small group of British politicians and intellectuals and it was not transmitted strongly to Australia, if at all. Australia simply could not develop the same appreciation as Japan of the Potemkin nature of British power after WWI until WWII brought it forcibly home. In modern terms, Australia was rather suddenly forced at the outbreak of WWII to confront the fact that she was living in a multi-polar world, one that no longer had a secure center of power.

Signs of British decline were there from the 1880s, when the American and German manufacturing economies began to draw even and then surpass that of Britain. They became, in retrospect, all too obvious in the early 1900s to Britain, America, Germany, and Japan. Joseph Chamberlain was a successful manufacturer and then Mayor of the Midlands manufacturing city of Birmingham before entering Parliament as a Free Trade Liberal. He was one of the founders of the Liberal Unionist Party that opposed Home Rule for Ireland and would eventually merge into the Conservative Party. He would eventually lead the Imperial Protectionist wing of the party. Chamberlain had seen firsthand Birmingham's export trade and thus its manufacturing start to collapse in the face of American competition from behind their tariff wall. British decline pushed him to argue for tariff reform and Imperial Protection by 1903. At the academic level a small number of educated Britons began to understand British decline from at least the publication of Mackinder's seminal "Pivot" paper of 1904 (Mackinder 1904). An increasingly confident America understood British decline in the deep cultural terms revealed in Edwardian boy's literature around the same period (Hugill 1999b). British decline was assumed in Germany's aims in WWI, notably in the demands for *Mittelafrika* and the Berlin to Baghdad Railway, the first at the expense of Britain's African colonies, the second at the expense of Britain's oil interests in Persia and, more particularly, Mesopotamia (Fischer 1967; 1974).

Japan was the most interesting case. Isolated like Australia by distance from the cockpit of Europe, her 1902 Naval Alliance with Britain seemed to indicate acceptance of British hegemony. The series of negative influences that pulled Japan away from Britain and toward Germany began with "The Gentleman's Agreement" of 1907 when America cut off Japanese migration to Hawai'i and California, and California's 1913 law against Japanese buying land in that state. They culminated in her inequitable treatment after WWI. In 1919 Britain and Australia refused, with American encouragement, to allow the Japanese to insert an "equality of nations" clause into the League of Nations Covenant (Meinig 2004, 314–315). In 1922 the Japanese, despite having been an ally in the war against Germany, were forced by the Washington Naval Treaty to accept an inferior fleet size to that of their co-victors, America and Britain (the 5-5-3 battleship ratio). America used this Treaty to effectively force Britain to side with America against Japan, which America saw as standing in the way of its ambitions in the eastern Pacific. On his trip to Japan before WWI the father of German *geopolitik*, Karl Haushofer, concluded that Germany had to ally with a maritime power if she was to rise against Britain and America. With Germany's loss of WWI, the scuttling of most of the High Seas Fleet of the Imperial German Navy at Scapa Flow in 1919, and the dishonoring of Japan by first American and then British, American, and Australian political action, Japan was the obvious ally for a Germany determined to redress the wrongs done it at Versailles (Haushofer 1925; Hayes 2004; Herwig 2016).

When Australia finally realized in late 1941 the truth of British decline her options were effectively down to two. Curtin's New Year message of 1942 sought them both. One was to eventuate in the American alliance, the other was to seek help from Russia. Although little remarked on by historians there are actually four appeals to Russia in the full text of the "without pangs" message and only three to America, although it is clear that the appeal to America was the stronger of the two (Prior 1987).

Without a model of hegemonic shifts it would have been almost impossible for any politician in the period of multi-polarity to understand the structure of the events they were living through. Given the events of the Depression of the 1930s and a lingering "America First" attitude, outside America it could not have been clear in 1941 that American hegemony was likely—that America, in the terms a politician of Curtin's era would have used, was destined to move beyond the hemispheric power it had achieved by the 1920s (Meinig 2004) to global power. At the time perhaps only Churchill and Roosevelt fully understood that it could and would: certainly Hitler did not. Curtin, Evatt, and Menzies were all men of their time, well able to remember and worry over American behavior of the 1920s: the refusal to ratify the League of Nations; the acceptance of continued British power at the Washington Naval Treaty of 1922; the noisome American domestic politics of isolationism during the 1920s; the autarkic Smoot-Hawley tariff of 1930 that forced Britain into Imperial Preference at Ottawa in 1931; and the disgraceful behavior of such "America Firsters" as Charles Lindbergh who openly embraced Nazi Germany. In such a light America hardly seemed a reliable or responsible security partner. Curtin, Evatt, and Menzies thus continued to look to Britain even after Curtin's "without pangs" message, albeit through different lenses. Curtin, encouraged by Evatt, hedged his bet on America against Japan with a side bet on Japan's other great Asian enemy, Russia, although he never called it in. Menzies, despite signing the ANZUS Treaty in 1951, abandoned his belief in British global power late, switching fully to America only in the 1960s, and with many a longing backward glance. Evatt pursued the most innovative strategy, demonstrating a belief that multi-polarity might not end. In world-system terms Evatt believed that a semi-peripheral polity such as Australia would be better served by acting in its own security interests if multi-polarity continued, or in concert with other lesser powers against the great powers if multi-polarity ended. To accomplish the first Evatt created the 1944 ANZAC pact with New Zealand to create a regional security zone without great power help. As for the second, he would be closely involved with the formation of the United Nations. Into the latter task he poured all his formidable skills and Australian historians rightly regard it as his greatest success as Minister of External Affairs.

The foreign policy of lesser powers in periods of hegemony—or unipolarity—and multi-polarity has to differ. Lesser powers always need security partners. In periods of hegemony it makes sense for lesser powers to ally with the hegemon, such as Japan did with Britain even though it had no cultural ties to Britain. In periods of multi-polarity lesser powers will want to hedge their bets. Australia, despite its massive cultural ties to Britain, classically hedged its bets in the period 1941 to 1945.

Boundary Problems

Seen through the lens of the three major types of boundary problems in the world-system: geopolitical, economic, and politico-military, Australia's problems become clearer.

Geopolitical Boundary Problems

These are greatly exacerbated when two polities with different geopolitical goals share a common territorial boundary. Until December 1941 Australia did not have a geopolitical boundary problem with any state, thereafter it was confronted with two: the first caused by the aggressive expansion of Japan to its very doorstep, the second, very much at the end of the period covered here, caused by the expansion of communism, especially Chinese communism, through Southeast Asia. Both produced the same response, a turn to the Asia-Pacific region's most significant military power, America, as a security partner. That does not mean that Australia had no geopolitical boundary problem before 1941, merely that its strong identification with the cultural core of Britain meant that it tended to accept Britain's geopolitical problems with Germany as its own. Of the Dominions, only New Zealand supported Britain at the Australian level by sending troops to fight in Europe and the Middle East in both WWI and WWII. Australia would later send troops to Korea and, most notably, be one of the few powers to send troops to support America in the Vietnam War.

If Japan posed an unrealized geopolitical boundary problem until 1941 it was because Japanese expansionism until that point had been driven by the policies of the Imperial Army and concerned with first Manchuria and then China proper. Two "Incidents" begun by aggressive elements in the Imperial Army in China, by the mid-1930s pretty much a law unto itself, brought this expansionism to a head and an end: the Chankufeng Incident of 1938 and the Nomonhan Incident of 1939. These, in reality, amounted to a serious, if undeclared, war. Japan claimed the inner Mongolian autonomous region of China as part of their puppet state in Manchuria: Manchukuo. The Mongolian People's Republic, backed by Russia following the signing of a mutual support pact in 1936, denied the claim.

In the summer of 1938 the Japanese Army and Army Air Force precipitated a series of relatively unresolved clashes with Mongolian and Russian forces at Changkufeng, where the Korean, Manchurian, and Siberian borders met (Sekigawa 1973, 246). The next summer, however, Russian corps commander Georgi Zhukov led a combined force of Mongolian and Siberian troops to a smashing victory over a Japanese Army attempting to occupy the Nomonhan Plateau, a region claimed by both parties. Zhukov would go on to a stellar career as the most important and effective of the Soviet Union's Marshals in WWII, arguably the greatest General of WWII. The Japanese Army lost some 17,000 troops in the Chankufeng and Nomonhan Incidents: in the latter the 23rd division alone lost 11,100 troops, 73 percent of its strength (Sekigawa 1973, 28). The Army Air Force fared better, never quite losing air superiority, but they could ill-afford the considerable attrition they suffered, losses that included some of their most experienced fighter leaders (Sekigawa 1973, 29).

The loss at Nomonhan was a marked loss of face for the Japanese army. At the geopolitical level this was furthered by the signing of the Non-Aggression Pact between Germany and Russia on August 23, 1939. This outraged Japan, a signatory with Germany to the anti-Comintern Pact of 1936, designed by both to contain Russian ambitions. In consequence of this and the loss at Nomonhan, on July 2, 1940, Japan also signed a Non-Aggression Pact with Russia. Shifting geopolitical sands, increasing American pressure on Japanese imports of iron and oil, and a turn of political power in Tokyo against the Imperial Army following its loss of face, brought the Imperial Navy center stage. The Japanese navy's plans to expand into the Pacific to acquire needed raw materials and markets began to take precedence over the Japanese army's plans to control China to the same end. Naval expansion into the Pacific, however, required preemptive strikes at the two main bases of opposing naval power, the American at Pearl Harbor and the British at Singapore. Both strikes were implemented in December 1941. Whereas Pearl Harbor survived the garrison at Singapore was forced to surrender in January of 1942.

Economic Boundary Problems

Australia's economic boundary problems were generally the same as Britain's. After the Ottawa Conference of 1930 Imperial Protection ruled the day, although Australia would maintain high tariffs far longer than Britain and is only now seeking Free Trade with America. The nature of Australian economic boundary problems with Japan in the 1930s was, in any case, more along the lines of increasing concerns about whether to ship Japan raw materials for its developing war economy, and such concern was not strong as long as Japan's geopolitical boundary was with Russia in China, not with America and Britain in the Pacific. As UAP Prime Minister in 1939, Menzies had intervened to ensure the flow of pig iron to Japan before WWII broke out and had resisted strengthening the British trade embargo against Japan (Day 1988, 43–44). ALP concerns over Japan were clearly greater, but Australia was at this time an almost totally peripheral economy, and reduction in exports of primary produce and raw materials would have been economically devastating. Almost all of Australia's massive export of agricultural surplus was going to Britain, half a world away. Such lines of communication could be easily disrupted by Mahanian commerce raiding and the Germans had demonstrated in WWI that their Uboats were extremely effective at such tactics.

Politico-military Boundary Problems

By the late 1930s it was becoming clear that politico-military boundary problems posed by far the greatest danger to Australia and, in Australia's case, would become closely linked to geopolitical boundary problems. The ongoing revolution in military technology that began in naval affairs in the 1890s by-passed Australia, except insofar as Australia was part of the British core. Most Australians believed they were largely outside this European politico-military ferment, although some of it had clearly filtered through to Curtin. But for most Australian politicians Japan did not seem much of a threat: "in early 1941 it was not foreseen that a holding war against Japan would be conducted from the very doorstep of Australia. Rather, given the poor opinion of Japanese abilities, it was thought that the limited defense could be conducted much closer to Japan itself" (Day 1988, 588). Most Australian politicians both deferred to Britain, believing in Britain's assessment of and response to the likely risk from Japan, and seemingly paid little attention to the rapid changes in military technology and capabilities that characterized the 1930s in particular.

Curtin saw it differently. As early as 1915, as ALP candidate for the Federal Parliament, Curtin noted in a speech the potential advantages of air as opposed to sea power, claiming that "instead of having to send warships to threatened ports, airships could be there in half the time and could be made quite as useful to Australia for defense purposes as the small naval boats now used" (quoted in Black 1995, 13). By 1937 Curtin's views were clearer and harder, almost straight out of Billy Mitchell, and he was well aware that the Japanese could threaten Australia: "it is our air-borne fleets that will most quickly arrive at the point of danger to combat enemy planes, to bomb his plane carriers and troop carriers, and put up first resistance until land-borne men and guns arrive" (quoted in Black 1995, 140). Given that in 1937 only Britain, America, France, and Japan possessed aircraft carriers he was clearly referring here to Japan.

By the late 1930s Curtin was not alone in these views. In 1939 the UAP government entered into an agreement with Britain's Bristol Aircraft Company to build the *Beaufort* torpedo-bomber under license, a plane designed for precisely the sort of missions Curtin had foreseen as early as 1915. By October 1939, however, the war in Europe was underway and supplies of the Bristol *Taurus* radial engines used in the *Beaufort* were shut off. At that point the Australian government shifted to American Pratt & Whitney *Twin Wasp* radials, already in production in Australia by General Motors' Holden subsidiary. Seven hundred *Beauforts* were built in Australia through January 1944: it was followed on Australia's production lines by 364 Bristol *Beaufighters*, effectively a lightened and hotted-up version of the *Beaufort* fitted with much more powerful Bristol *Hercules* engines, the war situation now allowing Britain to export engines. *Beaufighters*, called "whispering death" by the Japanese because their engines were so quiet, devastated Japanese shipping even more than its predecessor (Barnes 1988, 284, 287, 289, 303).

In late 1941 Curtin became prime minister, and part of his success derived from his belief in air power as part of a strong defense posture. Despite the UAP's investment in *Beauforts*, Australian voters had become increasingly distrustful of the defense policies of the UAP in the late 1930s. The UAP's twin policies of appeasement of Germany and dependence for defense on Britain and the Royal Navy worried many, and Curtin was a vocal and consistent critic of both UAP policies from the left. Many Australians wanted Australia to be better able to defend itself and attacks on the UAP came from the right as well as the left. Even well entrenched members of the UAP began to suffer if they were associated with appeasement of Germany. In the 1937 election Percy Spender, standing as an independent right-wing candidate on an anti-appeasement platform, scored a stunning upset when he knocked the sitting UAP Minister for Defence in the Lyons cabinet, Sir Archibald Parkhill, out of his very safe Sydney suburban seat (Spender 1972). Menzies would bring Spender into the UAP and appoint him Minister for the Army in 1941.

Throughout WWII, Curtin and Evatt both obsessed about Australia's lack of other forms of air power, demanding British Supermarine *Spitfires* for defense and, later, British Avro *Lancaster* bombers to carry the war to the enemy. In a barrage of memos both Curtin and Evatt harried Churchill and Roosevelt constantly for combat aircraft. Some of this enthusiasm for British airplanes arose from simple patriotism. Regrettably, much of it was technological naïveté. When the Pacific War began in December 1941 two combat airplanes were being built in Australia. One was the *Beaufort*, the other an advanced trainer and ground attack aircraft, the *Wirraway*. This was being built by the Commonwealth Aircraft Corporation (CAC) and was, essentially, a license built version of North American Aviation's *Texan* advanced trainer. CAC decided to marry the *Wirraway* to the Pratt & Whitney *Twin Wasp* engine being used in the *Beaufort* as the basis for a stopgap fighter although the eventual redesign, named *Boomerang*, was almost total. Curtin claimed the *Boomerang* was as good as the *Spitfire*, when it was almost as unsafe in skies infested with Japanese *Zero* fighters as the advanced trainer it was based on. CAC built some 200 and it found its niche in ground attack, its main advantages being that it was easy to fly, agile, had good range, and could lift heavy payloads.

Curtin and Evatt also made excessive claims to the Australian people for the abilities of the *Spitfire* over the Curtiss P-40, over 850 of which were supplied from America. The P-40 is generally under-rated, and it is true that it was "obsolete by European standards" even as it entered large-scale production in 1940 (Bowers 1976, 474). It was, however, easily available, reliable, easy to maintain, well-armed, long-ranging, and could absorb a great deal of battle damage and still get its pilot home. As American fighter leader Claire Chenault had shown in China, properly utilized and handled the P-40 could be a match for the heavily armed but very fragile *Zero*. The trick was for the P-40 pilot to make a fast diving pass at the *Zero* and not attempt to dogfight, the P-40 being much less maneuverable. In Australian conditions the P-40 was a better plane than the *Spitfire*, which was less well-armed, fragile, hard-to-maintain, and at its best as a fast-climbing, short-range, target defense interceptor. In addition, the *Spitfire* was designed to perform well in the temperate conditions of northern Europe, and outside northern Europe it needed to be "tropicalized," which invariably degraded its performance. The P-40 was designed from the outset for American conditions and Australia was much more like America in terms of climatic range than northern Europe.

In the event Australia got fewer *Spitfires* than P-40s, and no *Lancaster* bombers, although after the war they did build the *Lancaster's* successor, the *Lincoln*, under license. The first batch of some 250 *Spitfires* that made it to Australia were Mark Vs, and by the time they did so they had been replaced in Europe by the much faster Mark IXs. They were later replaced by some 400 more powerful Mark VIIIs, comparable to Mark IXs, and "tropicalized" to suit Australian conditions (Andrews & Morgan 1981, 376). Mark V *Spitfires* did see the sort of combat over Australia to which they were theoretically suited, fighting off the Japanese air attacks on the city of Darwin in 1943. Their generally poor performance underscored the need to modify them to suit Australian conditions (Cooper 2011).

The Geographic Problems of the Empire

In the most fundamental way one of Australia's most serious problems was "the tyranny of distance." It was (almost) the most distant part of the British Empire, and from gaining Dominion status in 1907 on had been a self-governing democracy in all but its foreign policy. In an era of relatively slow and expensive sea transportation and limited electronic communication there was limited intercourse between Australia and Britain, far less than existed between American and Britain across a North Atlantic served by numerous high-speed steamships and with the world's greatest density of submarine cables. Few Australians visited Britain and vice versa, so the British body politic was heavily dependent for its opinion of Australians on two things: the writings of Australian politicians and the reports of its professional diplomats.

In many ways the level of information and personnel flow between Britain and Australia in the late 1930s resembled that across the Atlantic at the end of the American Civil War. Whereas ships for the Atlantic run were built for speed and used powerful but fuel hungry high-speed steam turbines as early as 1906, those for the Australian and Pacific services focused on fuel economy. British yards built their first diesel passenger liners, such as the MV *Aorangi* of 1924, for precisely that sort of service. The first air service by Imperial Airways took about as long to reach Australia in the late 1930s as the steamships connecting New York with London in the late 1860s, some two weeks. Two submarine cables linked Australia to Britain, one via India, the other via Canada, although their capacity was bolstered in the 1930s by Marconi's beam wireless services. The first two submarine cables would open across the Atlantic in 1866, and their capacity and speed of transmission was still low and slow. In the 1860s it was usually the personal reading of American politicians by British professional diplomats that mattered most to British politicians when they had to make political judgments. There was no personal contact between prime ministers and presidents, secretaries of state for foreign affairs, or any other politicians from the different countries.

AUSTRALIA IN THE RUN-UP TO THE PACIFIC WAR

In order to analyze Australian relations with Britain up to and at the time of the Japanese entry into WWII it is therefore very necessary to look at agency: the personalities of the key Australian politicians of the period and the way those personalities were described to politicians in Britain and America. The three crucial Australian actors were the ALP Prime Minister at the time Japan entered the war, John Curtin; Herbert Evatt, the Curtin Government's Minister for External Affairs—the equivalent of Britain's Foreign Secretary or the American Secretary of State; and the UAP Prime Minister just before Japan entered the war, Robert Menzies. Curtin and Evatt played the major roles, Menzies an important subsidiary one, although he would eventually become the most important person in the shift of Australia out of Britain's geopolitical orbit and into America's with the creation of the ANZUS Treaty of 1951.

Australian historians tend to divide Australian politics into two main periods: before Curtin and after Curtin. Much of this has to do with the "without pangs" message, but Curtin also created a recognizable version of today's ALP. Although the ALP had been in effective existence since 1891, when it successfully fielded Trades Union candidates in the New South Wales state elections (McKinlay 1981, 8), it was wracked by internal dissension throughout the period between WWI and WWII. Like Labor parties everywhere it flirted with Socialism immediately following the success of the Bolshevik Revolution, although the ALP's specific origins as a Catholic and blue-collar movement caused it to reject any such revolutionary impulse by 1920. However, the party "continued to voice socialist aspirations, and retained for use the 'catch phrases of what was true international Socialism' which were to render it so electorally vulnerable to red scares" (Gordon 1979, 48). Despite Curtin's success in molding a reformed party from 1941 on, the ALP's opponents, especially Robert Menzies, continued to use "red scares" against it well into the McCarthyite 1950s and 1960s.

If the left-wing of Australian politics had problems finding its feet in the 1920s and 1930s the right wing did little better. In the period between WWI and WWII, the UAP was generally able to form a government only in coalition with the Country Party, a party that represented farm and ranch interests, supported "protection all around," balanced high secondary industry tariffs with subsidies for agricultural products, and had little concern with foreign affairs. In the late 1930s, however, such coalitions became increasingly fraught, in part over domestic policies, but also over rising voter concern that the UAP's support of the British Conservative Party's appeasement of Germany and Italy was actively dangerous.

> On European issues the documents [on Australian foreign policy] show Australia to have been consistently in favour of the appeasement of both Germany and Italy. The question of how to pacify Hitler runs as a theme through . . . volume [1], from the discussions at the Imperial Conference [of 1937] . . . through the confused consideration of the recognition of the *Anschluss* to the manifest relief of Prime Minister Lyons at the outcome of the Munich crisis. (DFAT I, xix)

In the 1937 election, however, the defeat of a sitting UAP Defense Minister in a supposedly safe seat by a right-wing opponent of appeasement indicated how much Australia's mood was changing. From then until late 1941 Australia went through coalition governments at an increasingly ferocious rate: the UAP/Country Party coalition finally failed when two nominally independent members, dissatisfied in part by UAP defense policies, crossed the aisle to sit with the ALP. In the aftermath of the virtual collapse of the UAP in the 1943 elections, won handsomely by the ALP, Menzies began the construction of a new right-wing party, the Liberal Party. In 1949 he led it to power, defeating the ALP decisively and holding office as Prime Minister for an unprecedented sixteen years.

One of the most interesting elements of Australian politics after Curtin is that both modern major political parties, Labor and Liberal, claim authorship of the American alliance, to the point that it has become almost a sacred cow. The Labor Party has a clear claim through Curtin's "without pangs" message "and subsequent Labor prime ministers . . . have depicted themselves as Curtin's disciples." The Liberal Party, however, considers its role in authoring the ANZUS alliance, formally signed in 1951 as a mutual defense past between Australia, New Zealand, and the United States of America, as the critical event. "This claim of joint ownership has been fundamental in instilling a structural disposition toward the U.S. into Australia's political culture" (Kelly 2004b, 18–19).

John Curtin

John Curtin's career in Australian politics was an interesting one. Despite his early interest in politics, his work as a labor agitator, and his opposition to conscription during WWI, which led to a brief jail term, he spent much of his life as a journalist, editing for many years the *Westralian Worker* in Perth, Western Australia. His informal education far outweighed his formal, but his several biographers point to the fact that he read widely and well (Chester 1943; Ross 1977; Day 1999). His speeches, writings, and letters have been selected and edited with some care (Black 1995; 2001). His *Backroom Briefings* of journalists during WWII, always a group he felt comfortable with, have also been published in edited form (Lloyd & Hall 1997). Visiting scholars at the John Curtin Prime Ministerial Library at the Curtin University of Technology, Perth, Western Australia have provided further specialized interpretation of different aspects of his career, including his role in shaping Australian foreign policy (John Edwards 2001; Paul Edwards 2001; Lake 2003; Kelly 2004a). There are several articles in academic books and journals on the same topic (Edwards 1979; Day 1988; Alcorn 1987; Prior 1987; Lee & Waters 1997). In a series of books dealing with Australian history during WWII David Day has paid particular attention to Curtin (Day 1988; 1992; 2002).

Given his background in organized labor, much of Curtin's early career and writing was spent on the basic issues of pay, working conditions, and union organization that are the daily stuff of left-wing politics at the practical level. On the basis of the body of scholarly attention noted above it is, however, easy to see the broad patterns in Curtin's thinking about the outside world. Almost all Curtin's career was within Australia and he left the country only twice. On the first occasion he represented Australia at the International Labor Organization (ILO) Conference organized under the auspices of the League of Nations in Geneva in 1924, traveling via Britain. On the second, in 1944, he traveled as Prime Minister to Britain via America to meet Roosevelt and Churchill.

At its inception, Curtin strongly supported the League of Nations, at least in the form he believed Woodrow Wilson intended it, but in his editorials in *Westralian Worker* he began to assail it as early as 1923 it for not living up to its promise: on June 1, 1923, he vilified the League as "the instrument for the dominant powers" and on August 17, 1923, as treating "nations other than the 'big five' (Britain, France, Italy, Japan, and the United States) as though they were of second or third rate importance." In the same year he described the treatment of Germany in the Treaty of Versailles as counterproductive and likely to cause more conflict in the future. On December 9, 1927, he described Russia as "the only powerful state which has given much tangible evidence of pacific intentions" (Black 1995, 73–75).

Curtin was first elected to the Australian Federal Parliament in 1928 and made his first serious foreign policy speech on August 7, 1930. Although he still supported the League he was increasingly worried about the American refusal to participate and argued that a League that included all the world's countries would better serve Australia's security. Despite his 1927 praise of Russia, his approach was that of a conservative internationalist, not a socialist one. He believed that "if the peace of the world is to be preserved, there must be ultimately half a dozen great world powers whose business it will be to preserve it" (quoted in Black 1995, 92). As Lake notes, he was not only an internationalist but also strongly anti-imperialist, although never able to reconcile this with his continued support of the "White Australia" policy that restricted migration to Australia. On the one hand, keeping out cheap colored labor would keep up the wages of the white Australian blue-collar workers he represented. On the other, he saw Australians as repressed by a colonizing power, not as herself a colonial power imposing its will on an indigenous people. In this he followed the general line of Australian reasoning at the time with regard to the aborigines: the massive impact of imported diseases meant that they were a "dying race" and the moral responsibility of the white man in Australia was "to ease the Black man's passing" (Lake 2003, 2–3).

Throughout the 1930s, because of "the deep divisions within the ranks of his own party among international socialists, pacifists, Catholics and other groups" Curtin generally avoided foreign policy statements (Black 1995, 136). He did, however, speak out on defense issues, arguing that Australia was by no means prepared, that the UAP was only too happy to depend dangerously on British security guarantees, and that Australian defense should be in the hands of an indigenous air force, not a Royal Navy based out of distant Singapore.

In a significant speech on the Australian Defense Estimates on November 5, 1936, Curtin noted Britain's decline: "with the development of air power and the re-emergence of Germany, Great Britain is becoming more and more a European and less and less a world power." When tackled by an opponent that "Great Britain has never failed us" Curtin showed an inkling of multipolarity: "history has had no experience of the situation I am visualizing." He was well aware that Japan was the likely aggressor, that the situation he envisioned was likely to be a global war, and that the Singapore naval base would prove useless as Australia's first line of defense.

> Great wars in which Australia's security is to be imperiled will not be European wars. They will be wars in the South Pacific. . . . That a naval unit based on Singapore which would obviously be weak in relation to Japan's naval strength would be a real deterrent to any hostile overseas intentions on Australia is contrary to all naval history. (quoted in Black 1995, 142)

Curtin's attitude to Russia was clearly based on Russian opposition to Japan in Manchuria and a belief that Russia might usefully restrain Japan from a naval push into the Pacific, as well as, one must presume, his socialist politics. On November 4, 1941, he cabled Lord Cranbourne, the British Secretary of State for Dominion Affairs, that the British should warn Japan that "any attack by Japan on Russia will be resisted by force by the British Commonwealth irrespective of the attitude of the United States of America" and that "a reciprocal guarantee should be sought from Russia that if Japan attacks in a southward direction Russia will declare war on Japan" (DFAT V, 97). In the event Russia, confronted by a European enemy that would be defeated by the Russian winter and the Red Army under General Zhukov, honored its Non-Aggression Pact with Japan until the last days of the war.

When Curtin found himself, shortly after he had become Prime Minister, forced to deal with the loss of Singapore to the Japanese, his reaction was one of righteous outrage. At the time, Churchill was pressing him to divert the Australian 7th division returning from the Middle East to the defense of Rangoon. Curtin summed up this period in a letter to his wife on January 5, 1942: "the war goes badly and I have a cable fight with Churchill almost every day" (quoted in Black 2001, 210). It got worse. Churchill dragged Roosevelt into the argument and on February 20, 1942, wrote Curtin one of his stiffest messages:

> Your greatest support in this hour of peril must be drawn from the United States. They alone can bring to Australia the necessary troops and air forces, and they appear ready to do so. As you know, the President attaches supreme importance to keeping open the connection with China, without which his bombing offensive cannot be started, and also most grievous results may follow in Asia if China is cut off from all allied help. . . . I am quite sure that if you refuse to allow your troops to stop this gap which [sic] are actually passing, and if in consequence the above evils affect the whole course of the war, a very grave effect will be produced upon the President and the Washington circle on whom you are so largely dependent. (BNA/FO954, 369–70)

Curtin did not give way and insisted, in the strongest terms, that the division be returned to Australia.

Part of the problem was that Churchill, at this period at the height of his powers and with America finally in the war, did not know if he could trust Curtin. It is true that Curtin had an impressive reputation in Australia as a totally honest and egalitarian politician. Interviewed by *Australian Women's Weekly* (*AWW*) in 1940 he was described in flattering terms: "calm, sensible . . . earnest, sincere . . . [and with an] attitude to women that is one of friendly equality without a trace of masculine superiority" (*AWW* September 28, 1940). But Curtin was a Labor politician and Churchill, for all his opposition to the appeasers in his party, a Conservative. The UAP governments of the 1930s had supported the appeasers of the British Conservative Party, who Churchill despised, but Curtin's ALP had never spoken out against appeasement, and Churchill had reason to believe they were complicit (Ross 1977, 255). Churchill had never met Curtin and the limited private information he had received conspired against Curtin, especially the unfavorable comments on Curtin made by Menzies to the British acting High Commissioner for Australia in July 1941. The whole situation could not have been helped by Curtin's reputation, which Menzies traded on: Curtin was a recovering alcoholic and subject to fits of depression.

The first positive view of Curtin, though not of his cabinet, did not reach Churchill until Viscount Cranbourne's long cable of January 9, 1942, by which time all were in crisis mode over Japan. Cranbourne, Britain's High Commissioner to Australia, provided Churchill his first balanced summary of Curtin's abilities. After the Japanese attack

> Parliament met, and the Prime Minister began and ended his appeal well, but for the rest read many pages of dreary departmentalese. In effect he failed to make use of the national rostrum on Australia's greatest parliamentary occasion in history. Such, however, is the lack of parliamentary tradition . . . that the Prime Minister's failure no doubt passed unnoticed amongst parliamentarians . . . Mr. Curtin made a further broadcast at the end of the year. This was excellent. The spirit and tone of leadership ran through the whole address. (BNA/FO954, 356)

Cranbourne's concluding sentence can have left no doubt in Churchill's mind that he could trust and depend on Curtin, although he would have retained Cranbourne's reservations about Curtin's "entourage," mainly Jack Beasley and Evatt.

> Mr. Curtin, contrary to all prognostications, seems to be standing the strain well. In my earliest acquaintance with him as Prime Minister I found that there was some little mistrust of a High Commissioner who so patently came out of a rival stable. I am glad to be able to record that our relations are now apparently cordial. I no longer apprehend any obstacle of a personal character, and conversation on both sides is friendly and frank. I find him likeable and capable, and whilst he is profiting greatly in prestige from the machinations of his entourage I esteem his character enough to doubt whether he has taken any part in giving character to false impressions. (BNA/FO954, 357)

In the event the "cable fights" continued until late March of 1942, when Churchill wanted to appoint Australia's Minister in America, Richard Casey, as British Minister to the Middle East. Curtin was not at all happy about this. Churchill pulled in the big guns: President Roosevelt and Viscount Halifax, the British ambassador to America, as well as others, all told Casey he should accept Churchill's offer. Eventually and reluctantly Curtin acceded. The "cable fight" over Casey ended with Roosevelt's telegram to Churchill of March 23, 1942 (BNA/FO954, 415). The appointment of Casey was part of a Cabinet reshuffle designed by Churchill to buy off some of his domestic and foreign opponents. Casey gained a seat in Britain's War Cabinet, thus blunting the constant criticism by Menzies of the lack of Dominion input to decisions about the course of the war. Evatt encouraged Casey to leave Washington because it assured Evatt far better access to the American government. Casey had reported direct to Curtin, not to Evatt and the Department of External Affairs.

One of Curtin's most serious and long-running cable fights with Churchill was over the "Beat Hitler First" strategy that Churchill and Roosevelt had agreed to at the "Arcadia" conference immediately America entered the war. This produced a top-secret document codenamed W.W.I. (Day 2003, 289). Curtin had no real understanding of Churchill's level of commitment to "Beat Hitler First" until Evatt, who had gone on to London from Washington, finally saw W.W.I. in May 1942 (Day 2003, 326). One of the reasons for Curtin's and Evatt's drift back toward Britain thereafter was that they realized that W.W.I. ensured that America was committed to more of a holding action in the Pacific than the aggressive response against Japan favored in Australia, at least until Hitler was beaten. A further reason was the behavior of the American General, Douglas MacArthur, sent to Australia after the fall of the Philippines as Commander-in-Chief of the American and Australian forces. Curtin received him as such, and this has been regarded as reflecting the "without pangs" frame of mind. But at the Prime Minister's War Conference of June 1, 1942, MacArthur made America's position quite clear to Curtin.

> Australia was part of the British Empire and it was related to Britain and the other Dominions by ties of blood, sentiment and allegiance to the Crown. The United States was an ally whose aim was to win the war, and it had no sovereign interest in the integrity of Australia. Its interest in Australia was from the strategical aspect of the utility of Australia as a base from which to attack and defeat the Japanese. As the British Empire was a Commonwealth of Nations, he presumed that one of its principal purposes was jointly to protect any part that might be threatened. (NAA/A5954/1/1)

MacArthur's statement was clearly throwing back in the Prime Minister's face Curtin's reference in the "without pangs" address "to 'traditional links

[and] kinship' with the United Kingdom." MacArthur "was asserting that it was to precisely those links and that kinship that Australia should look for its security" (Peter Edwards 2001, 4). Despite the "without pangs" message and his "cable fights" with Churchill, the combination of the implications of W.W.I. and MacArthur's nastily blunt statement ensured that Curtin had little choice but to turn back to Britain.

MacArthur was a problematic figure. His presidential aspirations, supported by the far right wing of the American Republican Party, caused Roosevelt to distrust him. Sending him to Australia got rid of him and silenced powerful domestic critics, who could scarcely complain that their standard bearer had been shut out of the war. Larrabee notes that Roosevelt thus used MacArthur both shamelessly and brilliantly to blunt the attacks of his right-wing opponents (Larrabee 1987, 351). Unsurprisingly, given his naval background, Roosevelt had agreed with "the strongest mind within the American Joint Chiefs of Staff" (Larrabee 1987, 153), Admiral Ernest J. King, that the way to beat Japan was through a naval war. The army came a distant fourth in influence in the Pacific war to the navy, the marines, and the USAAC, the last of which was nominally controlled by the army but effectively independent of it and certainly not under MacArthur's control.

MacArthur did not appreciate being banished and he justified his reputation for both nastiness and imperiousness: "he seemed not to regard American-Australian relations as worthy of his concern and rejected a proposal . . . for improving them. 'Shortly after I arrived in Australia,' said [Lieutenant General Robert] Eichelberger, 'General MacArthur ordered me to pay my respects to the Australians and then have nothing more to do with them'" (quoted in Larrabee 1987, 332). After MacArthur's bluntness at the meeting of June 1, 1942, Curtin may have cooled on him: he certainly did so on America (Lloyd & Hall 1997, 32), although he did not complain publicly. As he said in a press briefing of September 21, 1942, "I cannot moan to Roosevelt" (Lloyd & Hall 1997, 91): Australia lacked the "traditional links [and] kinship" that allow one partner to moan to another as Curtin certainly had to Churchill.

Australian historians argue that Curtin "retained a deep commitment to the British connection" (Black 1995, 237). The data can certainly be read that way, but Curtin's actions do not fully support a commitment to Britain *per se*. Curtin's commitment was more to the *idea* of Britain, and more specifically to a concert of British Dominions that would shape postwar policy and allow Britain and her Dominions to retrieve some of its declining Great Power status. In that regard Curtin was again showing how much he thought of Australia as a component of the British core, not an independent or semi-independent actor. This was at its clearest in the events of August 1943 to January 1944, especially in Curtin's "fourth empire" speech: "it has been said that we live in the third British Empire, and are in course of transition to the fourth." The third Empire had been composed of "autonomous communities within the British Empire, equal in status, in no way subordinate to one another": it began at the Imperial Conference of 1926 and, effectively, ended with the fall of Singapore. The fourth empire Curtin saw as an "association of independent sovereign peoples," the way to which was presaged by the ANZAC agreement between Australia and New Zealand, effectively a mutual defense pact, signed in February of 1944 (ANZAC Documents Online).

By the time of the "fourth empire" speech Curtin no longer believed in a Britain that could "manage the Empire on the basis of a government sitting in London." For Curtin Britain now needed the Empire in order to continue as a Great Power: "the power of Britain as a power for peace in the future will be strengthened in the world if the firm voice against potential aggressors comes from the Empire, and not merely London" (quoted in Black 1995, 237–239). Curtin pushed this approach at the Commonwealth Prime Ministers' Conference in London in 1944 but his "proposals failed to obtain serious attention. Curtin's opponents quietly pushed them aside. Churchill was unenthusiastic and suspicious; Sir Frederick Eggleston [Australian minister to China and a strong supporter of Evatt (*ADB* 8, 423)] walked out while Curtin was speaking" (Ross 1977, 360).

Curtin "was moody in England and very lonely" (Ross 1977, 359), hardly surprising when his proposals for a "fourth empire" were being shot down. Britain would seek to regain core status without the central participation of the Dominions. Upon his return to Australia in June 1944 Curtin's health deteriorated. In November he was hospitalized. He died on July 5, 1945.

Herbert Vere Evatt

Of the three key individuals, Herbert Evatt was the most complex and the most controversial: "a heroic and epic person who . . . attempted to change not only his country's way of life but that of the world" (Tennant 1972, 11). Evatt entered politics in 1925 as a member of the New South Wales legislature. He traveled much more than Curtin: in 1926 he attended a Congress on world migration in London and returned via Geneva, where he observed an ILO meeting. Back in Australia he had a successful career as a barrister and High Court judge. Like many who reshaped Australian politics in the late 1930s his distaste for appeasement made him enter Federal politics.

> Its little use talking about Chamberlain and his gang—they are really sympathetic with Fascism because their one fear is insecurity through Socialism. Money first and to hell with their own country. Even Palmerston's jingoism was better than their black treachery. (Evatt quoted in Buckley, Dale & Reynolds, 141)

In 1940 Evatt resigned from the bench and ran for office on the ALP ticket. An ambitious man, his star rose quickly, and he was appointed Attorney General and Minister for External Affairs when Curtin became Prime Minister in late 1941. Evatt was by no means the standard issue ALP politician: most such came from blue-collar origins and had little formal education. Evatt implied a more lowly family background than he had: he came from the middle-class, was well-educated, held an L.L.D., married into a wealthy American family, patronized the avant-garde arts, and wrote extensively on Australia's role in the world. He brought many "new men" from similar backgrounds into the ALP at External Affairs, notable among them John Burton, educated in Sydney but with a Ph.D. from the London School of Economics in 1941, who became Evatt's private secretary at External Affairs in 1942.

At his best Evatt refused to give way to traditional Great Power politics: at his worst he was "a generally poor leader whose political ability was best expressed when he was left to indulge perceived injustices free from party regulation. An arrogant inattention to ALP procedures aggravated his inadequate relations with parliamentary colleagues" (Crockett 1993, 8). He could, however, certainly charm when he wanted to. When interviewed for *AWW* before his election the interviewer described him as "as casual and as friendly as your own big brother," devoted to his wife, his two children, and the memory of his "adventurous, courageous" mother (*AWW* September 14, 1940). Much of his management of Australia's External Affairs was driven by his concern for international law and justice that came out of his legal background. In a speech of December 16, 1941, he presented his first "brief" against the Japanese.

> Japan's ultimate design against Russia can hardly be concealed. A typical instance occurred in July of 1938, when Japan was guilty of flagrant aggression against Russia in Manchuria. The claim of Japan was a dishonest one because a reference to the map annexed to the relevant treaty showed at once that the territory in dispute belonged to Russia. Russia resisted, the Japanese were forced to retire, and the [Chankufeng/Nomonhan] "incident" closed. But it was quite in keeping with Japan's long record of broken treaties even since 1931. . . . [I]t may be fairly said that Japan, although the last into the open in fighting the democracies, was really the first of the Axis aggressors. The truth is that Japan's conduct in 1931 turned out be the first step in the attempt to destroy the entire basis of international law and international justice. (Evatt 1945, 22)

One of Evatt's first acts in power was a War Cabinet Submission of November 4, 1941, for "direct trade and diplomatic relations" with Russia. One of his reasons for this was "common political interests, in particular the consideration that Soviet policy in respect to Japan . . . is of importance to Australia" (DFAT V, 96). Australia would use a considerable amount of time and effort attempting to cement this relationship, to no avail (DFAT V, 115, 212, 350). Evatt then lost little time in getting to Washington to plead Australia's case direct: "Evatt already had some standing on the world stage. From as far back as the thirties he had numbered Franklin D. Roosevelt, Clement Attlee, Stafford Cripps, Jawaharlal Nehru, and Harold Laski as his friends, and as a High Court judge with leanings to liberalism in law and socialism in politics, his erudition had earned him entrée to the salons of international politics" (Dalziel 1967, 21). This may be over-claiming by Dalziel, who was private secretary to Evatt in his role as Attorney General (*ADB* 13, 569), but it indicates that Evatt had far more standing outside Australia than any other ALP politician of his time. At the time of his visit to Washington in early 1942, however, Evatt had not seen W.W.I. and he believed MacArthur had some flexibility in interpreting his orders. MacArthur's statements certainly indicated that he believed in an aggressive response to Japan. At this time also Evatt had not met Churchill, and would not do so until he went to London from Washington in May 1942. Churchill had thus not been able to take Evatt's measure directly and had to rely on written reports of the man.

Churchill's concerns over Evatt's reliability continued much longer than his concerns over Curtin, and it may be this that contributed to his "cable fights" with Curtin extending into late March of 1942. On March 21, 1942, in the middle of the fight over Casey, Churchill cabled Halifax expressing concern about Evatt: "he is as you know reputed to be the least friendly of the Australian Ministers and most eager to throw himself into the arms of the United States" (BNA/FO954, 408). On March 22, 1942, Churchill cabled Roosevelt

> The [Casey] matter is complicated by Australian party politics which proceed with much bitterness and jealousy regardless of national danger. The present Labour Government in Australia, with a majority of one, contains various personalities, particularly Evatt and Beasley, who have made their way in local politics by showing hostility to Great Britain. . . . I shall be most interested to know your personal impressions of Evatt and how you get on with him. (BNA/FO954, 414)

Churchill's worries about Evatt's reliability were laid to rest by two cables from Halifax. In a cable of March 23, 1942, Halifax noted "I had a long talk with Evatt this afternoon." Halifax directly quoted Evatt as saying "we do not want MacArthur to be entirely running the war" and concluded that Evatt's "general attitude of mind, though he naturally sees the Australian side bigger than any other, was better than I expected; and having expected to dislike him, I found myself largely and rather pleasantly disappointed" (BNA/FO954, 416). On March 29, 1942, Halifax dashed off a second cable to Churchill: "Evatt addressed private meeting on March 26th. Influential people. One who was there tells me that after stressing Australian situation and needs he said 'but do not misunderstand me. If I had to choose whether Australia or Great Britain should go down I would wish to save Great Britain, for Great Britain is bastion of us all.' The audience much impressed" (BNA/FO954, 428). After that, Churchill's "cable fights" with Curtin were over.

Even before Evatt had seen W.W.I. he was clear in his support of Britain. After he had seen W.W.I, he supported it, noting in his address to the University of California in March 1945

> In the grave emergency of early 1942 it was arranged between President Roosevelt and Mr. Churchill that, broadly speaking, the strategic responsibility for the Pacific should lie with the United States. So we who were responsible in Australia bent almost the whole of our energies to forging links which would bind Australian manhood and Australian resources with whatever forces the United States could make available, having due regard to the undoubted fact that the American effort in the Pacific had for a time to be made subject to the over-riding conception of the accepted "Beat Hitler First" strategy. (Evatt 1946, 5)

Although Evatt began by supporting Curtin, as the war progressed "two rival centres of policy-making were being formed." Curtin placed his "emphasis on smoothing Australian relations with both the United Kingdom and the United States," Evatt "tended to assert Australian independence of the great powers" (Edwards 1979, 555). Evatt was at once by far the worst politician of the three key individuals, the best theoretician, and the most committed to Australian nationalism. Once he had realized the commitment of Roosevelt and Churchill to W.W.I. Evatt developed two solutions to the problem of Australia's role in international affairs, one based on the assumption that multi-polarity would continue and that Australia would not be able to depend on "great and powerful friends" for its security guarantees, the other on the assumption that multi-polarity would end and that Australia would need some sort of checks and balances to occasionally restrain such "great and powerful friends." Neither policy argued for dependence upon a hegemonic power. To the first end Evatt supported regional security guarantees such as ANZAC to control the region around the two countries. Technologically, air power would make this possible, hence his obsession with acquiring such as the means to make Australia relatively independent of its "great and powerful friends." This extended in the ANZAC Treaty to complete control by each country of "air transport services within its own national jurisdiction and its neighboring territories" (Evatt 1945, 156).

The second end Evatt proposed to reach through the United Nations (UN). Roosevelt and Churchill had made clear in the Atlantic Charter "the necessity for all free countries of the world to co-operate in the post-war period" (Evatt 1945, 12). Evatt strongly supported this, even before Pearl Harbor, specifically invoking ALP ideals to such ends in a speech of November 27, 1941:

> At the recent International Labour Organization Conference in New York, Australia was represented and a resolution sponsored by the American delegation was adopted. That resolution foreshadowed further conferences with the object of rebuilding international relationships on a basis of improved labour standards, liberal economic adjustments, and social security. (Evatt 1945, 12)

That support did not prevent reservations. Evatt read the history of the League of Nations as abysmal. In a speech of April 28, 1943, Evatt noted "unless the nations have the real *intention and will* of carrying out international obligations, little can be expected of international organizations" (Evatt 1945, 120). A UN with *intention and will* became a primary war aim of the western alliance, a UN that would owe little to the League. For Evatt "the main and immediate point of the UN for the world in general and Australia in particular was that it comprised a collective security system" (Hudson 1993, 153). However, the initial form of the UN Charter would have seen its power vested very heavily in a Security Council comprised of the victorious great powers. Evatt worked tirelessly at Dumbarton Oaks to ensure that no such intense concentration of authority came about. Although he was unable to reduce the centrality of the Security Council he was able to substantially increase the powers vested in the General Assembly, which gave the lesser powers at least some redress against high-handed actions by "great and powerful friends." His justification was straightforward:

> Countries that attempt blindly to follow a leader frequently find themselves in difficulties. The great powers are not always consistent in their own policies. . . . Nations which attempt to follow them are frequently humiliated by having to change simultaneously. (Evatt 1949, 36)

Robert Menzies

Of the three key individuals Robert Menzies is the simplest to understand. He saw the world in simple black and white terms. He opposed Bolshevism and Communism whatever the cost, and it cost him dearly in 1941, though not in the McArthyite 1950s and early 1960s. His readiness to sign the ANZUS Treaty in 1951 and bring Australia firmly into the geopolitical orbit of America was based on his strongly anti-communist stance. Despite his obvious preference for Britain in the 1930s and early 40s, he argued that Australia's need for "great and powerful friends" who would be "able and willing to support policies designed to further Australian interests" (Watt 1968, 109) outweighed all else, and that as Britain declined, the fact of which he realized very much later than Curtin or Evatt, Australia had no choice but to throw in its lot with America. In 1951 when Menzies and the Liberal government signed the ANZUS Treaty along with New Zealand and America this was seen at the time by America as the price America had to pay for a peace treaty with Japan about which Australia had very serious reservations. This treaty, plus Menzies' committing Australian troops to the wars in first Korea and then Vietnam, reinforced Menzies' anti-Communist credentials and continued to strengthen Australia's relationship with America.

Menzies was the son of a very small-town storekeeper who pulled himself up by his bootstraps. A good student, he was able to earn scholarships to good institutions and became, like Evatt, a successful barrister. He started in politics in the Legislative Council—the upper house—of the State Parliament of Victoria, probably to minimize the impact of politics on his professional life, but he moved up quickly and was acting Premier by 1933. In 1934 he entered Federal politics, being elected as a UAP member and being immediately taken into the UAP/Country Party coalition cabinet as Attorney General. His meteoric rise continued, and after the death of Joseph Lyons in office the UAP elected Menzies party leader, albeit in the face of strong opposition within the UAP, and he became Prime Minister in 1939. In 1941 the problems posed by his lack of real support from a divided UAP/Country Party coalition caused Menzies to leave his first term as Prime Minister under a cloud, effectively deposed by Arthur Fadden of the Country Party, who had been Menzies' acting Prime Minister while Menzies was on a long visit to London (Martin 1993, 61, 112, 122, 272, 382).

On that long visit to London, which stretched from January 24 to May 6, 1941, Menzies did two things. First, he sought more independence for Australia in industrial terms. In 1939 the UAP government had acquired advanced aircraft production facilities in the form of the license to build Bristol *Beauforts*. Menzies also pushed hard for the production of British tanks in Australia, a push that consumed much time and capital and produced no tanks. At one stage in his 1941 visit he left London "for a week long tour of Britain's industrial heartland" in the company of such British industrialists as William Rootes, President of the Society of British Motor Manufacturers. One of Menzies' stated goals on this trip was to establish in Australia a branch of the British automobile industry (Day 1993, 93). Several British companies would try to do this after WWII, and all would fail. It was the American giants, General Motors, Ford, and Chrysler, the first through their Holden subsidiary, that were successful. British automobiles were by no means suited to Australia's "colonial" conditions and poor roads, but the products of the great American companies indisputably were.

Menzies' second goal was to establish himself as a member of the British War Cabinet, then to become Churchill's successor. He made repeated efforts to enter the War Cabinet, which Churchill resisted strongly. In London he met William S. Robinson, the powerful and influential Australian mining baron, who suggested he meet David Lloyd George, one of Churchill's hidden but powerful opponents. Although Robinson was generally in favor of the ALP—he once described himself to Dalziel as a "Pink" (Dalziel 1967, 21) and after the ALP came to power he became an almost permanent presence in Evatt's entourage—Robinson also had a foot in both camps of the British Conservative Party. Before the ALP came to power Robinson seems to have believed that bringing Lloyd George and Menzies together would increase the chance of a negotiated peace and an early end to the war (Day 1993, 133–134). Menzies had plenty of contact with the Cliveden Set, hardly surprising given his predilection for appeasement: Lord Astor pushed the case hard for Menzies to enter the War Cabinet "to stand up to Churchill as his colleagues were not inclined to do" (quoted in Day 1993, 206).

Day argues that, had Menzies entered the British War Cabinet, he would have been able to lead a much more effective opposition to Churchill, interfere with Churchill's management of the war and, after the inevitable crisis, push Churchill aside in his own favor and that of the proponents of a settled peace who were supporting him: in other words, Menzies was a man whose over-weaning ambition at this point in his life was to be Prime Minister of Britain, not of Australia (Day 1993, 2). As Oliver Hardy, principal private secretary to Anthony Eden, then British Foreign Secretary, noted in his diary for June 2, 1941, "goodness knows where [any successor to Churchill] would lead the country to—to Berchesgarten [sic] probably" (quoted in Day 1993, 197). Churchill, whose political nose was sensitive, can hardly have missed Menzies' tendencies, whether to appeasement or to *putsch*. Churchill was glad to get Menzies back to Australia and, as Day makes clear, from that point on he resisted all Menzies' efforts to get back to London (Day 1993, 191–232). Menzies' only choice after the UAP disaster in the 1943 election was to get his nose back to the grindstone of Australian politics, which he did brilliantly in his invention of the Liberal Party, his guiding of that party to electoral victory in 1949, and his subsequent leadership of it as Prime Minister for sixteen years. In the aftermath of the war, "when Churchill and Menzies were both comfortably reinstated as Premiers of their respective countries, they sought to gloss over their previous conflict" (Day 1993, 4).

However, in the early part of WWII "Menzies . . . , in the 'phoney war' period, maintained his prewar desire for appeasement of Germany. Privately he hoped for a soft peace with Germany but he did not say this publicly for fear of opposition from the Australian press and some ministers of his own cabinet" (Buckley, Dale & Reynolds 1994, 139). His correspondence with London and Washington tells the true story as the Polish crisis devolved into the outbreak of war in Europe. On August 27, 1939, he cabled Neville Chamberlain, the British Prime Minister that "I would make it clear to Hitler that we regard merits of Danzig and the [Polish] Corridor as quite open to argument and that we would use our influence with Poland to procure some form of arbitration or adjustment" (DFAT II, 156). On September 1, 1939, Menzies cabled Chamberlain again that Australians "would not consider Danzig or the Corridor intrinsically worth a war" (DFAT II, 174). That same day he cabled S. M. Bruce, Australian High Commissioner in London, that the German proposals regarding Poland were "surprisingly reasonable (DFAT II, 176). Once the war had begun Menzies claimed he was committed to British victory: in his broadcast speech to the Australian people of September 3, 1939, he noted that if Hitler's ambition was allowed to go unchecked "there could be no security in Europe and there could be no just peace for the world" (DFAT II, 189). Yet on September 11, 1939, Menzies wrote to Bruce "I feel quite confident that Hitler has no desire for a first class war" (DFAT II, 218). He noted to Bruce on another occasion the desirability of a "new alignment of nations in which not only Great Britain and France, but Germany and Italy, combined to resist Bolshevism" (quoted in Buckley, Dale & Reynolds 1994, 141).

For his part, Bruce supported substantial appeasement of Japan. In a note to R. A. B. Butler, then British Parliamentary Under-Secretary of State for Foreign Affairs, of September 21, 1939, he argued that Britain should "facilitate Japan's acquiring necessary supplies of raw materials, including wool . . . [and] also should offer to remove restrictions against Japanese trade in the British Colonial Empire. . . . As part of a settlement it would probably be necessary for the Chinese Government to agree to recognise either permanently or for a period of years the existing status quo in Manchukuo" (DFAT II, 239). Given Butler's role in the coalition British War Cabinet after Churchill became Prime Minister it would be surprising if he had not appraised Churchill of Bruce's preference for appeasement.

As the situation in France deteriorated, on May 26, 1940, Menzies asked R. G. Casey, the Australian Minister to America, to arrange for an "immediate personal presentation to President Roosevelt on behalf of myself." This communication was in a much darker vein. In it he noted his concerns that America, to defend itself, should do all in its power to bolster the air defense of Britain, but that if Britain fell "I hope you will not find me unduly unconventional if I say to you, as the head of the greatest but most friendly neutral power, that the prevention of [the] British fleet falling into German hands in consequence of a defeat of Great Britain must be of vital importance to the United States" (DFAT III, 280). A British fleet in German hands would, of course, have been a nightmare for Australia.

Despite such grand strategic concerns, Menzies had plenty of energy left over for his political opponents at home. On July 7, 1941, R. R. Sedgwick, the Acting British High Commissioner for Australia, wrote Sir Eric Machtig in the Dominions Office on the subject of a possible ALP government. He reported a conversation in which Menzies was of the opinion that "neither mentally nor physically was Curtin strong enough to stand the strain" of being Prime Minister. Menzies had some but "not much use for Beasley, but none at all for Evatt whom he described as unquestionably an able lawyer but a man of no character, consumed with ambition, and hairy about the heel" (BNA/FO954). With such an evaluation on file in London it is scarcely surprising that Churchill should have had serious doubts about whether he could trust Curtin and Evatt, nor that the "cable fight" between Churchill and Curtin should have occurred. Menzies' character assassination of the Curtin Cabinet did not stop there. He next tried to infect Roosevelt. "In April 1942, the United States Consul-General in Melbourne, Dickover, reported to Washington on a conference between himself and R. G. Menzies, in which the latter referred to the Labor government as 'scum-positive scum.' In Menzies' view Curtin was 'reasonably safe,' but Ward and Evatt were 'positively menaces to Australia'" (Buckley, Dale & Reynolds 1994, 164). Before these attacks, Menzies had even gone so far as to claim in a letter dated September 11, 1939, to the Australian High Commissioner in London, Stanley Bruce, that Curtin did not want to be Prime Minister: "In point of fact, Curtin has privately made it clear to me that his people will not put me out to put the Country Party in, and that his own greatest ambition is to remain Leader of the Opposition for the duration of the war" (NAA/M103/1 1938/39).

Despite his meteoric rise in the UAP Menzies failed to impress other contemporaries, either with his behavior or his patriotism. Adele Shelton-Smith, the writer for *AWW* who also interviewed Curtin and Evatt to get a woman's perspective on the war, expressed an almost feminist reaction to Menzies' sexism, an unlikely reaction at the time in Australia: "he does seem to be a bit superior to women. His manner suggests that even if you don't want to be a helpless little thing or a hothouse flower, even if you don't want to stay home, keep house and do the mending, that's how you ought to feel." Shelton-Smith went on to comment that "Mr. Menzies' morning suits, homburgs, cigars, and sleek official cars tend to make us (some people say they make Mr. M. too) forget that he is still a 100 percent Australian, the son of a general storekeeper at Jeparit, Victoria" (*AWW* September 28, 1940).

SUMMARY AND CONCLUSIONS

State actions play out on the complex and often interrelated stages of international politics and warfare, and the character, motivation, and historical understanding of the various actors involved, in short, agency, can be highly relevant. Even the structural analysis of WWI and WWII presented concedes the huge impact of such as Churchill on the second of those wars. A war of hegemonic transition between America and Germany was pretty much a certainty by the early 1900s, and a fact by 1918, but America failed to take hegemony. *Pace* appeasement and the handing of hegemony to Germany by default, a second stage to that war was thus a certainty. Its precise course was, however, much less predictable, its outcome even less so.

Britain's role in all this was, of course, crucial. Britain had been locked in an economic struggle with America since 1861, but as a fellow trading state was most unlikely to convert that into any form of military struggle. Britain was also a politically much more complex actor than America: she had an Empire to consider, and many elements of that Empire were independently minded actors in the transition as it unfolded. Australia's role indicates many things about the British-American transition. First, even the best-educated and most intelligent politicians saw such transitions only darkly as they occurred. For example, only Keynes and his followers understood that it ought to be possible to manage business cycle induced swings and begin to tame capitalist economies. Keynes' contemporaries sought other solutions: International Socialism, National Socialism, and Communism were just the most prominent. The causes, even the existence of, longer economic swings in the capitalist world economy are still a subject of much academic discussion, with a substantial literature on Kondratiev waves, world leadership cycles and the like. None of the 1940s era actors discussed here could have had much if any intellectual understanding that they were in a multi-polar world caused by the hegemonic decline of Britain, although their discourse included notions of first, second, and third-rate powers, and they were clearly aware that Britain's power was declining and America's rising.

Australia's problem was compounded by Japan, which was an irritating sideshow in the war of hegemonic transition. Japan saw itself, like Britain and America, as a maritime power, had serious cultural obsessions with national honor, misunderstood the route to hegemony, and invested massively in a military its economy could not support. To make matters worse, it switched sides between WWI and WWII, although after the Washington Naval Treaty Japan had little choice given America's demand that Britain choose between them. The "Beat Hitler First" strategy of WWII owed its origins to both American and British understanding of the process of hegemonic transition, however ill defined. Defeating Germany mattered far more than defeating

Japan, a second-rate economy with a first rate navy, at least before the Battle of Midway.

Australia was thus caught in a classic "storm made in other men's worlds." Culturally Australians believed themselves part of the British core, although the actions of Curtin, Evatt, and Menzies were all moving Australia toward "independence." But when the Australian relation to the British core was revealed as hollow a series of very rapid shifts occurred in Australian thinking about External Affairs. Menzies, Evatt, and Curtin all had very different ideas about how these should work out.

Menzies was the simplest: he initially tried to bail out of what seemed a sinking Australian ship and failed to commit himself to his country until he had no other choice. If he could have exercised his considerable talents in the British War Cabinet he would have had a fighting chance of succeeding Churchill as Prime Minister and negotiating the peace his British supporters desired. Those supporters had supported appeasement in the late 1930s and were in disgrace once Churchill rose to power: Menzies, whose support for appeasement had been far less public, would have been their Trojan horse. Appeasement made perfect sense to those who believed Fascist Germany was better than Bolshevik Russia, although the likely price of appeasement was German hegemony. As Russia and then America entered the war, the idea of a negotiated peace faded, Menzies shifted back to a belief in the likely continuity of the British core, although his support for the ANZUS Treaty in 1951 showed him shifting to America. Not until some twenty years after the fact would Menzies complete, as Liberal Prime Minister, Australia's transfer of identification from a failed Britain to a rising America, from one "great and powerful friend" to another.

Evatt was a very different character, less a politician than a legal theoretician playing a politician. His great strength was in his attempt to plan for at least two possible futures. In the first possible future, multi-polarity would continue, necessitating a set of regional security guarantees crafted by Australia for an Australia without "great and powerful friends." Such a future also required a local armaments industry able to project power into at least the surrounding region. One of Evatt's legacies was that Menzies' Liberals were thus as committed as the ALP to an ability to project Australian power beyond Australia's boundaries.

In Evatt's second possible future, he saw that Australia might not always be able to trust its "great and powerful friends." After all, his experiences of 1941 and 1942 had taught him all about British and American *realpolitik*: Britain had withdrawn to the European war zone and abandoned Australia to its fate; America had treated Australia as an unsinkable aircraft carrier in Japan's backyard. Evatt's strong support of the UN, in particular his creation of a strong General Assembly, offered the possibility that groups of small powers would be able to deflect the "great and powerful" when their actions brought their friendship into question.

Curtin played perhaps the most complex role. Like Menzies he was first and foremost a practical politician. Like Menzies and Evatt he supported Australia's accelerating industrialization via high tariffs. As an ALP politician, industrialization, high tariffs, and the "white Australia" policy meant better pay and conditions for his core constituency, organized labor. Downplaying commodities would also have meant some lessening of the influence of the opposition Country Party. His obsession with aircraft production, a much higher technology item than automobiles or tanks, helped accelerate Australia's shift toward industrialization.

As with Evatt, it was in foreign policy that Curtin made his greatest contribution, but Curtin was more of a realist than Evatt. He always accepted that the world would be controlled by the "great and powerful" and that it would be better if they were Australia's "friends." Britain's decline gave him little choice but to embrace America once Japanese aggression shifted south into the Pacific in December 1941. The realization that America, as MacArthur so bluntly put it, "had no sovereign interest in the integrity of Australia," compelled Curtin to turn back to Britain after June 1942. By this time Curtin's "cable fights" with Churchill were over, he had no illusions over the commitment of Roosevelt and Churchill to W.W.I. and the "beat Hitler first "strategy, and Evatt was finally ensconced in Churchill's good graces. Despite his early praise of Wilsonian idealism, Curtin's career was that of a realist, never an idealist. The failure of a League of Nations unsupported by America was a valuable lesson. Long, hard work in the journalistic trenches also assured he was no theoretician, at least not in the sense that Evatt was. Neither of Evatt's options could have made much sense to him. Curtin thus failed to support the ANZAC Treaty to either Roosevelt or Churchill in 1944, and he pushed a very different worldview than that presented by Evatt in his proposals for the UN. Curtin's actions show that by 1944 he believed Britain might well recover some aspect of its global power, and that Australia's best interests lay in restoring the British core in its more expansive sense, as a Commonwealth of the Dominions in a "Fourth Empire." He knew, at any rate, that America would pursue its own interests and that Australia might or might not benefit. His death makes moot the question of whether, after it was clear that Britain was not about to accept the "Fourth Empire" idea, he might have come round to Evatt's solution of a powerful UN. In retrospect, however, something akin to a "Fourth Empire" was probably Britain's only chance of staying out of America's hegemonic shadow after 1944.

Conclusion

The simplest fact to remember about the British/American hegemonic transition is that it was far from simple, not least because so many other political actors became involved. The central transition was a classic Type II struggle between two trading states. However, the complex period of multi-polarity that began in the 1880s and the two German wars of the twentieth century ensured that both type I and type III struggles were also involved. Germany played, in general and despite Mackinder's comments on its trading polity in *Democratic Ideals and Reality* (1919), the role of a territorial state throughout. With the exception of France the other powers involved—China, Italy, Japan, and Russia acted as territorial states. Germany's struggle against Britain and America was thus a classic type I struggle. But Germany was also engaged in a type III struggle against Russia in WWI and the Soviet Union in WWII. As WWII developed Japan became embroiled in a type III struggle with China and then, on the Manchurian frontier, with the Soviet Union. Most histories look at the military actions of the twentieth century and conclude that these military struggles were what marked the transition, when they were really massive noise in the system that covered up the main action, which was the type II transition from British to American hegemony.

Another set of complications comes from political and military history, which tends to look for causation to agency rather than social and economic struggle. Certainly historical actors such as Churchill, Roosevelt, Hitler, and even many lesser actors, such as the Australians: Curtin, Evatt, and Menzies played important roles. Indeed, the first complication came from the failure of President Woodrow Wilson to persuade the United States Senate to sign off on the Treaty of Versailles and the slide into isolationism by the suite of largely incompetent and isolationist presidents who succeeded him. As a consequence

> A sense of betrayal permeated the British government. . . . When, promising isolation in foreign policy, the Republican Party under Warren Harding took the White House and both houses of Congress in elections in November 1920, the possibility of Anglo–American co-operation evaporated. As the 1920s unfolded, therefore, the British faced their external problems believing not only that the United States would probably remain aloof from political entanglements, but that its foreign policy was by nature unreliable. . . . [N]otions of retreating within the Empire could not be countenanced because of the peace treaties, earlier commitments like that concerning the Bosporus, and new responsibilities assumed by joining the League. (McKercher 1999, 14)

Wilson may also have played a more subtle and unstated role. As a Virginian he was the first president elected from the former Confederate States after the Civil War and a fervent segregationist. He headed the first Cabinet of Southern Democrats since secession. Segregation of Federal offices became normal in his administration. He believed fervently in the superiority of the white race and thus pushed America back to its mistakes of the 1770s and 1780s. As a white supremacist Britain's multiracial, multicultural empire must have sat poorly with him. As a political scientist he must have been struck by the ungovernability of such a construction. Tooze notes that the America that emerged from the Civil War was "something judged impossible by classical political theory. It was a consolidated federal republic of continental scale, a super-sized nation state" (Tooze 2014, 14). America flirted briefly with conventional imperial expansion in the 1890s, following the sense that its internal frontier had closed (Turner 1893). But America had realized quickly that such an approach was massively at odds with the political structure that had emerged from the Civil War: "the one thing that the US could well do without was a ragbag of ill-assorted, troublesome colonial possessions. On this simple but essential point there was a fundamental difference between the continental United States and the so-called 'liberal imperialism' of Great Britain" (Tooze 2014, 15).

In the rather narrow view that agency was all-important Wilson and the isolationist Republicans thus gave Britain no choice in the 1920s but to ignore America and cling to its Great Power role, despite the fact that it was increasingly obvious that the massive expenditure on the Empire and the military that this implied was economically unsustainable. Cost cutting was one possible way out, and certainly what Churchill had in mind in encouraging "air policing" of such fractious additions to the Empire as Iraq. But the primary emphasis of the 1920s had to be on containing the rising tide of Franco–German animosity over reparations and of the 1930s in attempting to contain a revitalized Germany and an increasingly belligerent Japan while trying to make sense of an unreliable, inward focused America and an inscrutable

Soviet Union— the last "a riddle wrapped in a mystery inside an enigma" as Churchill had it.

According to McKercher, from the viewpoint of political history "Britain's position as the leading Great Power was being eroded as 1941 ended" (1999, 310). From a world-system perspective that position had evaporated long before, beginning in the 1880s as the American and German economies overtook the British. In that regard Paul Kennedy's declinist views in *The Rise and Fall of the Great Powers* (1987) have relevance: Britain was massively overextended and clung to its sense of itself as a "Great Power" far longer than it should have, although part of that mistake can certainly be attributed to the American failure to assume hegemony after WWI.

CLARITY IN HISTORY

When one is searching for the forces underlying the hegemonic succession between Britain and America there is remarkable clarity to certain moments in history: the 1770s and 1780s; the 1860s; the late teens and early 1920s; and the mid-1940s. In the first of those moments America failed to set the scene for hegemony even though it was a potentially powerful actor. In the second it largely corrected the faults of the 1770s and 1780s and began to set the scene. In the third it failed again to take hegemony when it was "on the table." In the fourth and last it finally grasped it. Although social and economic forces were the main drivers, agency played a part.

The 1770s and 1780s were a period of multi-polarity in which France was clearly challenging a Britain that was at the end of its first world-leadership cycle (Modelski 1978). Lost in the general sense that France was strong was the fact that the new United States was also positioned to challenge a Britain seriously weakened by its loss of the American Revolutionary War and the massive economic impact of that war on the British national debt. In the 1780s all three states seemed to have major economic problems, though France certainly had military strength. France's taxation system was, however, massively inadequate and the accumulated costs of financing the American Revolution (Stinchcombe 1969) so strained it that King Louis XVI was forced to convene the estates general in an attempt to fix it. This would push the French state down the line to its own Revolution in 1789. France's credit worthiness suffered accordingly and Napoleon was unable to borrow to finance his wars. America's economic weakness and its dependence on loans from the Dutch and British once the Revolution was over was massively compounded by the fact that, despite the lofty statements of the Declaration of Independence, black men were apparently not created equal to white men and women, of course, did not count. An America that had not perpetrated slavery would have had a very different historical trajectory. Much of the refusal to embrace Hamilton's "Philadelphia System" of protection for American industry came from Southern slave-owners. An America of Jeffersonian yeoman farmers growing cotton across the South and fueling an American industrial revolution behind tariff walls and based on Samuel Slater's transfer of Arkwright's cotton spinning technology to New England in 1793 would have been a powerful challenger (Hugill 2010). Why focus on cotton? Textiles, and cotton was always the most important textile, were the driving force of industrialization around the globe. They were the leading sector of the British economy until 1870, of the American economy until 1930, of the Japanese economy until 1958, and of the Chinese economy until the mid-1990s (Hugill forthcoming 2020).

However, the 1770s and 1780s saw the conditions established for the hegemonic rise of a revitalized Britain by one major reform and "revolutions" in three arenas: political, economic understanding, and industrial organization. The reform was in taxation. As chancellor of the exchequer and prime minister, William Pitt the Younger undertook a massive overhaul of the British taxation system in the aftermath of the American Revolutionary War (1776–1783) to bring Britain's national debt under control and restore the nation's credit worthiness. Without such, Britain would not have been able to borrow heavily to finance the nearly continuous war with France that ran from 1792 to 1815 and that is generally known as the Napoleonic Wars.

In the political arena the revolt of the American colonies in 1776 was ultimately a victory for Britain in that it removed both a major source of expense in the administration of a distant and fractious territory and a massive source of political embarrassment for a British society that increasingly practiced democratic politics: "no taxation without representation" resonated as strongly in Britain as in the thirteen colonies. In the economic arena, that same year Adam Smith published his epochal *Wealth of Nations*, pushing Britain away from mercantilist economics and toward liberalism. Smith's work was complemented by that of David Ricardo: *Principles of Political Economy and Taxation* (1817) would affirm the doctrine of comparative advantage. Yet in many ways that doctrine was prefigured by the writings of Lord Sheffield, who made it clear as early as 1783 in the Sheffield doctrine that Britain would use "counting house imperialism," what Lenin would eventually label "financial imperialism," to take back control of the output of lands in America it had lost political control of to the Revolution (Kennedy 2013, 29). Sheffield recognized that Britain's comparative advantage here lay in extracting agricultural raw materials by finance, not colonial occupation. Finally, in the arena of industrial organization everything would change rapidly in Britain's favor after Arkwright developed his first water-powered cotton spinning mill at Cromford, Derbyshire, in 1771: such factories would prove to be a software innovation far superior in terms of labor management, capital efficiency, and concentration of energy to the old putting-out system that they would rapidly supersede (Hugill 1993, 73). The factory also allowed the use of more and more labor efficient machinery, especially in the production of cotton textiles: Arkwright's water frame for spinning cotton in 1771, Crompton's spinning "mule" in 1779, and Horrocks' power loom in 1802. In the aftermath of the Napoleonic Wars Arkwright's creation of the cotton textile industry was recognized as the economic engine that helped pay the immense costs of the long war with France. In 1816 Sir Robert Peel would testify before a select committee of the House of Commons in 1816 that Arkwright "has done more honour to the country than any man I know, not excepting our great military characters," by whom he meant Nelson and Wellington (quoted in Fitton 1989, 1). There were, of course, other legs to this first Industrial Revolution, principally wrought iron, pottery, woolen textiles, and steam power, but cotton textiles drove it hardest.

The second critical moment came in the 1860s. America had been held badly back in terms of progress to industrialization by two serious political failures of the 1770s and 1780s that had perpetrated what in the American South amounted to a dangerous *ancien régime*: the failure of the Declaration of Independence to define people of African origin as other than slaves followed by the enshrinement of that concept in the constitution. These failures were compounded by the Missouri Compromise of 1820 that ensured the Slave states would retain parity with the Free states in the United States Senate, and thus be able to veto the actions of the increasingly northern dominated House of Representatives and the president. A central focus of that veto power was to refuse the North the right to protect its infant industries, the slave South having little interest in industrialization and a vested interest in continued access to cheap British goods.

Those faults were righted by the passage of the Morrill Tariff in 1861 by the House of Representatives, just before secession and over southern opposition, followed by the Civil War and the resulting destruction of the slave system. The Morrill Tariff allowed American industry to develop free from British competition behind a wall of protective tariffs. It was, in effect, a declaration of economic war on Britain and a crucial move toward hegemony (Hugill 2009c). Once the Civil War was over the North pressed the British on the *Alabama* claims, demands that the British pay for the damage done to northern merchant shipping by Confederate cruisers built in British shipyards in violation of the Neutrality Acts. Initially the British refused, but when the USS *Miantonomoh* turned up in British waters and was recognized as the world's most powerful battleship the British allowed the case to go to arbitration in the first international court ever convened. When the court found against them the British paid up in 1872 affirming that they would, in future, abide by the Neutrality Acts. In the 1860s America thus reversed the political errors of the early Republic, established the economic grounds for successful industrialization, and faced down the world's greatest naval power in its chosen arena. In the aftermath they would develop a powerful cotton textile industry, overhauling Britain in cotton consumption by 1896.

The third critical moment came in the late teens and early 1920s when President Wilson recognized that, despite the Anglo–American rapprochement begun in the 1890s, Britain, its Empire enlarged at Versailles, was still a formidable competitor. He correctly identified three arenas of struggle, although his domestic policies did little to help, in part because he had poor technical advice. The protectionist Jones Act he signed into power discriminated in favor of American shipping, but in such a way that American shipyards effectively ignored a revolutionary new prime mover, Diesel's compression ignition internal combustion engine that allowed others to make a technological end run around America. Also in the transportation arena, in what seems to have been an example of the "not invented here" syndrome, American aviation failed initially to capitalize on the remarkable amount of advanced information about aerodynamics and metal aircraft structures that was translated from the German and transferred to every American Morrill Act University by the National Advisory Committee for Aeronautics (NACA) in the early 1920s. In the struggle over international communications, Wilson accepted the navy's advice on telecommunications at precisely the moment the British Marconi Company was making a vacuum tube generated short wave, high frequency, technological end run around the navy's Poulsen arc generated long wave, low frequency wireless system. As secretary of commerce Herbert Hoover compounded Wilson's error when, at the first National Radio Conference in 1922, he essentially gave away the valuable higher frequencies to the amateurs, retaining the lowest frequencies for the military and the midrange frequencies as the AM band for commercial wireless broadcast use (Hugill 1999a, 123). In addition to Wilson's three arenas of struggle, American banking, mired in small town attitudes and principles, failed to wrest control of global finance from London, and even more signally failed to correct the flaws inherent in the French demand for war reparations and the structural instability of a gold-backed currency that would push America far deeper into the Great Depression of the 1930s than any other country.

The fourth crucial moment came between 1944 and 1947 when America wrested fiscal control from Britain and Wilson's predictions finally matured, for the most part in America's favor.

PRECONDITIONS FOR AMERICAN SUCCESS

Six main factors were at work in America's rise and Britain's fall.

1. British weariness—the *Weary Titan* syndrome (Friedberg 1988). WWI was a massive drain on the British economy and psyche. Britain had to spend down its foreign reserves to pay for the war and went from creditor to debtor nation status (Ferguson 1999). Although Britain had fought its way through many imperial conflicts during the 1800s the death tolls had been moderate. The toll WWI took on young males, not just in deaths but also in disabling injury, often through gas attacks, was terrifying. And the British homeland had itself come under serious assault: the shelling by Admiral Hipper of the northern coastal resort of Scarborough, earning him the sobriquet of "baby-killer"; then widespread if inaccurate aerial attacks by Zeppelins; then focused aerial attacks by seaplanes on British coastal resorts; finally effective aerial attacks as far inland as London by heavier than air bombers. These last would persuade Parliament to merge the RFC and RNAS to create the RAF to act as a defensive air force as well as pursuing the RFC's role of spotting for the artillery on the Western Front and the RNAS role of strategic bombing and scouting for the fleet. But these constant attacks on the homeland combined with the toll taken by U-boat attacks on British merchant ships and its fishing fleet shattered Britain's long-standing self-image as an "island fortress." Although Britain would remain a serious military power as late as 1945 that power would come at an immense and ever-increasing financial cost. It was also clear that American economic might was vastly greater than that of any other power on the planet. There was simply no possible economic way Britain could keep up.
2. Part of the *Weary Titan* syndrome was a consequence of the increasing success of the decolonization movement within the Empire, exemplified by the success of Gandhi in India in promoting nonviolent resistance but also by Australia and New Zealand's disaffection with the way their troops had been used during WWI. The massive loss of Australian troops at Gallipoli in 1915 turned Australia in particular increasingly away from Britain. The economic costs of administering a sprawling global Empire were beginning to clearly exceed the returns, and American fiscal policy determined at Bretton Woods in June 1944 would drive a stake through the heart of imperial preference and effectively end the fiscal utility of the Empire.
3. The serious overextension of British power in the 1920s and 1930s brought on by the expansion of the Empire via the Versailles mandates and a sense that Britain had to retain its Great Power role in the absence

of continued support from America. American isolationism, refusal to sign the Versailles Treaty, and refusal to join the League of Nations took a heavy toll here.

4. The massive buildup of American military power in the late 1930s as America began to abandon its policy from the 1920s and 1930s of hemispheric defense (Meinig 2004) in favor of a more robust policy of direct involvement worldwide, in particular in opposition to the perceived global threat of a growing and successful Communist movement emanating from a resurgent Russia and a fascist Japan expanding aggressively into China. President Roosevelt's main advisor on China in the late 1930s, Owen Lattimore, was raised in China and an expert on inner Asia (Lattimore 1940a). Lattimore summarized America's geopolitical problems succinctly along with the need for a robust response:

> Above all, while we want to get Japan out of China, we do not want to let Russia in. Nor do we want to "drive Japan into the arms of Russia" . . . the savagery of the Japanese assault is doing more to spread Communism than the teaching of the Chinese Communists themselves or the influences of Russia. It supplies the pressure under which the detonative ideas can work. At the same time it destroys Chinese wealth of every kind—capital, trade, revenue from agricultural rent—thus weakening that side of Chinese society which is most antagonistic to Communism. (Lattimore 1940b, 162, 168)

   That robust response would begin in early 1941 when FDR adopted the plan of the leader of the American volunteer forces in China, Claire Chennault, to bomb Japan from bases in China with American supplied planes and pilots wearing Chinese markings and uniforms. The planes were shipped in October 1941 but arrived after the Japanese attack on Pearl Harbor and were diverted to Burma.

5. British failure in several of the arenas of struggle identified by Wilson in 1919, notably petroleum and international transportation.
6. American success in those same arenas. Of the three arenas of struggle identified by Wilson—international transportation, international communication, and petroleum—Britain did well in only one, international communication. America made serious mistakes in maritime transportation but did well in air transportation technology from the mid-1930s on and in air policy at the Bermuda Agreement of 1946. And America decisively secured petroleum through SOCAL's efforts in Saudi Arabia and then with FDR's visit with King Ibn Saud in 1945.

   Ironically, Wilson missed what was in many ways the main struggle and ultimately one of the main drivers of American hegemony, the shift of fiscal power from London to New York. By 1922 Britain owed America almost $4.5 billion, but as the British economist Alfred Keynes would note "the American delegation . . . arrived in Paris without a properly worked-out economic plan" (Tooze 2014, 302, 297, 439). In early 1923, however, Britain agreed to pay its debt over a sixty-two year period at 3.3 percent interest. The conditions were there for America to seize economic hegemony, and the Wilson government in 1919 indicated a willingness to extend the deferral of interest on British loans if Britain would "abstain from any future discrimination against America in the British Empire's trade policy." Britain simply rejected this and America made no attempt to pursue the issue at the time. Part of the failure came because of American fiscal incompetence, the role of the Federal Reserve Bank not yet being properly established, and part from political problems, notably the intransigence of the Senate in refusing to "cancel a cent of its claims on Europe" (Tooze 2014, 366, 373). The Bretton Woods Agreement of 1944 would reverse these failures of the 1920s. This fiscal struggle has not figured in this book since it was not part of Wilson's analysis, but it deserves a brief comment here.

BRETTON WOODS AND THE SWITCH TO AMERICAN FINANCIAL HEGEMONY

The Bretton Woods Conference opened on July 1, 1944. The opening speech by Henry Morgenthau, secretary of the treasury, laid out an agenda designed to destroy Britain economically by enforcing article VII of the 1942 Lend-Lease Bill. Article VII required the British, if they wanted the financial aid of Lend-Lease, to commit "to the elimination of all forms of discriminatory treatment in international commerce and to the reductions of tariffs and other trade barriers." Reading an early version of article VII in 1941 before the bill passed Congress Keynes had exploded in front of the state department's Dean Acheson, saying that it was a "lunatic proposal" by Secretary of State Cordell Hull. Article VII effectively destroyed imperial preference and Keynes recognized that in the postwar world Britain would be in no position to sell its exports in its colonies against American competition without such preference. "At every step to Bretton Woods the Americans had reminded them, in as brutal a manner as necessary, that there was no room in the new order for the remnants of British imperial glory" (Steil 2013, 14).

On July 15, 1946, President Harry Truman approved a British loan that "launched the pound sterling on an agonizing yearlong death march." The British goal had been to restore the pound sterling to health by making it fully convertible to the American dollar, but the rapid run up in the price of American imports and the forced commitment to free trade based on the enforcement of Article VII imposed an intolerable strain on sterling. Britain was thrust deeper and deeper into debt to America. On August 20, 1947, the British government was thus forced to suspend convertibility and "the dream of reestablishing sterling as a world currency was shattered" (Steil 2013, 309–310). The consequence was the displacement of the pound sterling as global reserve currency by the dollar and thus, finally, the establishment of American financial hegemony.

WILSON'S PREDICTIONS

Wilson's predictions about what we can see as the final stage in the struggle over the hegemonic succession between a declining Britain and a rising America that was clearly underway by 1919 were a mixed bag, although the results were clear. By 1947 British power was largely spent and America clearly the successor power on the world stage.

The key arena was petroleum. In 1919 America had the advantage of being the world's largest producer, and subsequent major domestic strikes in East Texas and the Texas Permian Basin, as well as a host of lesser strikes along the Gulf Coast in Oklahoma and Louisiana ensured that advantage remained. The problem was keeping Britain out of the main oil reserves of the Middle East, especially Saudi Arabia. Not much of this can be put down to policy. President Roosevelt was willing to deal with King Ibn Saud only in 1945, very late in the day. Much of American success here thus came from the technological proficiency of SOCAL, which was willing to accept the possibility of salt dome oil deposits in Saudi Arabia and made the appropriate deals with King Ibn Saud to gain access to them in the 1930s, setting the stage for American dominance of the oil output of the region after 1945. But the technological blindness of both Shell and Standard Oil of New Jersey, which looked for oil trapped in porous rocks such as sandstones and limestones by anticlines and synclines rather than in salt domes, ensured their failure in the development of Saudi oil. Both Shell and Standard Oil had sought to use the Red-Line Agreement to restrict the access of companies such as SOCAL to what they saw as the more likely oiliferous areas in Iran and Iraq, notably in the Asmari limestone deposits.

The second key to American success was in the arena of international transportation, though not in its maritime component. Bad policy—the Jones Act—dissuaded American shipyards from embracing the crucial technological shift to the diesel engine for merchant ships. There were simply no really long hauls for American ships within the American controlled region of the globe—the longest by far being to Hawai'i and the Philippines in the Pacific—where the diesel engine could come into its own. American ships were always within range of oil bunkering using domestically produced fuel, and for a while oil fired steamships held their own on efficiency grounds. For some of the longest imperial routes British yards had started to build diesel ships in the early 1920s, but stayed wedded to coal and steam too long. The market for new, efficient, diesel merchant ships slipped out of both America and Britain's grasp.

In aviation it was another story. Although America came late to the table in its acceptance of Lanchester's aerodynamic theory in the early 1930s it then fully embraced it along with German structural engineering. Here policy played several roles. First, American banks needed airmail subsidized by the Federal Government to reduce the lost interest caused by "float." Most of this was between New York and Chicago, by far the two most important banking cities in the American economy. Legal fiscal transfers required paper evidence that they had occurred. If trains moved that paper the banks lost a full day's interest on very large sums. Although the New York Central's crack Twentieth Century Limited cut the time for its 958 mile route between the two cities to twenty hours when it began service in 1902 and cut it again to sixteen hours in 1938 it was nowhere close to airmail. Overnight airmail cut the "float" to zero (Hugill 1993, 275). Initially the airmail service used converted WWI bombers capable of carrying only the mail, landing a time or two to refuel, and taking about ten hours. But by 1933 Boeing's 247 airliner was capable of such service carrying passengers as well as the mail, and it would be rapidly followed by the Douglas DC-2 and DC-3, the last of which is usually regarded as the first truly profitable commercial airplane. Such planes could fly the just over 800 mile air route nonstop in under five hours.

The second policy area was domestic. All through the 1930s America developed a complex web of policies at the federal, state, and local level to encourage domestic aviation. The Federal Government paid for an increasingly dense and complex navigational network. As cities saw that without air service they might decline, they began to invest in airports. All this paid off in the market place by creating a network of airports, navigational aids, and airlines that enjoyed airmail subsidies (Dobson 2011).

The third policy area was geopolitical—the obsession with "hemisphere defense" and the creation at military expense of very large four engine airplanes such as Boeing's B-17 *Flying Fortress* (1934) and B-15 (1937) was the genesis for equally large four engine airliners. The wing and engines of the B-15 were used direct on the Boeing 314 *Clipper* flying boat, the first commercial airplane with the range to reach both Britain and Hawai'i from America carrying a useful payload. The wings from the B-17 were used on Boeing's 377 *Stratoliner* which had the potential to fly the New York–Los Angeles service nonstop. By 1939 America thus had far more experience with long-distance airmail, commercial air service, and aircraft with the ability to provide that service than any other state. America parlayed that ability into commercial hegemony by persuading Britain in 1942 to discontinue the development of transport aircraft in favor of bombers and fighters during wartime, then by pushing for what amounted to American control of the international airways via *cabotage* at the Chicago conference of 1944 and the Bermuda Agreement of 1946 (Dobson 1991). The British tried hard to recover, establishing the Brabazon Committee in 1943 to formulate postwar commercial airplane designs. Although it had some successes it misread the market for long-distance air travel badly, seeing the future in relatively slow but luxurious first class style travel rather than in the faster airplanes aimed at the mass market pioneered by American plane makers (Phipp 2007). The committee's type 1 long range designs, the Bristol *Brabazon* landplane and the Saunders-Roe *Princess* flying boat, were large, lumbering, inefficient designs that could in no way compete with the Boeing *Stratoliner*, Douglas DC-4, or Lockheed *Constellation*. American airplane designers had made the mistake of one too large, inefficient, four engine airplane, the first DC-4, retroactively called the DC-4E, sold it to the Japanese who redesigned it as an equally unsuccessful long range bomber, and never repeated their error. In this arena American technology and policy were fully aligned and successful in ensuring that Britain, especially when it made serious technological errors and was unable to control policy, would have a hard time recovering.

In international communications the results were mixed. The British had some success in pursuing a model of broadcasting that was at once commercial and yet stressed the need for accuracy in providing information—in other words the BBC avoided propagandizing news while commercial stations in Europe beamed British made commercial entertainment to British shores. Although American radio was initially highly commercial it adopted an essentially British view that it should not be an outlet for state supported political propaganda. At the level of international telecommunications Britain scored its lone success at the Atlantic City Conference of 1947, preventing the move of the International Telecommunications Union (ITU) to New York City and retaining it under essentially British leadership in Geneva. Even so the American Federal Communications Commission essentially exempted American companies from ITU control, rendering it a somewhat hollow victory.

THE CURRENT DEBATE: UNIPOLARITY OR MULTI-POLARITY?

Currently there is debate over whether America is enjoying a unipolar moment or the world is in a condition of multi-polarity. The argument for unipolarity is military and is derived in some ways from the navalist model of world-leadership cycles in Modelski & Thompson's *Seapower in Global Politics* (1988). This would suggest the continuation of American hegemony since the American fleet clearly continues to outweigh the fleets of the rest of the world by a large margin. The American air force enjoys a similar advantage. Alternatively, from the perspective of Wallerstein's World-System theory America is clearly not hegemonic.

Wallerstein's way of looking at hegemony is simple and useful. It requires the intersection of supremacy in three arenas: agro-industrial; commercial; and financial (Wallerstein 1984). Agro-industrial supremacy in many ways describes the organic world-economy of the period up to around 1900 where political or economic control of large swaths of territory crossing over many different climatic regions where different crops could be grown was a crucial precondition. In the late 1800s, however, the German development of the science of organic chemistry began to shift much of what had been agro-industrial production out of the organic world into a world based on hydrocarbons, hence the central importance of oil to modern geopolitics. Since WWI all major industrial polities have adopted such technologies. Two examples are characteristic: the shift from cotton textiles to polyester, and from natural rubber automobile tires to synthetic ones. Most textiles in the modern world come from hydrocarbons, as does what we still call "rubber." They require capital and petroleum, not land. Cotton used to need control of subtropical lands on which to grow it, rubber tropical. One can easily substitute "techno-industrial" for agro-industrial supremacy, but technology has a nasty way of diffusing very rapidly among the major contenders for hegemony, so no one polity ends up with anywhere near as much of an advantage as was conferred by the old agro-industrial supremacy which involved control over large swaths of earth space. One can argue that adoption environment can very well play a role—in the past ideology could certainly play a destructive role in technology adoption, as for example in the bizarre Nazi argument about what were the appropriate Aryan frequencies for airborne radar, a row that held the Nazis back a crucial year in radar development during WWII (Hugill 1999a, 203). But for the most part in the major polities today ideological blinders have been removed and all such states treat technology for what it is, a method of wealth enrichment and power.

The current period seems to be one of deep and very likely continuing multi-polarity in which no one polity can readily gain a serious advantage in at least two of the three arenas: techno-industrial and commercial. The financial arena seems currently fairly fixed in New York and London by historical inertia, despite the rise of new banking centers and stock markets. This plus the use of the American dollar as the world reserve currency gives America a substantial but not a hegemonic advantage. Britain's ill-considered vote to leave the European Union—Brexit—will probably massively reduce

London's advantage, focusing world finance more strongly in New York to America's great advantage.

America acquired agro-industrial and partial commercial supremacy in the aftermath of WWI, but this was at the moment when agro-industrial was seguing into techno-industrial. Deep down, and however much he and his advisers did not and could not understand it, much of what Wilson was concerned about was America's techno-industrial struggle with Britain in the 1920s and 1930s, with the struggle over petroleum by far the most important. But America failed to displace gold, the pound sterling, and the City of London to create financial supremacy until Bretton Woods in 1944, at which time the dollar became the world reserve currency. This was then underscored with complete victory in the military war of transition in 1945. America emerged from that war as indisputably the world military and economic superpower, that is, the hegemonic power. But hegemony is always brief. Techno-industrial supremacy slipped away with the recovery of the European and Japanese economies by the early 1970s and has slipped further with the recent rise of China. Commercial supremacy began to slip with the inability to control the world's oil market after the formation of the Organization of Petroleum Exporting Countries (OPEC), an organization ironically based on the Texas Railroad Commission that controls the production and transportation of oil and gas in the State of Texas. Financial supremacy began to slip with the creation of the Euro, whatever its flaws, as an alternative world reserve currency.

How deep is the current multi-polarity? In simplistic terms, out of a total of 230 states the world's three largest economies by far in 2015 were China, the European Union, and America: China was at $19.51 trillion; the European Union at $19.18 trillion; America at $17.97 trillion. Fourth highest was India at just over $8 trillion. But total output is not the only important measure of economic power. In terms of Purchasing Power Parity (PPP) per capita the ranking among the top three is reversed: in 2015 America ranked nineteenth in the world at $56,300; the European Union forty-fourth at $37,800; China one hundred and twelfth at $15,300. India makes the list at number one hundred fifty-eight (CIA Factbook). At one level this certainly reads as evidence of economic multi-polarity. But it does not compare well to the much more even multi-polarity evident at the previous hegemonic shift point. In 1890 the five largest economies were those of China at just over $218 billion, America at nearly $215 billion, India at just over $163 billion, Britain at just over $150 billion, and Germany at over $115 billion. Although America's population was just over 63 million at the time it was not radically larger than Germany at nearly 48 million or Britain at just over 37 million. In terms of PPP per capita, however, Britain ranked first in the world at $4009, America third at $3392, and Germany eighth at $2428. India and China came in well down the list at $584 and $540 respectively (Maddison 2003).

The American neoconservative argument that America is currently enjoying a unipolar moment thus cannot be an argument for American hegemony, merely for an American military supremacy that is undeniable at the level of conventional warfare, naval power, and nuclear weaponry—though obviously challenged by its difficulties responding to insurgencies and asymmetric warfare, for which it is not well prepared. Kagan, for example, takes the view that, at the end of the Cold War Europe, believing that a world was emerging where economic power would matter more than military power, cashed in its "peace dividend" and emerged as "an economic power of the first rank. . . . But the end of the Cold War did not reduce the salience of military power, and Europeans discovered that economic power did not necessarily translate into strategic and geopolitical power" (Kagan 2003, 21–22). From a neoliberal perspective, Kupchan argues that "unipolarity is here, but it will not last long" (Kupchan 2003, 61) and that multi-polarity will re-exert itself, in a normal and predictable historical pattern. In Kupchan's analysis, Europe has the economic power that would allow it, if sufficiently provoked by America, to become a superpower very quickly, and to argue simplistically that Europe is mired in a "postmodern paradise," or that the Euro crisis and a possible breakup of the Union may happen is to miss the point. Rapid transitions have occurred in the past: Japan's following the Meiji Restoration; Germany's expansion in the Wilhelmine period; Germany's recovery from WWI under the Nazi Party; Soviet expansion under Stalin. Such transitions can and will occur again. Kagan's point is that Europe currently lacks the will to such power, but the point of European political scientists is that Europe is becoming, under primarily German leadership, a very new sort of global superpower that fits no existing model (Maull 1993; 2004).

Neither Kagan, Kupchan, nor the many who are contributing to the current policy arguments in America from the point of view of international affairs scholars, can be described as following a world-system argument. Kagan's book is simply an essay devoid of references, although he clearly views history from an evolutionary perspective. Kupchan's work has an extensive bibliography in which perhaps the main theoretical focus is provided by Paul Kennedy, in particular his arguments for imperial overextension and subsequent decline (Kennedy 1983; 1987). Kupchan notes, however, that his work "views history as having a cyclical as well as an evolutionary character" (Kupchan 2003, 304), which implies acceptance of something akin to a world-system perspective.

THE "LESSONS" OF HISTORY

At the theoretical level Halford Mackinder argued in his "Pivot" paper of 1904 that the day of the naval powers was ending, which implied that all the technology forcing in naval armaments of the period was relatively pointless. As Mackinder noted in 1919, the range of the guns carried by the largest battleships meant that naval power could not reach more than thirty miles or so into the continental interiors, and the new technologies of railroads, telegraphs, telephones, and bureaucratic organization meant that a great state could now be created safe from naval power in the continental interior of Eurasia. From there it would be eventually able to erupt as the world's greatest economic power, able to build the world's greatest fleet. The father of German *geopolitik*, Karl Haushofer, adopted Mackinder's arguments after WWI.

Both Mackinder and Haushofer enjoyed substantial influence on important political figures in the first half of the 1900s. Mackinder's came through his colleagues in the Co-Efficients, a "think tank" founded in London in 1902 that included such luminaries as Beatrice and Sydney Webb, the founders of the intellectual wing of the British Labour Party, H. G. Wells and, briefly, Bertrand Russell. More to the point, the Co-Efficients also included two powerful and influential members of the Asquith cabinet that would lead Britain into war in 1914, Viscount Haldane, secretary of state for war 1905–1912 and Lord Chancellor 1912–1915, and Sir Edward Grey, foreign secretary 1905–1916 (Blouet 1987, 116–117).

Karl Haushofer visited Japan as a military adviser from 1908 to 1910, became fluent in Japanese, and in 1914 completed his Ph.D. dissertation in geography on Japan at the University of Munich. Its title translates as *The German Share in the Geographical Opening up of Japan and the sub-Japanese Earth Space, and its advancement through the influence of War and Defense Politics*. In WWI Haushofer rose to the rank of major general in the Bavarian army. Like many Germans he was appalled at the treatment meted out to Germany by the Entente Powers at Versailles. After WWI he returned to the University of Munich as professor of geography and came to regard maritime Japan as Germany's natural geopolitical ally in a next war he saw as inevitable, a war that would redress the wrongs done Germany at Versailles. He began to support Hitler and in 1924 founded *Zeitschrift für Geopolitik* at the University of Munich to create the intellectual blueprints for German and Japanese expansion. Haushofer's most famous student was Rudolph Hess and Haushofer visited both Hess and Hitler in prison in Munich following the failed "Beer-Hall *Putsch*" of 1923. Hitler's *Mein Kampf* was clearly influenced by Hess and Haushofer (Herwig 2016). In 1936 Haushofer used his good offices with the Japanese to arrange for the signing of the Anti-Comintern Pact between Germany and Japan at his home in Munich (Hayes 2004).

Politico-military problems began to change character in the period surrounding WWI: technologically, air power came of age. Geopolitical theory lagged, still mired in navalist thinking. At the geostrategic level, revolutionary military thinkers, codified in the book *The Apostles of Mobility* (Carver 1979), began to realize by WWI's end that the new military technologies developed during the war would allow a return to mobility in land warfare, using advances by aircraft and tanks coordinated by wireless. Given the technology of the period such advances would have been relatively slow. Had the war gone into 1919 British plans called for exactly such an approach. In WWII Germany would use faster tanks and aircraft but the same plans, renamed as *Blitzkrieg*.

Soon after WWI, American brigadier-General Billy Mitchell demonstrated that land-based bombing planes were able to destroy capital ships within their range by sinking the former Imperial German navy battleship, *Ostfriesland*, although many critics of Mitchell argued that he was mistaken, since *Ostfriesland* was a stationary target and not shooting back. Despite such criticism the USAAC would place a heavy bet on the Boeing B-17 in the 1930s, not as a strategic bomber but as a bomber capable of intercepting an enemy battlefleet well out to sea and acting as an extension of America's "moat defensive," hence its name of *Flying Fortress*. A fleet of B-17s was on its way to defend Pearl Harbor just as the Japanese attacked in December 1941. The Japanese sinking with land-based bombers of the Royal Navy's *Repulse* and *Prince of Wales* in open sea off Singapore in December 1941 demonstrated that Mitchell and the USAAC were correct. At sea, carrier advocates argued that blue water naval battles would no longer be between battleships but would become battles between carrier based aircraft for control of the air. The battles of Coral Sea and, especially, Midway, confirmed this analysis in 1942. Once control of the air had been achieved by the carriers' fighter planes, torpedo and dive-bombers would easily destroy the opponent's carriers. At Midway the Americans did not gain air superiority so much as the Japanese lost it when their fighter planes, having savaged the American torpedo bombers, landed to refuel, allowing the American dive-bombers to destroy four carriers. All of this geostrategy had been thought through and the designs begun for the necessary weaponry with the ink hardly dry on the Washington Naval Treaty of 1922.

In terms of agency, of the actors discussed, Churchill and Roosevelt were the most aware of the transition. FDR made no bones about his desire to ultimately curb the power of the British Empire, but when it came to waging a war of hegemonic transition he equally did not doubt that his primary enemy was Germany. America and Britain, always close economically, underwent a political and, perhaps most importantly, a cultural rapprochement in the early 1900s that culminated in America seeing itself as Britain's logical successor, despite Wilson and the isolationist Republicans who followed him. From Britain's perspective, if the "divine spirit" were to shift, better it shifted to an America admired by many in Britain for purifying itself of many of old England's shortcomings. Allying with America would help stave off or minimize the decline academics such as Mackinder were sure was underway by the early 1900s. As a major consequence America and Britain fought both German wars as allies, America to ensure its hegemonic rise and Britain to minimize the costs of its hegemonic decline.

The truly nightmare problem for the maritime powers would have been what the architect of German *geopolitik*, Karl Haushofer, always wanted: an agreement between Germany, Russia, and Japan to share the world-island. Despite the Anti-Comintern Pact, for a brief time after the signing of German and Japanese non-aggression pacts with Russia, that was the case. Fortunately Hitler's race hatred trumped academic analysis and he invaded Russia. Japan's non-aggression pact with Russia was signed in 1940 after the humiliation of the Imperial Japanese Army (IJA) at the hands of then-cavalry commander Georgy Zhukov at the Battle of Khalkin Gol. Zhukov, promoted in 1940 and eventually to be regarded as Russia's greatest general, regarded his experiences at Khalkin Gol, in particular with the flanking use of tanks and strike aircraft, as crucial to the Soviet victory over Nazi Germany. Following the defeat of the IJA at Khalkin Gol Japanese domestic politics thereafter favored the Imperial Japanese Navy (IJN) and maritime over territorial expansion, which brought Japan into conflict with America and Britain instead of Russia, a trial of strength long wished for by the IJN, but which would ultimately be just as destructive to Japan as the IJA's preferred conflict with Russia.

Britain's world-economy as hegemon was based on maritime trade in organic materials, cotton in particular, which in many ways played a role in the global economy somewhat akin to that of oil today. Britain's trade was organized through global capitalism and later through control over submarine cable telecommunications, not by control over territory (Hugill 2009b). America's rise was greatly helped by being a capitalist power in possession of a politically unified and secure territorial base in the mid-section of the North American continent with excellent resources and energy availability. Once the *ancien régime* of the slave South was curbed America was on the road to hegemony. Throughout the 1800s and even well into the 1900s it was the world's major source of cotton and cotton was the country's major source of income. Behind the wall erected in 1861 by the Morrill Tariff cotton textile industrialization fueled a significant part of America's massive post-Civil War industrial growth. The world-economy Japan briefly attempted to establish was based on an imperfectly understood British model, with military power substituted for trade. Japan's geographical expansion occasioned particular heartache in the Pacific: America, Australia, and China were particularly affected, the last for the cotton Japan needed to help fuel its industrial growth. As America moved toward hegemony Australia would make a deliberate and calculated shift out of the British sphere of influence into the American sphere, and China, after brutal Japanese occupation, would, in the Maoist period, attempt to completely secede from the capitalist world-economy although it is now returning to it.

America and, later, Russia operated a territorial version of the British world-economy at a continental rather than a global scale, using railroads (and telegraphs) instead of ships to politically control territory. Germany, in the absence of tropical and subtropical colonies, developed a radically new world-economy that would depend on the synthesis of resources previously only available from the organic world-economy. Because of its technological fixation, Germany sought to replace agro-industrial production based on control of territory through politics, trade, or military power with techno-industrial production based on basic research in science and technology. This led Germany to heavy emphasis on the practical applications of chemistry to produce dyestuffs, in electrochemistry to produce explosives, and in physics to better understand such forces as aerodynamics. German superiority in these three areas allowed Germany to fight a protracted WWI. The two losers in the first stage of the hegemonic struggle, Russia and Germany, succumbed to ideological utopianism and totalitarianism after the war. German scientists and technologists migrated wherever possible and after the rise of the Nazi Party anti-Semitism increased the loss of such key personnel. German science and technology had begun to diffuse to American corporate laboratories before WWI. In the period immediately after WWI, as the German economy collapsed and what was effectively an internal civil war raged, American laboratories and universities became massive beneficiaries of skilled German migration (Hugill & Bachmann 2005). Such new technologies were relatively easily adopted, especially when the personnel who had developed them migrated. Importantly, American superiority was in software rather than in hardware technologies, in the technologies of management developed by Taylor and Gilbreth and implemented by Carnegie and Ford rather than in those of resource substitution and production (Hugill 1993). These software technologies promoted superior productivity and did not diffuse easily because they required a very flexible social structure, which America enjoyed. America was, however, rather slow to adopt new telecommunication technologies, an area of generally British superiority (Hugill 1999a).

WHERE NEXT?

The hegemony achieved by America in 1945 had five foundations:

1. the establishment of the dollar as the global reserve currency;
2. secure domestic markets via protectionism, begun with the Morrill Tariff of 1861;
3. abundant domestic resources, including oil;
4. control of the Middle Eastern oil spigot for the future; and
5. general scientific and technological superiority.

This last was based in part on the improvements in the American higher educational system that came about as a result of WWI, which had made plain the superiority of German chemistry, and in part on the immigration during the 1920s and 1930s of a large number of German scientists. Jews fleeing Nazi Germany were only part of this picture, although perhaps the one most commented upon. In the 1920s as American chemical companies rebuilt themselves on the basis of German organic chemistry they offered German scientist handsome salaries to relocate. The same thing happened in aviation as American companies saw the advantages of German metal structures and aerodynamic knowledge in the early 1930s.

These conditions do not all prevail in the current multi-polar period, although some are less susceptible to challenge than others. Of the five foundations only one clearly remains, the dollar as the global reserve currency. America's domestic markets are open to massive and often successful competition. A great deal of America's natural resource endowment has been burned through, although improved recovery technology has recently allowed a return to substantial domestic production of oil. OPEC has, however, massively reduced the ability of American oil companies to control the Middle Eastern oil spigot economically. Finally, American scientific and technological superiority has begun to slip. The American inability to reproduce a native technocracy and the need to "import" STEM (Science Technology Engineering Mathematics) workers via H1-B visas is a significant failing in a world-economy driven heavily by technology, although many such workers find ways to stay in America rather than return home when their visas expire. If America shifts to a more restrictive immigration policy this advantage could, however, be reduced.

The evidence from the last two hegemonic transitions (Britain I to Britain II and Britain II to America) is that they were driven heavily by the adoption by the hegemonic successor of new technologies. Simply put, the rate of return in older technologies drops away as new companies enter the market and take a lower rate of profit—Marx's "law of the tendency of the rate of profit to fall." Sensible investors remove their capital from declining technologies—causing a downturn in the Kondratiev wave—and invest in new, more profitable ones—producing an upturn. This transition is accompanied by what Schumpeter described as a "gale of creative destruction" as the new economy emerges. The world-leadership cycles identified by Modelski contain two Kondratiev waves (Hugill 1993, 9–12). If America is not to stay in a long period of multi-polarity it is important to identify a "carrier technology" (the term is from Hall and Preston 1998) that may possibly drive the next upturn.

The three most likely carrier technologies for the next world-leadership cycle, should a hegemon emerge from the current period of multi-polarity, are electronics, biotechnology, and environmental management leading to sustainable development. Electronics technology is hardly new. Hall and Preston (1998) note that it was a carrier technology in the period 1910 to 1920 or so with wireless. It was then elaborated through the 1930s by television and radar, and reached a point of dominance with the creation of the first electronic computer, *Colossus*, by 1944. The development of solid-state electronics markedly improved all of this, as have microcomputers, smart phones, and the development of the internet. But this now has all the hallmarks of a mature technology (1999a). Certainly there will be massive refinements, but electronics technology is now globally ubiquitous, not a carrier technology. Biotechnology certainly has massive promise, although its main contribution seems likely to be in the area of prolonging human life expectancy and, hopefully, human health into old age, simply because that is what most humans will want and be willing to pay for. It also has promise for generating the more productive and disease-resistant crops that will be needed to feed a human population that continues to increase. In addition it may allow the production of hydrocarbon fuels in a sustainable way.

A serious problem for the contemporary world-economy is that longer human life expectancy and more food translates into more consumers, and in some ways that can be counterproductive. Each hegemonic world-economy thus far has laid the seeds of its own destruction and the succeeding hegemon has had to resolve the inherent contradictions of the previous world-economy. American economic hegemony was based on flexible accumulation strategies that very successfully promoted mass consumption of energy and resources. The political geographer Peter Taylor argues that American style hegemony is doomed because of the environmental stress it places on the planet. None of the alternatives recounted by Taylor are very palatable and all argue against the continued success of capitalism. "Eco-fascism" negates humanistic liberalism. "Conspicuous asceticism" and a "deep-green world-system" are both centered on reducing human impacts on the environment in draconian ways, overall, or in favor of other species (Taylor 1996, 216–223). All such solutions would require massive belief shifts against capitalism that seem culturally unlikely. Logically, the only way for capitalism to survive the environmental pressures brought on by mass consumption is to seek technologies that allow environmentally sustainable productivity increases. The current phase of the world-economy is one where productivity increases are coming far more easily and cheaply than production increases. Combined with Taylor's analysis this would suggest that productivity increases in the carrier technology of environmental management are likely to be a major key to sustainable development of the world-economy and thus to any possible continuation of American hegemony. Even then, such technologies would easily and rapidly diffuse through the global economy and even if an American hegemony could be restored it would be necessarily brief. It would seem the world-system is likely to be multi-polar for the foreseeable future.

Bibliography

PRIMARY SOURCES

British National Archives (BNA): materials from Admiralty (ADM); Cabinet (CAB); Foreign Office (FO); and Treasury (T) files.

John Curtin Prime Ministerial Library, Curtin University of Technology, Perth, Western Australia (JCPMS).

National Air and Space Museum, Washington D.C. (NASM).

National Archives of Australia, Canberra (NAA).

GOVERNMENT DOCUMENTS

Department of Foreign Affairs and Trade (DFAT). *Documents on Australian Foreign Policy, 1937–1949*. Canberra: Australian Government Printing Office. Sixteen Volumes. Also available online at http://www.info.dfat.gov.au/info/historical

U.S. Department of Commerce (USDC), 1960. *Historical Statistics of the United States: Colonial Times to 1957*. Washington, DC: U.S. Government Printing Office.

WEBSITES

100 Years of Anheuser Busch, accessed January 16, 2013. http://www.subvetpaul.com/Diesel_Engines.html

ANZAC Documents, accessed April 23, 2013. www.dfat.gov.au/geo/new_zealand/01-ANZAC.pdf

Chester on Armenia, accessed February 20, 2018. www.tallarmeniantale.com/Chester-dad.htm

The Electro-Motive Story, accessed February 19, 2018. http://history.gmheritage-center.com/wiki/index.php/The_Electro-Motive_Story

History of NASA, accessed May 27, 2014. http://history.nasa.gov/SP-4103/ch4.htm

Louisiana State University, Department of Geology, accessed January 16, 2013. http://geology.lsu.edu/AboutLSUGeology/ProgramHistory/item32703.html

Old Radio World, accessed March 13, 2013. English version marked closed 19 February 2018. http://www.oldradioworld.de/volks.htm

Radio Luxembourg Wikipedia Entry, accessed March 13, 2013. http://en.wikipedia.org/wiki/Radio_Luxembourg_%28English%29

Sir Edwin Pears Wikipedia Entry, accessed February 19, 2018. https://en.wikipedia.org/wiki/Edwin_Pears

Smithsonian Time and Navigation, accessed February 19, 2018. http://timeandnavigation.si.edu/multimedia-asset/pioneer-earth-inductor-compass

Sterling Times, accessed December 16, 2014. http://www.sterlingtimes.co.uk/radio_sponsorship.htm

Texas Railroad Commission, accessed January 16, 2013. http://www.rrc.state.tx.us/about/history/chronological/chronhistory01.php

Texas State Handbook, entry on Permian Basin, accessed January 16, 2013. http://www.tshaonline.org/handbook/online/articles/ryp02

JOURNALS ETC.

Australian Dictionary of Biography (ADB). Melbourne: Melbourne University Press. Sixteen Volumes.

Australian Women's Weekly (AWW), Sydney.

Hunt's Merchants' Magazine and Commercial Review, 1839–1870.

PUBLISHED MATERIALS

Adams, Ralph J. Q. *British Politics and Foreign Policy in the Age of Appeasement, 1935–39*. Stanford, CA: Stanford University Press, 1993.

Agnew, John. *Geopolitics: Re-Visioning World Politics*. Second Edition. New York: Routledge, 2003.

———. *Hegemony: The New Shape of Global Power*. Philadelphia: Temple University Press, 2005.

Ahamed, Liaquat. *Lords of Finance: The Bankers who Broke the World*. New York: Penguin, 2009.

Albion, Robert Greenhalgh. *Forests and Sea Power. The Timber Problem of the Royal Navy, 1652–1862*. Cambridge: Harvard University Press, 1926.

Alexander, Fred. *From Curtin to Menzies and after: Continuity or Confrontation?* Melbourne: Nelson, 1973.

Alexanderson, Ernst. Transoceanic Radio Communication. In *Proceedings of the IEEE* 72, 1984, 626–33. Originally published in *Transactions of the American Institute of Electrical Engineers* 38, 1919, 1269–85.

Alomes, Stephen. *A Nation at Last? The Changing Character of Australian Nationalism 1880–1988*. North Ryde, NSW: Angus & Robertson, 1988.

Anderson, John D., Jr. *A History of Aerodynamics*. New York: Cambridge University Press, 1997.

Andersson, Lennart. *Soviet Aircraft and Aviation, 1917–1941*. Annapolis, MD: Naval Institute Press, 1994.

Andrews, C. F., & E. B. Morgan. *Vickers Aircraft since 1908*. London: Putnam, 1969.

———. *Supermarine Aircraft since 1914*. Annapolis, MD: Naval Institute Press, 1987.

Ardagh, Colonel. The Red Sea Petroleum Deposits. In *Proceedings of the Royal Geographical Society* 8:8, 1886, 502–7.

Austin, B. A. Wireless in the Boer War. In IEE, *International Conference on 100 Years of Radio*, 1995, 44–50.

Ayrton, Frederick. *Railways in Egypt: Communication with India*. London: Ridgway, 1857.

Baker, W. A., & Tre Tryckare. *The Engine Powered Vessel*. New York: Crescent, 1965.

Barker, A. J. *The Bastard War. The Mesopotamian Campaign of 1914–1918*. New York: Dial, 1967.

Barnes, C. H. *Handley Page Aircraft since 1907*. Second Edition. London: Putnam, 1987.

———. *Bristol Aircraft since 1910*. London: Putnam, 1988.

———. *Shorts Aircraft since 1900*. Revised Edition. London: Putnam, 1989.

Beauchamp, Ken. *History of Telegraphy*. London: Institution of Electrical Engineers, 2001.

Black, David, ed. *In His Own Words. John Curtin's Speeches and Writings*. Perth, WA: Paradigm Books/Curtin University, 1995.

———. *'Friendship is a Sheltering Tree': John Curtin's Letters 1907 to 1945*. Perth, WA: John Curtin Prime Ministerial Library/Curtin University, 2001.

Black, Edwin. *Internal Combustion: How Corporations and Governments Addicted the World to Oil and Derailed the Alternatives*. New York: St. Martin's, 2006.

Blouet, Brian W. *Geopolitics and Globalization in the Twentieth Century*. Second Edition, revised. London: Reaktion, 2011.

Bloxham, Donald. *The Great Game of Genocide: Imperialism, Nationalism, and the Destruction of the Ottoman Armenians*. New York: Oxford University Press, 2005.

Bogert, J. J. The New Oil Engines. In *The New York Times*, September 26, 1912, 4.

Bolton, Geoffrey. *The Oxford History of Australia*. Volume 5, *1942–1988: The Middle Way*. Melbourne: Oxford University Press, 1990.

Bordewich, Fergus M. *America's Great Debate: Henry Clay, Stephen A. Douglas, and the Compromise That Preserved the Union*. New York: Simon & Schuster, 2012.

Bowers, Peter M. *Curtiss Aircraft 1907–1947*. London: Putnam, 1979.

———. *Boeing Aircraft since 1916*. Annapolis, MD: Naval Institute Press, 1989.

Bowyer, M. J. F. The Bomber Revolution. In Philip Jarrett, ed., *Aircraft of the Second World War. The Development of the Warplane, 1939–1945*. London: Putnam, 1997, 91–125.

Brandon, Ruth. *The Dollar Princesses: Sagas of Upward Mobility, 1870–1914*. New York: Knopf, 1980.

Breit, Gregory, & Merle A. Tuve. A Test of the Existence of the Conducting Layer. In *Physical Review* 28:3, 1926, 554–75.

Briggs, Asa. *The History of Broadcasting in the United Kingdom.* Volume I, *The Birth of Broadcasting*. London: Oxford University Press, 1961.

———. *The History of Broadcasting in the United Kingdom.* Volume II, *The Golden Age of Wireless*. London: Oxford University Press, 1965.

———. *The BBC: The First Fifty Years*. New York: Oxford University Press, 1985.

Bright, Charles. *Submarine Telegraphs: Their History, Construction, and Working*. New York: Arno, 1974 [1898].

Bromley, Simon. *American Hegemony and World Oil: The Industry, the State System and the World Economy*. State College, PA: Pennsylvania State University Press, 1991.

Buckley, Kenneth D., Barbara Dale, & Wayne Reynolds. *Doc Evatt: Patriot, Internationalist, Fighter and Scholar*. Melbourne: Longman Cheshire, 1994.

Buehler, E. C. *American versus British System of Radio Control*. New York: H. W. Wilson, 1933.

Bunting, Sir John. *R. G. Menzies: A Portrait*. Sydney: Allen & Unwin, 1988.

Burns, Russell W. *Communications: An International History of the Formative Years*. London: Institution of Electrical Engineers, 2004.

Burton, John. *Labour in Transition*. Kingston, A.C.T.: Morgan's Publications, n.d.

———. *The Alternative (a dynamic approach to our relations with Asia)*. Sydney: Morgans Publications, 1954.

———. A Human Component: The Failure of the Liberal Tradition. In David Lee & Christopher Waters. *Evatt to Evans: The Labor Tradition in Australian Foreign Policy*. St. Leonards, NSW: Allen & Unwin, 1997, 23–32.

Butler, David L. Technogeopolitics and the Struggle for Control of World Air Routes, 1910–1928. In *Political Geography* 20, 2001, 635–58.

Cadman, John. Middle East Geography in Relation to Petroleum. In *The Geographical Journal* 84:3, 1934, 201–12.

Carter, Charles Frederick. The Advent of the Motor Ship. In *The American Review of Reviews* 44:5, 1911, 591–4.

Carver, Field Marshall Lord. *The Apostles of Mobility: The Theory and Practice of Armoured Warfare*. New York: Holmes and Meier, 1979.

Catherwood, Christopher. *Churchill's Folly. How Winston Churchill Created Modern Iraq*. New York: Barnes & Noble, 2004.

Chapman, Lawrence B. *The Marine Power Plant*. New York: McGraw Hill, 1924.

Chester, Alan. *John Curtin*. Sydney: Angus & Robertson, 1943.

Chester, Colby M. Turkey Reinterpreted. In *Current History* 16:6, 1922, 939–47.

Choudhury, Deep Kanta Lahiri. *Telegraphic Imperialism: Crisis and Panic in the Indian Empire c. 1830–1920*. New York: Palgrave Macmillan, 2010.

Clay, Christopher G. A. *Gold for the Sultan: Western bankers and Ottoman finance 1856–1881, a Contribution to Ottoman and to International Financial History*. London: Tauris, 2000.

Coates, Vary T., & Bernard Finn. *A Retrospective Technology Assessment: Submarine Telegraphy. The Transatlantic Cable of 1866*. San Francisco, CA: San Francisco Press, 1979.

Codding, George Arthur Jr. *The International Telecommunication Union: An Experiment in International Cooperation*. New York: Arno, 1972 [1952].

Codding, George Arthur Jr., & Anthony M. Rutkowski. *The International Telecommunication Union in a Changing World*. Dedham, MA: Artech House, 1982.

Cohen, Eliot A. *Supreme Command. Soldiers, Statesmen, and Leadership in Wartime*. New York: Simon & Schuster, 2002.

Constant, Edward W. H. *The Origins of the Turbojet Revolution*. Baltimore, MD: The Johns Hopkins University Press, 1980.

Cooper, Anthony. *Darwin Spitfires: The Real Battle for Australia*. Sydney: University of New South Wales Press, 2011.

Corum, James S. Defeat of the Luftwaffe, 1935–1945. In Robin Higham & Stephen J. Harris, eds. *Why Air Forces Fail: The Anatomy of Defeat*. Lexington, KY: University Press of Kentucky, 2006, 203–26.

Craig, Robin. *The Ship*. Volume 5, *Steam Tramps and Cargo Liners 1850–1950*. London: Her Majesty's Stationery Office/National Maritime Museum, 1980.

Crockett, Peter. *Evatt: A Life*. Melbourne: Oxford University Press, 1993.

Cumings, Bruce. *Dominion from Sea to Sea. Pacific Ascendancy and American Power*. New Haven, CT: Yale University Press, 2009.

Curzon, George N. *Russia in Central Asia and the Anglo–Russian Question*. London: Longman, Green, 1889.

Dalziel, Allan. *Evatt the Enigma*. Melbourne: Landsdowne, 1967.

Daveluy, René. *The Genius of Naval Warfare*. Volume 1, *Strategy*. Annapolis MD: United States Naval Institute, 1910.

Davenport, E. H., and Sidney R. Cooke. *The Oil Trusts and Anglo–American Relations*. New York: Macmillan, 1924.

Davies, R. E. G. *British Airways. An Airline and its Aircraft*. Volume 1, *1919–1939, The Imperial Years*. McLean, VA: Paladwr, 2005.

Day, David. *The Great Betrayal: Britain, Australia & the Onset of the Pacific War 1939–42*. New York: Norton, 1988.

———. *Reluctant Nation: Australia and the Allied Defeat of Japan 1942–45*. Melbourne: Oxford University Press, 1992.

———. *Menzies and Churchill at War*. New York: Oxford University Press, 1993.

———. *John Curtin: A Life*. Sydney: Harper Collins, 1999.

———. *The Politics of War*. Sydney: Harper Collins, 2003. Revised and combined edition of *The Great Betrayal* and *Reluctant Nation*.

Deffeyes, Kenneth S. *Beyond Oil: The View from Hubbert's Peak*. New York: Hill & Wang, 2006.

Delaisi, Francis. *Oil: Its Influence on International Politics*. London: Allen & Unwin, 1922.

De la Tramerye, Pierre L'Espagnol. *The World–Struggle for Oil*. New York: Knopf, 1924.

Delgado, James P. *Gold Rush Port. The Maritime Archeology of San Francisco's Waterfront*. Berkeley, CA: University of California Press, 2009.

Dening, Greg. *Mr. Bligh's Bad Language: Passion, Power, and Theatre on the Bounty*. New York: Cambridge University Press, 1992.

Denny, Reginald. *We Fight For Oil*. New York: Knopf, 1928.

———. *America Conquers Britain: A Record of Economic War*. New York: Knopf, 1930.

DeNovo, John A. The Movement for an Aggressive American Oil Policy Abroad, 1918–1920. In *American Historical Review* 61:4, 1956, 854–76.

———. *American Interests and Policies in the Middle East, 1900–1939*. Minneapolis, MN: University of Minnesota, 1963.

Dewey, John. *German Philosophy and Politics*. New York: Henry Holt, 1915.

Dibner, Bern. *The Atlantic Cable*. Norwalk, CT: Burndy Library, 1959.

Dobson, Alan. *Peaceful Air Warfare: The United States, Britain, and the Politics of International Aviation*. Oxford: Clarendon, 1991.

———. *FDR and Civil Aviation: Flying Strong, Flying Free*. New York: Palgrave Macmillan, 2011.

Eade, Charles, ed. *Secret Session Speeches by the Right Hon. Winston S. Churchill O.M., C.H., M.P.* London: Cassell, 1946.

Earle, Carville. The Myth of the Southern Soil Miner: Macrohistory, Agricultural Innovation, and Environmental Change. In D. Worster, ed., *The Ends of the Earth: Perspectives on Modern Environmental History*, New York: Cambridge University Press, 1988, 175–210.

Earle, Edward Mead. *Turkey, The Great Powers, and the Bagdad Railway: A Study in Imperialism*. New York: Macmillan, 1923.

———. The Turkish Petroleum Company—a Study in Oleaginous Diplomacy. In *Political Science Quarterly* 39:2, 1924, 265–79.

——— ed. *Makers of Modern Strategy: Military Thought from Machiavelli to Hitler*. Princeton, NJ: Princeton University Press, 1943.

Edgar, E. Mackay. Sees Oil Control Passing to Britain. In *The New York Times*, 21 September 1919.

Edgerton, David. *Warfare State: Britain 1920–1970*. Cambridge: Cambridge University Press, 2006.

Edwards, John. In Search of John Curtin. 2001. http://espace.library.curtin.edu.au/R/8M3SNPNTM2A4F9EC146HJAYK8QK35SI388MMLVMVAM4VIK2XD5-02514?func=results-jump-full&set_entry=000002&set_number=000237&base=ERA01JCPML accessed 26 February 2013.

Edwards, Peter G. Evatt and the Americans. In *Historical Studies* 18, 1979, 546–60.

———. From Curtin to Beazley: Labor leaders and the American alliance, 2001. http://espace.library.curtin.edu.au/R/8M3SNPNTM2A4F9EC146HJAYK8QK35SI388MMLVMVAM4VIK2XD5-01189 accessed 26 February 2013.

Elkind, Sarah S. Los Angeles, the Energy Capital of Southern California. In Joseph A. Pratt, Martin V. Melosi, & Kathleen A. Brosnan, eds. *Energy Capitals: Local Impact, Global Influence*. Pittsburgh, University of Pittsburgh Press, 2014, 77–93.

Engdahl, William. *A Century of War: Anglo-American Oil Politics and the New World Order*. London: Pluto, 2004.

Engel, Jeffrey A. *Cold War at 30,000 Feet: The Anglo-American Fight for Aviation Supremacy*. Cambridge, MA: Harvard University Press, 2007.

Engelbrecht, Helmuth C., & Frank C. Hanighen. *Merchants of Death*. New York: Dodd, Mead, 1934.

Evatt, Herbert Vere. *Foreign Policy of Australia: Speeches*. Sydney: Angus & Robertson, 1945.

———. *Australia in World Affairs*. Sydney: Angus & Robertson, 1946.

———. *The Task of Nations*. Westport, CT: Greenwood, 1972 [1949].

Fanning, Leonard M. *American Oil Operations Abroad*. New York: McGraw Hill, 1947.

Farnie, D. A. *The English Cotton Industry and the World Market, 1815–1896*. Oxford: Clarendon, 1979.

Ferguson, Niall. *The Pity of War: Explaining World War One*. New York: Basic, 1999.

———. *The War of the World: History's Age of Hatred*. New York: Allen Lane, 2006.

———. *Civilization: The West and the Rest*. New York: Penguin, 2011.

Fischer, F. *Germany's Aims in the First World War*. New York: Norton, 1967.

———. *World Power or Decline. The Controversy over Germany's Aims in the First World War*. New York: Norton, 1974.

Fischer, Louis. *Oil Imperialism: The International Struggle for Petroleum*. New York: International, 1926.

Fisk, Robert. *The Great War of Civilization: The Conquest of the Middle East*. New York: Knopf, 2005.

Fitton, R. S. *The Arkwrights: Spinners of Fortune*. Manchester: Manchester University Press, 1988.

Fitzgerald, Edward Peter. The Iraq Petroleum Company, Standard Oil of California, and the Contest for Eastern Arabia, 1930–1933. In *International History Review* 13:3, 1991, 441–65.

———. Business Diplomacy: Walter Teagle, Jersey Standard, and the Anglo-French Pipeline Conflict in the Middle East, 1930–1931. In *Business History Review* 67:2, 1993, 207–45.

———. France's Middle East Ambitions, the Sykes-Picot Negotiations, and the Oil-Fields of Mosul, 1915–1918. In *Journal of Modern History* 66:4, 1994, 697–725.

Foreman-Peck, James. *A History of the World-Economy: International Economic Relations since 1850*. New York: Harvester-Wheatsheaf, 1995.

Fox, Edward Whiting. *History in Geographic Perspective: The Other France*. New York: Norton, 1971.

Francillon, René J. *McDonnell Douglas Aircraft since 1920*. Volume I. Annapolis, MD: Naval Institute Press, 1988.

Freris, L. L. A Re-Appraisal of the Goldschmidt Generator. In IEE. *International Conference on 100 Years of Radio*. London: Institution of Electrical Engineers, 1995, 71–5.

Friedberg, Aaron. *The Weary Titan: Britain and the Experience of Relative Decline, 1895–1905*. Princeton, NJ: Princeton University Press, 1988.

Fromkin, David. *A Peace to End All Peace. The Fall of the Ottoman Empire and the Creation of the Modern Middle East*. New York: Avon, 1989.

Fuller, Howard J. *Clad in Iron: The American Civil War and the Challenge of British Naval Power*. Westport, CT: Praeger, 2008.

Ganz, A. Harding. The German Navy in the Far East and Pacific: The Seizure of Kiautschou and After. In John A. Moses & Paul M. Kennedy, eds. *Germany in the Pacific and the Far East, 1870–1914*. Brisbane: University of Queensland Press, 1977, 115–36.

Gardner, Brian. *The East India Company: A History*. New York: Barnes & Noble, 1971.

Gardner, Lloyd C. *Safe for Democracy. The Anglo-American Response to Revolution, 1913–1923*. New York: Oxford University Press, 1984.

Gathorne-Hardy, G. M. Review of *Kampen en Oljen* by Anton Mohr. In *Journal of the British Institute of International Affairs* 5:1, 1926, 52–3.

Gibbons, Tony. *The Complete Encyclopedia of Battleships and Battlecruisers*. London: Salamander, 1983.

Gibbs-Smith, Charles H. *Aviation. An Historical Survey from Its Origins to the End of World War II*. London: Science Museum, H.M.S.O., 1970.

Gilbreth, Frank W. *Motion Study: A Method for Increasing the Efficiency of the Workman*. New York: Van Nostrand, 1911.

Glauert, Hermann. *The Elements of Airfoil and Airscrew Theory*. Cambridge: Cambridge University Press, 1926.

Goebbels, Joseph. Der Rundfunk als achte Großmacht [Radio as the Eighth Great Power] *Signale der neuen Zeit. 25 ausgewählte Reden von Dr. Joseph Goebbels*. Munich: Zentralverlag der NSDAP, 1938, 197–207. http://www.calvin.edu/academic/cas/gpa/goeb56.htm accessed 16 August 2014.

Goldsborough, Brice. The Earth Inductor Compass. In *Aero Digest* 10:6, 1927, 542–4.

Goldstein, Joshua S. *Long Cycles. Prosperity and War in the Modern Age*. New Haven, CT: Yale University Press, 1988.

Goldstein, Laurence. *The Flying Machine and Modern Literature*. Bloomington, IN: Indiana University Press, 1986.

Gordon, Amanda. The Conservative Press and the Russian Revolution. In Cameron Hazlehur ed. *Australian Conservatism: Essays in Twentieth Century Political History*. Canberra: Australian National University Press, 1979, 29–50.

Gourdon, Édouard. *Histoire du Congrés de Paris*. Paris: Librairie Nouvelle, 1857.

Gray, Peter, & Owen Thetford. *German Aircraft of the First World War*. London: Putnam, 1962.

Green, William. *Warplanes of the Third Reich*. New York: Galahad, 1970.

Gunston, Bill. *Aircraft of the Soviet Union*. London: Osprey, 1983.

———. *World Encyclopedia of Aero Engines*. Cambridge: Patrick Stephens, 1989.

———. *Development of Piston Aero Engines*. Second Edition. Sparkford, UK: Patrick Stephens, 1999.

Haddow, G. W., & Peter M. Grosz. *The German Giants. The German R–Planes 1914–1918*. Second Edition. New York: Funk & Wagnalls, 1969.

Hall, Peter. *Cities in Civilization*. New York: Pantheon, 1998.

Hamilton, B., & Robert J. Falkiner. Motor Gasoline. In George E. Totten, Steven R. Vestbrook, & Rajesh H. Shah, eds., *Fuels and Lubricants Handbook: Technology, Properties, Performance, and Testing*. West Conshohocken, PA: ASTM, 2003, 61–88.

Hanighen, Frank C. *The Secret War*. New York: John Day, 1934.

Hanle, Paul A. *Bringing Aerodynamics to America*. Cambridge, MA: MIT Press, 1982.

Hardy, A. C. *Motorships. An Investigation into the Character of Mercantile Vessels Powered by Internal Combustion Engines*. New York: Van Nostrand, 1926.

———. *Oil Ships and Sea Transport. A Story of Oil in Relation to its Effect on Sea Transportation*. London: Routledge, 1931.

Hare, Paul R. *The Royal Aircraft Factory*. London: Putnam, 1990.

Hart, P. T. *Saudi Arabia and the United States: Birth of a Security Partnership*. Bloomington, IN: Indiana University Press, 1998.

Harvey, David. *The Condition of Postmodernity*. Oxford: Blackwell, 1990.

———. *The New Imperialism*. New York: Oxford, 2003.

Hassell, Patrick. Advances in Aerodynamics. In Philip Jarrett, ed., *Biplane to Monoplane: Aircraft Development 1918–39*. London: Putnam, 1997, 140–54.

Haushofer, Karl. *Geopolitics of the Pacific Ocean: Studies on the Relationship between Geography and History*. Lewiston, NY: Edwin Mellen, 2002 [1925].

Hayes, Tracey. Karl Haushofer and his Impact on the German–Japanese Alliance. Unpublished M.S. Thesis, Department of Geography, Texas A&M University, 2004.

Hazlehurst, Cameron. *Menzies Observed*. Hornsby, NSW: Allen & Unwin, 1979.

——— ed. *Australian Conservatism: Essays in Twentieth Century Political History*. Canberra: Australian National University Press, 1979.

Headrick, Daniel, R. *The Invisible Weapon: Telecommunications and International Politics, 1850–1940*. New York: Oxford University Press, 1991.

Heffer, Jean. *The United States and the Pacific. History of a Frontier*. Translated from the French by W. Donald Wilson. Notre Dame, IN: Notre Dame University Press, 2002.

Henning, Graydon R., & Keith Trace. Britain and the Motorship: A Case of the Delayed Adoption of New Technology. In *Journal of Economic History* 35:2, 1975, 353–85.

Henry, J. D. *Oil Fuel and the Empire*. London: Bradbury, Agnew & Co., 1908.

Herrera, Geoffrey L. *Technology and International Transformation. The Railroad, the Atom Bomb, and the Politics of Technological Change*. Albany, NY: State University of New York Press, 2006.

Herris, Jack. *Germany's Fighter Competitions of 1918*. Lexington, KY: Aeronaut, 2013.

Herwig, Holger H. *The Politics of Frustration: The United States in German Naval Planning, 1889–1941*. Boston: Little, Brown, 1976.

———. *The Demon of Geopolitics: How Karl Haushofer "Educated" Hitler and Hess*. Lanham, MD: Rowman & Littlefield, 2016.

Heymann, Hans Jr. Oil in Soviet–Western Relations in the Inter-War Years. In *American Slavic and East European Review* 7:4, 1948, 303–16.

Higham, Robin, & Stephen J. Harris, eds. *Why Air Forces Fail: The Anatomy of Defeat*. Lexington, KY: University Press of Kentucky, 2006.

Hills, Jill. *The Struggle for Control of Global Communications: The Formative Century*. Urbana, IL: University of Illinois Press, 2002.

———. *Telecommunications and Empire*. Urbana, IL: University of Illinois Press, 2007.

Hilmes, Michele. *Radio Voices: American Broadcasting, 1922–1952*. Minneapolis, MN: University of Minnesota Press, 1997.

———. *Network Nations: A Transnational History of British and American Broadcasting*. New York: Routledge, 2012.

Hong, Sungook. *Wireless: From Marconi's Black-box to the Audion*. Cambridge, MA: MIT Press, 2001.

Hough, Richard. *The Great War at Sea, 1914–1918*. New York: Oxford University Press, 1983.

Howeth, L. S. *History of Communications-Electronics in the United States Navy*. Washington, D.C.: GPO, 1963.

Hudson, John C. *Across This Land: A Regional Geography of the United States and Canada*. Baltimore, MD: The Johns Hopkins University Press, 2002.

Hudson, William James. *Australia and the New World Order: Evatt in San Francisco 1945*. Canberra: Australian Foreign Policy Publications Program, Australian National University, 1993.

Hugill, Peter J. Good Roads and the Automobile in the United States, 1880–1929. In *Geographical Review* 72:3, 1982, 327–49.

———. The Rediscovery of America. Elite Automobile Touring Early This Century. In *Annals of Tourism Research* 12:3, 1985, 435–47.

———. English Landscape Tastes in the United States. In *Geographical Review* 76:4, 1986, 408–23.

———. Technology Diffusion in the World Automobile Industry, 1885–1985. In Peter J. Hugill & D. Bruce Dickson, eds., *The Transfer and Transformation of Ideas and Material Culture*. College Station, TX: Texas A&M University Press, 1988a, 110–42.

———. Structural Changes in the Core Regions of the World–Economy. In *Journal of Historical Geography* 14, 1988b, 111–27.

———. Home and Class among an American landed elite. In John Agnew & James S. Duncan, eds., *The Power of Place. Bringing Together Geographical and Sociological Imaginations*. Boston: Unwin Hyman, 1989, 66–80.

———. *World Trade since 1431: Geography, Technology, and Capitalism*. Baltimore, MD: The Johns Hopkins University Press, 1993.

———. *Upstate Arcadia. Landscape, Aesthetics, and the Triumph of Social Differentiation in America*. Lanham, MD: Rowman & Littlefield, 1995.

———. *Global Communications since 1844: Geopolitics and Technology*. Baltimore, MD: The Johns Hopkins University Press, 1999a.

———. Imperialism in Edwardian Boys' Novels. In *Ecumene* 6:3, 1999b, 318–40.

———. Technology, its Innovation and Diffusion as the Motor of Capitalism. In *Comparative Technology Transfer and Society* 1:1, 2003, 89–113.

———. Trading States, Territorial States, and Technology: Mackinder's Contribution to the Discourse on States and Polities. In Brian W. Blouet, ed., *Global Geostrategy: Mackinder and the Defence of the West*. London: Cass, 2005, 107–124.

———. The Geostrategy of Global Business: Wal-Mart and its Historical Forbears. In Stanley R. Brunn, ed., *Wal-Mart in the World-Economy*. New York: Routledge, 2006, 3–14.

———. Transitions in Hegemony: A Theory Based on State Type and Technology. In William R. Thompson, ed., *Systemic Transitions: Past, Present, and Future*. New York: Palgrave Macmillan, 2009a, 31–54.

———. The Geopolitical Implications of Communications under the Sea. In Bernard Finn & Daqing Yang, eds., *Communication Under the Seas. The Evolving Cable Network and Its Implications*. Cambridge, MA: M.I.T. Press, 2009b, 257–77.

———. The American Challenge to British Hegemony. In *Geographical Review* 99:3, 2009c, 403–25.

———. The Shaping of the American Empire. In *Journal of Historical Geography* 36:3, 2010, 261–5.

———. Re-Making America: Soil Mechanics, Earth Moving, Highways, and Dams. In Stanley Brunn, ed., *Engineering Earth: The Impacts of Megaengineering Projects*. New York: Springer, 2011, 1395–408.

———. *Cotton in the World-Economy: Geopolitics and Globalization since 1771*. Staunton VA: George F. Thompson Publishing, forthcoming 2020.

Hugill, Peter J., & D. Bruce Dickson, eds. *The Transfer and Transformation of Ideas and Material Culture*. College Station, TX: Texas A&M University Press, 1988.

Hugill, Peter J., & Veit Bachmann. The Route to the Techno-Industrial World Economy and the Transfer of German Organic Chemistry to America Before, During, and Immediately After World War One. In *Comparative Technology Transfer and Society* 3:2, 2005, 159–86.

Huntington, Samuel P. *The Clash of Civilizations and the Remaking of World Order*. New York: Simon & Schuster, 1996.

Huntington, S. Transnational Organizations in World Politics. In *World Politics* 25:3, 1973, 333–68.

IEE. *International Conference on 100 Years of Radio*. London: Institution of Electrical Engineers, 1995.

Innis, Harold A. *Empire and Communications*. Oxford: Clarendon, 1950.

———. *The Bias of Communication*. Toronto: University of Toronto Press, 1951.

———. *Empire and Communications*. Revised by Mary Q. Innis with a foreword by Marshall McLuhan. Toronto: University of Toronto Press, 1972.

———. *Empire and Communications*. With a foreword by Alexander John Watson. Lanham, MD: Rowman & Littlefield, 2007.

Jackson, A. J. *De Havilland Aircraft since 1909*. Annapolis, MD: Naval Institute Press, 1987.

———. *Blackburn Aircraft since 1909*. Annapolis, MD: Naval Institute Press, 1989.

———. *Avro Aircraft since 1908*. London: Putnam, 1990.

Johnson, George. *The All-Red Line: The Annals and Aims of the Pacific Cable Project*. Ottawa: James Hope & Sons, 1903.

Jones, Lloyd S. *U.S. Fighters, Army Air Force 1925 to 1980s*. Fallbrook, CA: Aero, 1975.

———. *U.S. Naval Fighters, Navy/Marine Corps 1922 to 1980s*. Fallbrook, CA: Aero, 1977.

———. *U.S. Bombers 1928 to 1980s*. Third Edition. Fallbrook, CA: Aero, 1980.

Jones, B. Melvill. The Streamline Aeroplane. In *Aeronautical Journal* 33, 1929, 358–85.

Josephson, Matthew. *Empire of the Air. Juan Trippe and the Struggle for World Airways*. New York: Harcourt, Brace & Co., 1944.

Kagan, Robert. *Of Paradise and Power: America and Europe in the New World Order*. New York: Knopf, 2003.

Kandel, Eric R. *The Age of Insight: The Quest to Understand the Unconscious in Art, Mind, and Brain, from Vienna 1900 to the Present*. New York: Random House, 2012.

Kay, Antony L. *Junkers Aircraft and Engines, 1913–1945*. London: Putnam, 2004.

Kelly, Paul. An Australian Vision: Foreign Policy Challenges in the 21st Century, 2004a. http://espace.library.curtin.edu.au/R/8M3SNPNTM2A4F9EC146HJAYK8QK35SI388MMLVMVAM4VIK2XD5-01313 accessed 26 February 2013.

———. *Shipwrecked in Arcadia: The Australia Experiment*. Cambridge, MA: Harvard University Weatherhead Center for International Affairs, 2004b.

Kennedy, Paul M. *The Samoan Tangle: A Study in Anglo-German-American Relations 1878–1900*. New York: Harper & Row, 1974.

———. *The Rise and Fall of British Naval Mastery*. London: Macmillan, 1983a.

———. The Tradition of Appeasement in British Foreign Policy, 1865–1939. In Paul M. Kennedy, ed., *Strategy and Diplomacy 1870–1945: Eight Studies*. London: Allen & Unwin, 1983b, 15–39.

———. *The Rise and Fall of the Great Powers: Economic Change and Military Conflict from 1500 to 2000*. New York: Random House, 1987.

Kennedy, Roger G. *Cotton and Conquest: How the Plantation System Acquired Texas*. Norman, OK: University of Oklahoma Press, 2013.

Kent, Marian. *Oil and Empire. British Policy and Mesopotamian Oil, 1900–1920*. London: Macmillan, 1976.

Kieve, Jeffrey. *The Electric Telegraph in the U.K.: A Social and Economic History*. Newton Abbott, Devon: David & Charles, 1973.

Knaack, Marcelle S. *Encyclopedia of U.S. Air Force Aircraft and Missile Systems*. Volume II, *Post-World War II Bombers, 1945–1973*. Washington, D.C.: Office of Air Force History, 1988.

Kosin, Rüdiger. *The German Fighter since 1915*. London: Putnam, 1988.

Kupchan, Charles A. *The End of the American Era: U.S. Foreign Policy and the Geopolitics of the Twenty-First Century*. New York: Knopf, 2003.

Lake, Marilyn. John Curtin: Internationalist, 2003. www.http://john.curtin.edu.au/events/speeches/lake.html accessed 26 February 2013.

Landahn, Dr. Krupp's Big Battleship Diesel Engine. In *Journal of the American Society for Naval Engineers* 37:1, 1925, 168–73.

Larrabee, Eric. *Commander-in-Chief. Franklin Delano Roosevelt, his Lieutenants, and their War*. Annapolis, MD: Naval Institute Press, 1987.

Lattimore, Owen. *Inner Asian Frontiers of China*. New York: American Geographical Society, 1940a.

———. American Responsibilities in the Far East. In *Virginia Quarterly Review* 16:2, 1940b, 161–74.

Lee, David. The Curtin and Chifley Governments, Liberal Internationalism and World Order. In David Lee & Christopher Waters. *Evatt to Evans: The Labor Tradition in Australian Foreign Policy*. St. Leonards, NSW: Allen & Unwin, 1997, 48–61.

Lee, David, & Christopher Waters. *Evatt to Evans: The Labor Tradition in Australian Foreign Policy*. St. Leonards, NSW: Allen & Unwin, 1997.

Lee, Stephen J. *Gladstone and Disraeli (Questions and Analysis)*. Abingdon: Routledge, 2005.

Levine, Bruce. *The Fall of the House of Dixie: War and the Social Revolution that Transformed the South*. New York: Random House, 2013.

Lewis, Peter. *The British Fighter since 1912*. Fourth Edition. London: Putnam, 1979.

Lewis, Tom. *Empire of the Air: The Men Who Made Radio*. New York: Harper Collins, 1991.

Lindsey, William Schaw. *History of Merchant Shipping and Ancient Commerce*. Volume 3. London: Sampson Low, 1876.

Lippman, Thomas W. The Day FDR Met Saudi Arabia's Ibn Saud. In *The Link* 38:2, 2005, 1–12.

Lloyd, Clem, & Richard Hall, eds. *Backroom Briefings: John Curtin's War*. Canberra: National Library of Australia, 1997.

London, Peter. *Saunders and Saro Aircraft since 1917*. London: Putnam, 1988.

Lucke, Charles Edward. Internal Combustion Engines in Marine Service, Parts I & II. In *Journal of the Franklin Institute* 192:1, 1921, 11–45 & 192:2, 1921, 203–32.

MacKeand, J. C. B., & M. A. Cross. Wide–Band High Frequency Signals from Poldhu? In IEE, *International Conference on 100 Years of Radio*, London: Institution of Electrical Engineers, 1995, 26–31.

Mackinder, Halford J. The Geographical Pivot of History. In *Geographical Journal* 23, 1904, 421–44.

———. *Democratic Ideas and Reality: A Study in the Politics of Reconstruction*. New York: Henry Holt, 1919 [1942].

Maclennan, Graeme. Marine Propulsion. In Robert Gardiner, ed., *The Golden Age of Shipping: The Classic Merchant Ship, 1900–1960*. Annapolis, MD: Naval Institute Press, 1994, 152–63.

MacMillan, Margaret. *Paris 1919. Six Months that Changed the World*. New York: Random House, 2001.

Maddison, Angus. *The World Economy: Historical Statistics*. Paris: OECD, 2003.

Mahan, A. T. *The Influence of Sea Power upon History, 1660–1783*. Boston: Little, Brown & Company, 1890.

Mann, F. J. Federal Telephone and Radio Corporation. A Historical Review: 1909–1946. In *Electrical Communication* 23:4, 1946, 377–405.

Mann, Michael. *The Sources of Social Power*. Volume 2, *The Rise of Classes and Nation-States, 1760–1914*. New York: Cambridge University Press, 1993.

Marcosson, Isaac F. *The Black Golconda. The Romance of Petroleum*. New York: Harper, 1924.

Martin, A. W. *Robert Menzies: A Life*. Volume 1, *1894–1943*. Melbourne: Melbourne University Press, 1993.

———. *Robert Menzies: A Life*. Volume 2, *1944–1978*. Melbourne: Melbourne University Press, 1999.

Massey, Robert K. *Dreadnought. Britain, Germany, and the Coming of the Great War*. New York: Ballantine, 1991.

Maull, Hanns W. Civilian Power: The Concept and Its Relevance for Security Issues. In Lidija Babic & Bo Huldt, eds., *Mapping the Unknown, Towards a New World Order*. Stockholm, Sweden: Swedish Institute of International Affairs, 1993, 115–31.

Maull, Hanns W. Germany and the Use of Force: Still a Civilian Power? In Saori N. Katada, Hanns W. Maull, & Takashi Inoguchi, eds., *Global Governance. Germany and Japan in the International System*. Aldershot, UK: Ashgate, 2004, 89–109.

Maunsell, Captain F. E. Kurdistan. In *The Geographical Journal* 3:2, 1894, 81–92.

———. The Mesopotamian Petroleum Field. In *The Geographical Journal* 9:5, 1897, 528–32.

Mawdsley, Ewan. *The Russian Civil War*. Boston: Allen & Unwin, 1987.

McFarland, Stephen L. *America's Pursuit of Precision Bombing, 1910–1945*. Washington, D.C.: Smithsonian, 1995.

McKercher, B. J. C. *Transition of Power: Britain's Loss of Global Pre-eminence to the United States, 1930–1945*. Cambridge: Cambridge University Press, 1999.

McKinlay, Brian. *The ALP. A Short History of the Australian Labor Party*. Richmond, Victoria: Heinemann, 1981.

Meinig, Donald W. *The Shaping of America*, Volume 4. *Global America, 1915–2000*. New Haven, CT: Yale University Press, 2004.

Mejcher, Helmut. *The Struggle for a New Middle East in the 20th Century. Studies in Imperial Design and National Politics*. Berlin: Lit, 2007.

Melosi, Martin V., & Joseph A. Pratt. The Energy Capital of the World? Oil-Led Development in Twentieth–Century Houston. In Joseph A. Pratt, Martin V. Melosi, & Kathleen A. Brosnan, eds. *Energy Capitals: Local Impact, Global Influence*. Pittsburgh: University of Pittsburgh Press, 2014, 30–57.

Messere, Fritz. Regulation. In Christopher H. Sterling, ed., *Encyclopedia of Radio*. New York: Fitzroy Dearborn, 2004, 1201–07.

Meyer, Karl E., & Shareen Blair Brysac. *Tournament of Shadows: The Great Game and the Race for Empire in Central Asia*. Washington, DC: Counterpoint, 1999.

Mikesh, Robert C., & Shorzoe Abe. *Japanese Aircraft 1910–1941*. London: Putnam, 1990.

Mirchandani, Rajesh, & Abdullahi Tasiu Abubakar. *Britain's International Broadcasting*. Los Angeles, CA: Figueroa Press, 2014.

Miyake, Masaki. German Cultural and Political Influence in Japan, 1870–1914. In John A. Moses & Paul M. Kennedy, eds. *Germany in the Pacific and the Far East, 1870–1914*. Brisbane: University of Queensland Press, 1977, 156–81.

Modelski, George. The Long Cycle of Global Politics and the Nation-State. In *Comparative Studies in Society and History* 20, 1978, 214–35.

Modelski, George, & William R. Thompson. *Seapower in Global Politics, 1494–1993*. Seattle: University of Washington Press, 1988.

———. *Leading Sectors and World Powers*. Columbia, SC: University of South Carolina Press, 1996.

Mohr, Anton. *The Oil War*. New York: Harcourt, Brace, 1926.

Montgomery, Maureen. *"Gilded Prostitution:" Status, Money, and Transatlantic Marriages, 1870–1914*. New York: Routledge, 1989.

Moses, John A., & Paul M. Kennedy, eds. *Germany in the Pacific and the Far East, 1870–1914*. Brisbane: University of Queensland Press, 1977.

Mowat, R. B. *The Life of Lord Pauncefote: First Ambassador to the United States*. Boston: Houghton Mifflin, 1929.

Müller, Simone M. *The Social and Cultural Creation of Global Telegraph Networks*. NewYork: Columbia University Press, 2016.

Mumford, Lewis. *Technics and Civilization*. New York: Harcourt Brace, 1934.

Munson, Kenneth. The Biplane's Fall from Favour. In Philip Jarrett, ed., *Biplane to Monoplane: Aircraft Development 1918–39*. London: Putnam, 1997, 11–27.

Nickles, David Paull. *Under the Wire. How the Telegraph Changed Diplomacy*. Cambridge, MA: Harvard University Press, 2003.

Norrell, Tracey Hayes. *For the Honor of our Fatherland: German Jews on the Eastern Front During the Great War*. Lanham, MD: Lexington, 2018.

Offer, Avner. *The First World War: An Agrarian Interpretation*. Oxford: Clarendon, 1989.

O'Hara, J. G., & W. Pricha. *Hertz and the Maxwellians. A Study and Documentation of the Discovery of Electromagnetic Wave Radiation, 1873–1894*. London: Peter Peregrinus, 1987.

Omissi, David E. *Air Power and Colonial Control: The Royal Air Force, 1919–1939*. Manchester: Manchester University Press, 1990.

Oren, Michael B. *Power, Faith, and Fantasy. America in the Middle East, 1776 to the Present*. New York: Norton, 2007.

Paris, Michael. *Winged Warfare. The Literature and Theory of Aerial Warfare in Britain, 1859–1917*. Manchester, UK: Manchester University Press, 1992.

Pegg, Mark. *Broadcasting and Society, 1918–1939*. London: Croom Helm, 1983.

Pelletier, Alain J. *Bell Aircraft since 1935*. Annapolis, MD: Naval Institute Press, 1992.

Phipp, Mike. *The Brabazon Committee and British Airliners, 1945–1960*. Stroud, Gloucestershire: Tempus, 2007.

Podobnik, Bruce. *Global Energy Shifts: Fostering Sustainability in a Turbulent Age*. Temple University Press, 2006.

Pool, Ithiel de Sola. *Technologies of Freedom*. Cambridge, MA: Belknap, 1983.

———. *Technologies without Boundaries: On Telecommunications in a Global Age*. Cambridge, MA: Harvard University Press, 1990.

Postma, Thijs. *Fokker: Aircraft Builders to the World*. London: Jane's, 1979.

Pratt, Joseph A., Martin V. Melosi, & Kathleen A. Brosnan, eds. *Energy Capitals: Local Impact, Global Influence*. Pittsburgh, University of Pittsburgh Press, 2014.

Prior, James. Without Pangs: John Curtin and the Soviet Union. Paper presented at the Australian War Memorial History Conference, 6–10 July 1987.

Ramsbottom, C. F. The Era of the Home Wireless Constructor. In IEE, *International Conference on 100 Years of Radio*, London: Institution of Electrical Engineers, 1995, 114–9.

Ransom, Stephen, & Robert Fairclough. *English Electric Aircraft and their Predecessors*. London: Putnam, 1987.

Read, Donald. *The Power of News. The History of Reuters 1849–1989*. Oxford: Oxford University Press, 1992.

Reid, J. D. *The Telegraph in America*. New York: Derby Bros., 1879.

Rennstich, Joachim K. *The Making of a Digital World. The Evolution of Technological Change and How It Shaped Our World*. New York: Palgrave Macmillan, 2009.

Ricardo, David. *On the Principles of Political Economy and Taxation*. London: John Murray, 1817.

Ricardo, Harry R. *The High-Speed Internal Combustion Engine*. Glasgow: Blackie. 1931 [1923].

Rosen, P. The Development of the Domestic Radio Receiver. In IEE, *International Conference on 100 Years of Radio*, London: Institution of Electrical Engineers, 1995, 128–33.

Rosenberg, Emily S. *Spreading the American Dream: American Economic and Cultural Expansion, 1890–1945*. New York: Hill & Wang, 1982.

Ross, Lloyd. *John Curtin: A Biography*. Melbourne: Melbourne University Press, 1996 [1977].

Russ, William Adam. *The Hawaiian Republic (1894–98) and Its Struggle to Win Annexation*. Selinsgrove, PA: Susquehanna University Press, 1992.

Sandler, Stanley. *The Emergence of the Modern Capital Ship*. Newark: University of Delaware Press, 1979.

Sassi, Mohammed. The Emergence of the French Oil Industry Between the Two Wars. In *Business and Economic History On-Line* 1, 2003, 1–25.

Sekigawa, Eiichiro. The Undeclared Air War. In *Air Enthusiast* 4, 1973, 247–50, 294–6 & *Air Enthusiast* 5, 1973, 26–29, 47.

Shi, David E. Seward's Attempt to Annex British Columbia, 1865–1869. In *Pacific Historical Review* 47:2, 1978, 217–38.

Shulman, Mark Russell. *Navalism and the Emergence of American Sea Power, 1882–1893*. Annapolis, MD: Naval Institute Press, 1995.

Slade, Edmond. Review of *Oil Imperialism: The International Struggle for Petroleum* by Louis Fischer and *The Oil War* by Anton Mohr. In *Journal of the Royal Institute of International Affairs* 7:2, 1928, 148–9.

Sloan, Alfred P. *My Years with General Motors*. Garden City, NY: Doubleday, 1964.

Smil, Vaclav. *Prime Movers of Globalization: The History and Impact of Diesel Engines and Gas Turbines*. Cambridge, MA: MIT Press, 2010.

Smith, Adam. *An Inquiry into the Nature and Causes of the Wealth of Nations*. Dublin: Whitestone, 1776.

Smith, Anthony. *The Shadow in the Cave. The Broadcaster, His Audience, and the State*. Urbana, IL: University of Illinois Press, 1973.

Smith, Edgar C. *A Short History of Marine Engineering*. Cambridge: Cambridge University Press, 1938.

Smith, F. E. John Cadman, Baron Cadman. 1877–1941. In *Obituary Notices of Fellows of the Royal Society* 3:10, 1941, 915–28.

Smith, Herschel. *A History of Aircraft Piston Engines*. Manhattan, KS: Sunflower, 1981.

Sobel, Robert. *RCA*. New York: Stein & Day, 1986.

Sondhaus, Lawrence. *The Great War at Sea. A Naval History of the First World War*. Cambridge: Cambridge University Press, 2014.

Spender, Percy. *Politics and a Man*. Sydney: Collins, 1972.

Sprout, Harold, & Margaret Sprout. *The Rise of American Naval Power, 1776–1918*. Princeton, NJ: Princeton University Press, 1939.

Standage, Tom. *The Victorian Internet. The Remarkable Story of the Telegraph and the Nineteenth Century's On–line Pioneers*. New York: Berkley, 1998.

Starosielski, Nicole. *The Undersea Network*. Durham, NC: Duke University Press, 2015.

Steil, Ben. *The Battle of Bretton Woods: John Maynard Keynes, Harry Dexter White, and the Making of a New World Order*. Princeton, NJ: Princeton University Press, 2013.

Sterling, Christopher H., & Michael Keith, eds. *The Museum of Broadcast Communications Encyclopedia of Radio*. In Three Volumes. New York: Fitzroy Dearborn, 2004.

Stewart, Colonel C. E. The Herat Valley and the Persian Border, from the Hari-rud to Sistan. In *Proceedings of the Royal Geographical Society* 8:3, 1886, 137–56.

Stilgoe, John R. *Metropolitan Corridor. Railroads and the American Scene*. New Haven, CT: Yale University Press, 1983.

Stinchcombe, William C. *The American Revolution and the French Alliance*. Syracuse NY: Syracuse University Press. 1969.

Stinton, Darrol. The Structural Revolution. In Philip Jarrett, ed., *Biplane to Monoplane: Aircraft Development 1918–39*. London: Putnam, 1997, 126–39.

Stivers, William. *Supremacy and Oil: Iraq, Turkey, and the Anglo-American World Order, 1918–1930*. Ithaca, NY: Cornell University Press, 1982.

Strauss, Kurt. Aviation Fuels. In George E. Totten, Steven R. Vestbrook, & Rajesh H. Shah, eds., *Fuels and Lubricants Handbook: Technology, Properties, Performance, and Testing*. West Conshohocken, PA: ASTM, 2003, 89–114.

Street, Séan. *Crossing the Ether: Pre–War Public Service Radio and Commercial Competition in the UK*. Bloomington: Indiana University Press, 2006.

Swanborough, Gordon, & Peter M. Bowers. *United States Navy Aircraft since 1911*. Annapolis MD: Naval Institute Press, 1976.

Tapper, Oliver. *Armstrong Whitworth Aircraft since 1913*. London: Putnam, 1973.

Taylor, Frederick W. *The Principles of Scientific Management*. New York: Harper, 1911.

Taylor, Peter J. *Political Geography: World–Economy, Nation-State and Locality*. London: Longman, 1985.

Taylor, Peter J., & Colin Flint. *Political Geography: World-Economy, Nation-State and Locality*. Fourth Edition. Englewood Cliff, NJ: Prentice Hall, 2000.

Templeton, Viscount [Samuel Hoare]. *Empire of the Air: The Advent of the Air Age, 1922–1929*. London: Collins, 1957.

Tennant, Kylie. *Evatt: Politics and Justice*. Second Edition, revised. Sydney: Angus & Robertson, 1972.

Turner, Frederick Jackson. The Significance of the Frontier in American History. In *Annual Report of the American Historical Association*, 1893, 199–227.

Tilly, Charles. *Coercion, Capital, and European States, AD 990–1990*. Cambridge, MA: Basil Blackwell, 1990.

Tooze, Adam. *The Deluge. The Great War, America and the Remaking of the Global Order, 1916–1931*. New York: Penguin, 2014.

Townshend, Charles. *When God Made Hell. The British Invasion of Mesopotamia and the Creation of Iraq, 1914–1921*. London: Faber & Faber, 2010.

Tuma, E. H. Technology Transfer and Economic Development: Lessons in History. In *Journal of Developing Areas* 21, 1987, 403–28.

Vance, James E., Jr. *Capturing the Horizon: The Historical Geography of Transportation Since the Transportation Revolution of the Sixteenth Century*. Baltimore, MD: The Johns Hopkins University Press, 1990.

Venn, Fiona. *Oil Diplomacy in the Twentieth Century*. New York: St. Martin's, 1986.

Vyvyan, R. N. *Wireless over Thirty Years*. London: Routledge, 1933. Re-issued as *Marconi and Wireless*. Wakefield: EP, 1974.

Wallerstein, Immanuel. *The Modern World-System.*Volume 1, *Capitalist Agriculture and the Origins of the European World-Economy on the Sixteenth Century*. New York: Academic, 1974.

———. The Three Instances of Hegemony in the History of the Capitalist World-Economy. In *The Politics of the World-Economy*. New York: Cambridge University Press, 1984, 37–46.

Warf, Barney. *Time-Space Compression: Historical Geographies*. London: Routledge, 2008.

Waters, Christopher. Creating a Tradition: The Foreign Policy of the Curtin and Chifley Labor Governments. In David Lee & Christopher Waters. *Evatt to Evans: The Labor Tradition in Australian Foreign Policy*. St. Leonards, NSW: Allen & Unwin, 1997, 35–47.

Watt, Alan Stewart. *The Evolution of Australian Foreign Policy, 1938–1965*. Cambridge: Cambridge University Press, 1967.

Weedon, K. The Alexanderson Alternator, A "Near Perfect" System of W/T Transmission. In IEE, *International Conference on 100 Years of Radio*, 1995, 69–70.

Wehl, A. R. *Fokker: The Creative Years*. London: Putnam, 1965.

Wells, Herbert George. *The War in the Air*. London: Bell, 1908.

Wilkins, Mira, & Frank E. Hill. *American Business Abroad: Ford on Six Continents*. Detroit, MI: Wayne State University Press, 1964.

Wilson, Woodrow. *Congressional Government: A Study in American Politics*. Cleveland, OH: Meridian, 1956 [1885].

Winchester, Simon. *Krakatoa. The Day the World Exploded: August 27, 1883*. New York: Harper Collins, 2003.

Winkler, Jonathan Reed. *Nexus: Strategic Communications and American Security in World War I*. Cambridge, MA: Harvard University Press, 2008.

Winkler, Matthew F. Introduction to Marine Petroleum Fuels. In George E. Totten, Steven R. Vestbrook, & Rajesh H. Shah, eds., *Fuels and Lubricants Handbook: Technology, Properties, Performance, and Testing*. West Conshohocken, PA: ASTM, 2003, 145–68.

Winseck, Dwayne R., & Robert M. Pike. *Communication and Empire: Media, Markets, and Globalization, 1860–1930*. Durham, NC: Duke University Press, 2007.

Woodhouse, Henry. American Oil Claims in Turkey. In *Current History* 15:6, 1922, 953–9.

Woods, H. Charles. The Baghdad Railway and Its Tributaries. In *The Geographical Journal* 50:1, 1917, 32–56. Discussion 56–7.

Wright, Gavin. *The Political Economy of the Cotton South*. New York: Barnes & Noble, 1978.

Yergin, Daniel. *The Prize: The Epic Quest for Oil, Money, and Power*. New York: Simon & Schuster, 1991.

Younger, R. M. *A Concise History of Australia and the Australians*. Richmond, Victoria: Hutchinson, 1982.

Zelinsky, Wilbur. *The Cultural Geography of the United States*. Englewood Cliffs, NJ: Prentice-Hall, 1973.

Zischka, Anton. *La Guerre Secrete pour le Pétrole*. Paris: Payot, 1933.

———. *Der Kampf um die Weltmacht Öl*. Leipzig: Wilhelm Goldmann, 1934.

———. *Ölkrieg: Wandlung der Weltmacht*. Leipzig: Wilhelm Goldmann, 1939.

Index

About the Author

Peter J. Hugill was born in York, England, and holds degrees from British, Canadian, and American universities. He is currently professor emeritus of geography at Texas A&M University, where he taught from 1978 through 2016, and a fellow of the Scowcroft Institute in the Bush School of Government and Public Service. He has published five books. His major books are on the relationship between technology, geography, and the world-system. Two are published by Johns Hopkins University Press: *World Trade since 1431* in 1993 and *Global Communication since 1844* in 1999. His third major book, *Cotton in the World Economy since 1771*, is due to be published by George F. Thompson Publishing in 2020. His interests focus on the hegemonic struggle between Britain and America in the early twentieth century, an economic struggle between two trading states obscured by the period of multi-polarity and the larger, often military, struggles that occupied much of the 1900s and that we call the two world wars. He has also published numerous articles, book chapters, and many book reviews. He has served as president of both Texas Conference of the American Association of University Professors and Texas Association of College Teachers.

